Linear System Theory

陆　军　编著

哈尔滨工程大学出版社
Harbin Engineering University Press

内容简介

线性系统理论是控制科学领域的一门重要的基础课程。本书以线性系统为研究对象，对线性系统理论做了全面论述。本书的主要内容包括线性系统的数学模型、连续时间系统的运动分析、线性系统的能控性和能观性测、线性系统运动的稳定性和系统的状态反馈和状态观测等。本书内容丰富，理论严谨，深入浅出地阐述了线性系统的基础理论和基本方法，并配有丰富的例题和习题，帮助读者理解书中所阐述的内容。

本书可作为来华留学生线性系统理论课程的教材，也可作为控制科学与工程专业、系统工程专业和电子信息类专业等相关专业的高年级本科生和研究生的专业英语教材；还可供科学工作者和工程技术人员学习与参考。

图书在版编目（CIP）数据

线性系统理论 = Linear System Theory ：英文/陆军编著.—哈尔滨 ：哈尔滨工程大学出版社，2023.3
ISBN 978-7-5661-3889-7

Ⅰ.①线… Ⅱ.①陆… Ⅲ.①线性系统理论-英文 Ⅳ.①O231

中国国家版本馆 CIP 数据核字（2023）第 054379 号

选题策划 唐欢欢
责任编辑 张 彦 田雨虹
封面设计 李海波

出版发行 哈尔滨工程大学出版社
社　　址 哈尔滨市南岗区南通大街 145 号
邮政编码 150001
发行电话 0451-82519328
传　　真 0451-82519699
经　　销 新华书店
印　　刷 哈尔滨午阳印刷有限公司
开　　本 787 mm×1 092 mm 1/16
印　　张 15
字　　数 498 千字
版　　次 2023 年 3 月第 1 版
印　　次 2023 年 3 月第 1 次印刷
定　　价 59.00 元
http://www.hrbeupress.com
E-mail:heupress@hrbeu.edu.cn

前　言

线性系统是系统和控制领域研究中最基本的研究对象，它伴随着航空航天、过程控制、最优控制、通信、电路和系统等众多学科的发展而日益成熟，已形成十分完整和成熟的线性系统理论。线性系统理论的概念、方法、原理和结论，对于系统和控制理论的许多分支，诸如最优控制、非线性控制、系统辨识、随机控制、智能控制、信号检测与估计等都具有重要的作用。国内外许多大学将线性系统理论列为系统和控制科学课程方面的一门基础课程。线性系统理论方面的教材和专著有很多，比较著名的有陈啟宗教授著的《线性系统理论与设计》和清华大学郑大钟教授编著的《线性系统理论》。

随着控制科学与工程学科来华留学生人数的不断增加，迫切需要一本结合国内课程设置的线性系统理论方面的教材。本书是编著者结合二十多年“线性系统理论”课程教学经验并参考国内外优秀教材而编撰的，系统地阐述了分析和综合线性多变量系统的时域理论和方法。本书主要内容包括：线性系统的数学模型，重点介绍了状态空间模型，并论述了系统状态空间描述和输入输出描述在建模能力上的异同；线性连续时间系统的运动分析；线性系统的能控性和能观测性，通过结构分解原理，阐述了状态空间描述在系统建模方面比输入输出描述更为全面；线性系统运动的稳定性；线性系统的状态反馈和状态观测，阐述了状态反馈的基本方法和一些应用，包括状态反馈解耦、镇定问题、状态观测器等。本书还适当地加入了系统分析和设计的 Matlab 实现，使读者能够运用 Matlab 工具完成系统的建模、分析和设计；章后的习题，方便读者对知识点的掌握。

由于编著者水平有限，书中难免存在错误、疏漏和不妥之处，恳请广大读者批评指正。

编著者
2022 年 12 月于哈尔滨工程大学

Contents

Chapter 1 Mathematical Descriptions of Systems

In order to analyze and synthesize systems, the first step is to establish the mathematical models of systems, that is, to establish the mathematical relationship between the variables in the system. The mathematical descriptions of systems include the input-output description and state-space description in general. The input-output description, also known as the external description, describes the characteristics of systems by the mathematical relationships between the inputs and outputs of systems. In classical control theory, the transfer function and ordinary differential equations belong to the external descriptions of systems. In the state-space description which is also called internal description, some internal variables, known as state variables, are chosen to describe the behavior of the system perfectly. The state-space description describes the behavior of the system by establishing the mathematical relationship between the state and the input-output variables of the system. The external description can't describe all characteristics of systems in general, while the state-space description is a perfect description of the behavior of systems.

A system with only one input variable and only one output variable is called a single-input and single-output (SISO) system or a single-variable system. A system with two or more input variables and/or two or more output variables is called a multivariable system. More specifically, we can call a system a multi-input multi-output (MIMO) system if it has two or more input variables and output variables. The research object of this book is extended to multi-input and multi-output linear time-varying (LTV) systems from single-input single-output linear time-invariant (LTI) systems in classical linear control theory. This chapter first introduces the external description of systems and then focuses on the internal description of systems.

1.1 System's input-output descriptions

Input-output descriptions of systems reveal mathematical relationships between the inputs and outputs of systems. When we establish the input-output description of a system, the system is viewed as a "black box" and there is not prior knowledge about the internal structure of the system. The input-output description is established based on external exciting inputs and corresponding responses. The input-output descriptions of systems reflect or determine the intrinsic nature of systems based on the external performance of systems. Therefore, the input-output descriptions of systems are also called the external descriptions of systems. The transfer function and differential equation of a single-variable system are typical examples of input-output descriptions. Firstly, the definition of the linear system is introduced by input-output descriptions.

1.1.1 Linear systems

Linear system theory mainly studies the related theory about multi-input multi-output linear systems. The following concepts are given.

1. Number field $\mathscr{F}$

Definition 1.1.1 (Number field $\mathscr{F}$) The number field $\mathscr{F}$ is the aggregate of scales of a given type (as integers, irrationals, complex numbers) which can be combined by addition or multiplication to obtain a result of the same type. The operation addition or multiplication must satisfy the following conditions:

(1) $\forall \alpha \in \mathscr{F}, \beta \in \mathscr{F}$, we have

$$\alpha + \beta \in \mathscr{F} \text{ and } \alpha \cdot \beta \in \mathscr{F} \tag{1.1.1}$$

(2) Addition and multiplication are both commutative:

$\forall \alpha \in \mathscr{F}, \beta \in \mathscr{F}$, we have

$$\alpha + \beta = \beta + \alpha \text{ and } \alpha \cdot \beta = \beta \cdot \alpha \tag{1.1.2}$$

(3) Addition and multiplication are both associative:

$\forall \alpha, \beta, \gamma \in \mathscr{F}$, we have

$$\alpha + (\beta + \gamma) = (\alpha + \beta) + \gamma \text{ and } (\alpha \cdot \beta) \cdot \gamma = \alpha \cdot (\beta \cdot \gamma) \tag{1.1.3}$$

(4) Multiplication is assignable with respect to addition:

$\forall \alpha, \beta, \gamma \in \mathscr{F}$

$$\alpha \cdot (\beta + \gamma) = \alpha \cdot \beta + \alpha \cdot \gamma \tag{1.1.4}$$

(5) For any element α in the $\mathscr{F}$, there must exist the elements 0 and 1 in the $\mathscr{F}$ that satisfy the following conditions:

$$\alpha + 0 = \alpha \text{ and } 1 \cdot \alpha = \alpha \tag{1.1.5}$$

(6) For any element α in the $\mathscr{F}$, there must exist an element β that satisfies the following condition:

$$\alpha + \beta = 0 \tag{1.1.6}$$

The element β is called the additive inverse of α.

(7) For any non zero element α in the $\mathscr{F}$, there must exist an element γ that satisfies the following condition:

$$\alpha \cdot \gamma = 1 \tag{1.1.7}$$

The element γ is called the multiplicative inverse of α.

Example 1.1.1 The set $\{\mathbf{0}, \mathbf{1}\}$ consisted of **0** and **1** is not a number field if we use common addition or multiplication. Because $1 + 1 = 2 \notin \{\mathbf{0}, \mathbf{1}\}$. But if the operation addition and multiplication are adopted as the following

$$\mathbf{0}+\mathbf{0}=\mathbf{1}+\mathbf{1}=\mathbf{0},\quad \mathbf{1}+\mathbf{0}=\mathbf{1},\quad \mathbf{0}\cdot\mathbf{0}=\mathbf{0}\cdot\mathbf{1}=\mathbf{0},\quad \mathbf{1}\cdot\mathbf{1}=\mathbf{1} \tag{1.1.8}$$

then the set $\{0,1\}$ satisfies the all conditions of the number field. Now, the set $\{0,1\}$ is a number field and it is called binary number field. In general, the symbol **R** represents real number field and $R(s)$ represents rational function field.

2. Linear space $\mathscr{X}$ over a number field $\mathscr{F}$

Definition 1. 1. 2 (Linear space) A linear space (also called a vector space) is a nonempty set of objects, called vectors, which may be added together and multiplied ("scaled") by numbers which be called scalars. A linear space over a field $\mathscr{F}$ is a set $\mathscr{X}$ together with two operations (vector addition and scalar multiplication) that satisfy the following conditions:

(1) Vector addition is closed: $\forall\ \boldsymbol{x}_1,\ \boldsymbol{x}_2\in\mathscr{X}$, we have

$$\boldsymbol{x}_1+\boldsymbol{x}_2\in\mathscr{X} \tag{1.1.9}$$

(2) Vector addition is commutative: $\forall \boldsymbol{x}_1,\ \boldsymbol{x}_2\in\mathscr{X}$, we have

$$\boldsymbol{x}_1+\boldsymbol{x}_2=\boldsymbol{x}_2+\boldsymbol{x}_1 \tag{1.1.10}$$

(3) Vector addition is associative: $\forall \boldsymbol{x}_1,\ \boldsymbol{x}_2,\boldsymbol{x}_3\in\mathscr{X}$, we have

$$(\boldsymbol{x}_1+\boldsymbol{x}_2)+\boldsymbol{x}_3=\boldsymbol{x}_1+(\boldsymbol{x}_2+\boldsymbol{x}_3) \tag{1.1.11}$$

(4) There is a vector $\mathbf{0}$ in the set $\mathscr{X}$: $\forall \boldsymbol{x}\in\mathscr{X}$, we have

$$\mathbf{0}+\boldsymbol{x}=\boldsymbol{x} \tag{1.1.12}$$

The vector $\mathbf{0}$ is called a zero vector or original point.

(5) $\forall \boldsymbol{x}\in\mathscr{X}$, there is a vector $\bar{\boldsymbol{x}}$ in the set $\mathscr{X}$, we have

$$\boldsymbol{x}+\bar{\boldsymbol{x}}=\mathbf{0} \tag{1.1.13}$$

The $\bar{x}$ is called the additive inverse of $\boldsymbol{x}$.

(6) $\forall \alpha\in\mathscr{F}$ and $\boldsymbol{x}\in\mathscr{X}$, we have

$$\alpha\boldsymbol{x}\in\mathscr{X} \tag{1.1.14}$$

$\alpha\boldsymbol{x}$ is scalar multiplication of α and $\boldsymbol{x}$.

(7) The scalar multiplication with field multiplication is compatible: $\forall \alpha,\beta\in\mathscr{F},\boldsymbol{x}\in\mathscr{X}$, we have

$$\alpha(\beta\boldsymbol{x})=(\alpha\beta)\boldsymbol{x} \tag{1.1.15}$$

(8) The scalar multiplication with respect to vector addition is distributive:

$\forall \alpha\in\mathscr{F}$ and $\boldsymbol{x}_1,\boldsymbol{x}_2\in\mathscr{X}$, we have

$$\alpha(\boldsymbol{x}_1+\boldsymbol{x}_2)=\alpha\boldsymbol{x}_1+\alpha\boldsymbol{x}_2 \tag{1.1.16}$$

(9) The scalar multiplication with respect to field addition is distributive:

$\forall \alpha,\beta\in\mathscr{F},\ \boldsymbol{x}\in\mathscr{X}$, we have

$$(\alpha+\beta)\boldsymbol{x}=\alpha\boldsymbol{x}+\beta\boldsymbol{x} \tag{1.1.17}$$

(10) Identity element of scalar multiplication: $\forall \boldsymbol{x}\in\mathscr{X}$, we have

$$1\boldsymbol{x}=\boldsymbol{x} \tag{1.1.18}$$

where 1 denotes the multiplicative identity in $\mathscr{F}$.

Example 1.1.2 Given a number field $\mathscr{F}$, $\mathscr{F}^n$ denotes a set composed by the n-dimensional vector $\boldsymbol{x}_i$, where

$$\boldsymbol{x}_i = \begin{bmatrix} x_{1i} \\ x_{2i} \\ \vdots \\ x_{ni} \end{bmatrix}$$

$x_{ji} \in \mathscr{F}, j = 1, 2, \cdots, n$.

In this example, the vector addition is defined as

$$\boldsymbol{x}_i + \boldsymbol{y}_j = \begin{bmatrix} x_{1i} + y_{1j} \\ x_{2i} + y_{2j} \\ \vdots \\ x_{ni} + y_{nj} \end{bmatrix}$$

and the scalar multiplication is

$$\alpha \boldsymbol{x}_i = \begin{bmatrix} \alpha x_{1i} \\ \alpha x_{2i} \\ \vdots \\ \alpha x_{ni} \end{bmatrix}$$

So, $(\mathscr{F}^n, \mathscr{F})$ is a linear space.

If $\mathscr{F} = \mathbf{R}$, call $(\mathbf{R}^n, \mathbf{R})$ an n-dimensional real vector space.

Definition 1.1.3 (Linear mapping) For the vector spaces $\mathscr{U}$ and $\mathscr{Y}$ over a number field $\mathscr{F}$, a mapping $L: \mathscr{U} \to \mathscr{Y}$ is called a linear mapping if

$$L(a_1 \boldsymbol{u}_1 + a_2 \boldsymbol{u}_2) = a_1 L(\boldsymbol{u}_1) + a_2 L(\boldsymbol{u}_2), \forall \boldsymbol{u}_1, \boldsymbol{u}_2 \in \mathscr{U}, \forall a_1, a_2 \in \mathscr{F} \quad (1.1.19)$$

The eq. (1.1.19) is also called superposition principle or superposition theorem. It is equivalent to the following two equations:

$$\begin{cases} L(\boldsymbol{u}_1 + \boldsymbol{u}_2) = L(\boldsymbol{u}_1) + L(\boldsymbol{u}_2) \\ L(a\boldsymbol{u}) = aL(\boldsymbol{u}) \end{cases} \quad (1.1.20)$$

where $\boldsymbol{u}_1, \boldsymbol{u}_2, \boldsymbol{u} \in \mathscr{U}, a \in \mathscr{F}$. The two equations is called additivity and homogeneity properties, respectively.

Definition 1.1.4 (Linear system) Under zero initial conditions, if the relationship between input and output of a system is a linear mapping, the system is linear system, i. e.

$$\boldsymbol{y} = L(\boldsymbol{u}) \quad (1.1.21)$$

where L is linear mapping, $\boldsymbol{y} = \begin{bmatrix} y_1 \\ y_2 \\ \vdots \\ y_q \end{bmatrix} \in \mathscr{Y}, \boldsymbol{u} = \begin{bmatrix} u_1 \\ u_2 \\ \vdots \\ u_p \end{bmatrix} \in \mathscr{U}$, $\mathscr{U}$ is a input linear space, $\mathscr{Y}$ is a output linear space.

The zero initial condition means the system doesn't have energy storage at the initial time.

1.1.2 Non-zero initial condition and impulse input

First look at the circuit example in the Figure 1.1.1.

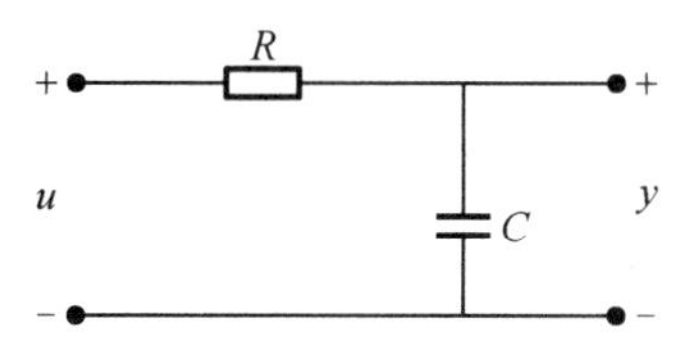

Figure 1.1.1 *RC* circuit

According to circuit laws, we have

$$\begin{cases} u = iR + y \\ i = C\dfrac{\mathrm{d}y}{\mathrm{d}t} \end{cases} \tag{1.1.22}$$

From eq. (1.1.22), we have

$$\dot{y}(t) = -\frac{1}{RC}y(t) + \frac{1}{RC}u(t) \tag{1.1.23}$$

Let $t = t_0, y = y_0$, we have

$$y(t) = \mathrm{e}^{-\frac{1}{RC}(t-t_0)}y_0 + \int_{t_0}^{t} \mathrm{e}^{-\frac{1}{RC}(t-\tau)}\frac{1}{RC}u(\tau)\mathrm{d}\tau, t \geqslant t_0 \tag{1.1.24}$$

The symbol L represent the mapping between u and y, i. e.

$$y = L(u)$$

Let $u_1(t), u_2(t), t \geqslant t_0, y_1(t) = L(u_1(t)),\ y_2(t) = L(u_2(t)), y_{12}(t) = L(u_1(t) + u_2(t)),\ a \in \mathbf{R}$

If $y_0 = 0$, we have

$$y_{12}(t) = y_1(t) + y_2(t)$$
$$ay_1(t) = L(au_1(t))$$

That means the mapping L is linear.

If $y_0 \neq 0$, we have

$$y_{12}(t) \neq y_1(t) + y_2(t)$$

It indicates the mapping L is not linear. Meantime, from the eq. (1.1.24), we can get that, for same input $u(t)$, outputs $y(t)$ are different for different y_0. Thus we can't obtain a unique relationship between input and output for different initial conditions.

1. Zero initial conditions

When we establish an input-output description of a system, we must assume the initial conditions of the system are zero, in order to uniquely determine the relationship between the input and output of the system.

Here, the initial conditions of systems are zero means systems don't have energy storage at the initial moment.

In fact, the initial conditions of systems are often not zero. So that, input-output descriptions can't deal with this situation. In order to describe the effects of non-zero initial conditions, we can view the non-zero initial conditions as a impulse inputs at the initial moment. We will discuss it in detail in the following. First the unit impulse function is introduced.

2. Unit impulse function(δ function)

Let

$$\delta_\Delta(t-t_1)=\begin{cases}0, & t<t_1\\ \dfrac{1}{\Delta}, & t_1\leqslant t\leqslant t_1+\Delta\\ 0, & t>t_1+\Delta\end{cases}\tag{1.1.25}$$

Its graphical expression is shown in Figure 1.1.2.

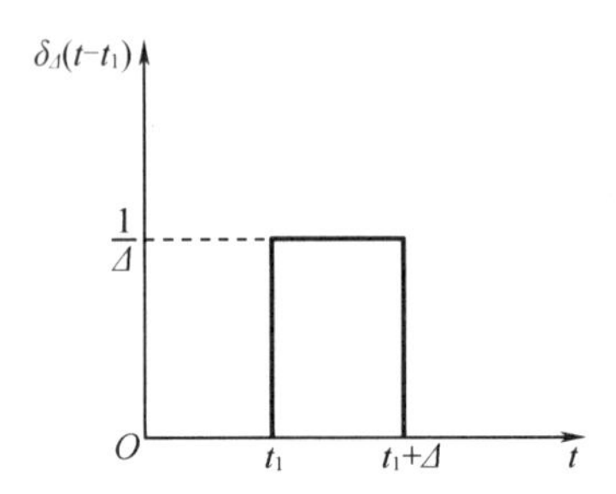

Figure 1.1.2 Function $\delta_\Delta(t-t_1)$

When $\Delta\to 0$, the limit of $\delta_\Delta(t-t_1)$ is defined as δ function, i. e.

$$\delta(t-t_1)\triangleq\lim_{\Delta\to 0}\delta_\Delta(t-t_1)\tag{1.1.26}$$

$\delta(t-t_1)$ is called unit impulse function or δ function.

The properties of δ function are

(1) $\forall\varepsilon>0$, we have

$$\int_{-\infty}^{+\infty}\delta(t-t_1)\,\mathrm{d}t=\int_{t_1-\varepsilon}^{t_1+\varepsilon}\delta(t-t_1)\,\mathrm{d}t=1\tag{1.1.27}$$

(2) For any function $f(t)$ which is continuous at the moment t_1,

$$\int_{-\infty}^{+\infty}f(t)\delta(t-t_1)\,\mathrm{d}t=f(t_1)\tag{1.1.28}$$

3. Non-zero initial conditions and equivalent impulse inputs

The non-zero initial conditions can be equivalent to the impulse inputs of systems at the initial time.

Conclusion 1. 1. 1 The system's output excited by non-zero initial conditions equals response excited by a impulse input at the initial moment.

Consider the following two systems

$$\begin{cases} \dot{y}(t) = f(y(t),t) + \varphi(t)\delta(t-t_0) \\ y(t_0) = 0 \end{cases} \tag{1.1.29}$$

and

$$\begin{cases} \dot{y}(t) = f(y(t),t) \\ y(t_0) = \varphi(t_0) \end{cases} \tag{1.1.30}$$

The solutions of above two systems are same as

$$y(t) = \varphi(t_0) + \int_{t_0}^{t} f(y(\tau),\tau)\mathrm{d}\tau$$

It indicates that the initial energy of a system can be the result of previous accumulation ($y_0 \neq 0$) or it can be established by instantaneous impulse input. That is, the non-zero initial condition can be equivalent to an impulse input at the initial moment with a zero initial condition.

In the future, the initial conditions of systems are assumed to be zero when we establish input-output descriptions of systems.

1.1.3 Unit impulse response of linear systems

The unit impulse response of a system is a form of input-output description of the system. First, the unit impulse response of linear systems is discussed by taking single-variable linear systems as an example, and then the results are extended to multiple input-multiple output systems. For a single-variable system, under zero initial conditions, the system output excited by the unit impulse function is called the unit impulse response of the system.

Let the input-output relation of a single-variable LTI (linear time-invariant) system be expressed as:

$$y = L(u) \tag{1.1.31}$$

The unit impulse response of the system is denoted by the symbol $g(t,\tau)$, i. e.

$$g(t,\tau) \triangleq L(\delta(t-\tau)) \tag{1.1.32}$$

It should be noted that $g(t,\tau)$ is a bivariate function, τ represents the moment when the δ function acts on the system, and t is the moment when the response of the system is observed. For any input, the corresponding system output can be expressed by using the unit impulse response, and the corresponding conclusions are as follows:

Conclusion 1.1.2 The unit impulse response representation of linear systems. For a SISO (single input single output) linear system, under the zero condition, the output response of the system is

$$y(t) = \int_{-\infty}^{+\infty} g(t,\tau)u(\tau)\mathrm{d}\tau \tag{1.1.33}$$

where $u(t)$ is the system's input variable and $g(t,\tau)$ is the system's unit impulse response.

Proof The single-variable linear system eq. (1.1.31) is taken as an example to prove the conclusion. As shown in Figure 1.1.3, an arbitrarily time-continuous input $u(t)$ can be approximated by a series of impulse functions, i. e.

$$u(t) \approx \sum_{i=-\infty}^{+\infty} u(t_i)\delta_\Delta(t-t_i)\Delta \tag{1.1.34}$$

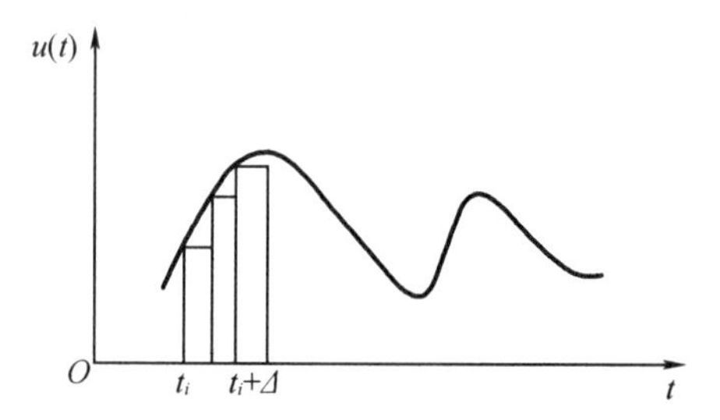

Figure 1.1.3 Approximate representation of $u(t)$

The output of the system excited by input $u(t)$ is

$$\begin{aligned} y &= L(u) \\ &\approx L\left(\sum_{i=-\infty}^{+\infty} u(t_i)\delta_\Delta(t-t_i)\Delta\right) \\ &= \sum_{i=-\infty}^{+\infty} L(u(t_i)\delta_\Delta(t-t_i)\Delta) \\ &= \sum_{i=-\infty}^{+\infty} L(\delta_\Delta(t-t_i)) \cdot u(t_i)\Delta \end{aligned} \tag{1.1.35}$$

When $\Delta \to 0$, eq. (1.1.35) becomes

$$y(t) = \int_{-\infty}^{+\infty} L(\delta(t-\tau))u(\tau)\mathrm{d}\tau = \int_{-\infty}^{+\infty} g(t,\tau)u(\tau)\mathrm{d}\tau$$

Conclusion 1.1.2 shows that the input-output description of a linear system can be expressed by the unit impulse response of the system under zero initial conditions. That is, the unit impulse response of linear systems can be used as a form of external description of systems.

Definition 1.1.5 (Linear time-invariant (LTI) system) If the linear mapping L between input and output of a linear system satisfies the following equations:

$$\begin{cases} y(t+t_1) = L(u(t+t_1)) \\ y(t) = L(u(t)) \end{cases} \tag{1.1.36}$$

That is, the linear mapping L does not change with time, then the system is called a LTI system.

For a LTI system, its unit impulse response satisfies the following conclusions:

Conclusion 1.1.3 The unit impulse response of a LTI system depends only on the difference between the time at which the response is observed and the time at which the function acts on the system, i. e.

$$g(t,\tau)=g(t-\tau) \tag{1.1.37}$$

Proof For a LTI system, there are

$$g(t+t_1,\tau)=L(\delta(t+t_1-\tau))=L(\delta(t-(\tau-t_1)))=g(t,\tau-t_1) \tag{1.1.38}$$

Let $\tau_1=\tau-t_1$, from eq. (1.1.38), we can get

$$g(t+t_1,\tau_1+t_1)=g(t,\tau_1) \tag{1.1.39}$$

Because eq. (1.1.39) is true for any t,t_1,τ_1. So choose $t_1=-\tau_1$, for any t and τ_1, we can get

$$g(t,\tau_1)=g(t-\tau_1,0)=g(t-\tau_1) \tag{1.1.40}$$

The equation indicates that the unit impulse response of a LTI system only depends on the difference between the observed time of the response and the time when the function acts on the system, i. e. eq. (1.1.37) is true.

From conclusion 1.1.3 and eq. (1.1.33), the following conclusions can be drawn.

Conclusion 1.1.4 For any input $u(t)$, the response of a LTI system can be expressed as

$$y(t)=\int_{-\infty}^{+\infty}g(t-\tau)u(\tau)\mathrm{d}\tau \tag{1.1.41}$$

or

$$y(t)=\int_{-\infty}^{+\infty}g(\tau)u(t-\tau)\mathrm{d}\tau \tag{1.1.42}$$

where $g(\cdot)$ is the unit impulse response of the LTI system.

Definition 1.1.6 (Causal system) A system is called a causal system if its current output depends on the past and current inputs but not on the future inputs. If a system is not causal, then its current output will depend on the future inputs.

Conclusion 1.1.5 For any input $u(t)$, the response of a causal LTI system can be expressed as

$$y(t)=\int_{-\infty}^{t}g(t-\tau)u(\tau)\mathrm{d}\tau \tag{1.1.43}$$

Proof From eq. (1.1.41), we can get

$$y(t_1)=\int_{-\infty}^{t_1}g(t_1-\tau)u(\tau)\mathrm{d}\tau+\int_{t_1}^{+\infty}g(t_1-\tau)u(\tau)\mathrm{d}\tau$$

For the causal system, the second term of above equation is zero. So

$$g(t_1-\tau)=0,\ \text{when } t_1<\tau$$

Because t_1 is arbitrary in $(-\infty,+\infty)$, so

$$g(t-\tau)=0,\ \text{when } t<\tau$$

or

$$g(t)=0,\ \text{when } t<0 \tag{1.1.44}$$

So for a causal LTI system,

$$y(t)=\int_{-\infty}^{t}g(t-\tau)u(\tau)\mathrm{d}\tau$$

Next, we extend the conclusion of the single-variable systems to the MIMO systems, and obtain the unit impulse response matrix of the MIMO LTI system. The conclusion is:

Conclusion 1.1.6 The output expression of a MIMO LTI system using the unit impulse response matrix.

For a p-input q-output system, $u(t)$ is input vector, i. e. $\boldsymbol{u} = [u_1 \quad u_2 \quad \cdots \quad u_p]^{\mathrm{T}}$, $\boldsymbol{y}(t)$ is output vector, i. e. $\boldsymbol{y} = [y_1 \quad y_2 \quad \cdots \quad y_q]^{\mathrm{T}}$, under zero initial conditions, the response of the system is

$$\boldsymbol{y}(t) = \int_{-\infty}^{+\infty} \boldsymbol{G}(t,\tau)\boldsymbol{u}(\tau)\mathrm{d}\tau \tag{1.1.45}$$

where

$$\boldsymbol{G}(t,\tau) = \begin{bmatrix} g_{11}(t,\tau) & g_{12}(t,\tau) & \cdots & g_{1p}(t,\tau) \\ g_{21}(t,\tau) & g_{22}(t,\tau) & \cdots & g_{2p}(t,\tau) \\ \vdots & \vdots & & \vdots \\ g_{q1}(t,\tau) & g_{q2}(t,\tau) & \cdots & g_{qp}(t,\tau) \end{bmatrix} \tag{1.1.46}$$

and $g_{ij}(t,\tau)\,(i=1,2,\cdots,q;j=1,2,\cdots,p)$ is the response at time t at i^{th} output terminal due to an unit impulse applied at time τ at j^{th} input terminal, the inputs at other terminals being identically zero. The $q \times p$ matrix $\boldsymbol{G}(t,\tau)$ is called unit impulse response matrix of the system.

Proof Assume symbol L_1 is the mapping between input vector $\boldsymbol{u}$ and first output variable y_1, i. e.

$$\begin{aligned} y_1 &= L_1\left(\begin{bmatrix} u_1 \\ u_2 \\ \vdots \\ u_p \end{bmatrix}\right) \\ &= L_1\left(\begin{bmatrix} u_1 \\ 0 \\ 0 \\ \vdots \\ 0 \end{bmatrix} + \begin{bmatrix} 0 \\ u_2 \\ 0 \\ \vdots \\ 0 \end{bmatrix} + \cdots + \begin{bmatrix} 0 \\ \vdots \\ 0 \\ 0 \\ u_p \end{bmatrix}\right) \\ &= L_1\left(\begin{bmatrix} u_1 \\ 0 \\ 0 \\ \vdots \\ 0 \end{bmatrix}\right) + L_1\left(\begin{bmatrix} 0 \\ u_2 \\ 0 \\ \vdots \\ 0 \end{bmatrix}\right) + \cdots + L_1\left(\begin{bmatrix} 0 \\ \vdots \\ 0 \\ 0 \\ u_p \end{bmatrix}\right) \\ &= L_{11}(u_1) + L_{12}(u_2) + \cdots + L_{1p}(u_p) \end{aligned} \tag{1.1.47}$$

$$y_1(t) = \int_{-\infty}^{+\infty} g_{11}(t,\tau)u_1(\tau)\mathrm{d}\tau + \int_{-\infty}^{+\infty} g_{12}(t,\tau)u_2(\tau)\mathrm{d}\tau + \cdots + \int_{-\infty}^{+\infty} g_{1p}(t,\tau)u_p(\tau)\mathrm{d}\tau \tag{1.1.48}$$

where $g_{1i}(t,\tau) = L_{1i}(\delta(t-\tau))$, $i=1,2,\cdots,p$. In a similar way, we can get

$$y_2(t) = \int_{-\infty}^{+\infty} g_{21}(t,\tau)u_1(\tau)\mathrm{d}\tau + \int_{-\infty}^{+\infty} g_{22}(t,\tau)u_2(\tau)\mathrm{d}\tau + \cdots + \int_{-\infty}^{+\infty} g_{2p}(t,\tau)u_p(\tau)\mathrm{d}\tau$$

$$\vdots$$

$$y_q(t) = \int_{-\infty}^{+\infty} g_{q1}(t,\tau)u_1(\tau)\mathrm{d}\tau + \int_{-\infty}^{+\infty} g_{q2}(t,\tau)u_2(\tau)\mathrm{d}\tau + \cdots + \int_{-\infty}^{+\infty} g_{qp}(t,\tau)u_p(\tau)\mathrm{d}\tau$$

Write the above forms in matrix form, and get

$$\boldsymbol{y}(t) = \int_{-\infty}^{+\infty} \boldsymbol{G}(t,\tau)\boldsymbol{u}(\tau)\mathrm{d}\tau$$

For a LTI system,

$$\boldsymbol{G}(t,\tau) = \boldsymbol{G}(t-\tau) \tag{1.1.49}$$

For a causal LTI system with zero initial conditions,

$$\boldsymbol{y}(t) = \int_{t_0}^{t} \boldsymbol{G}(t-\tau)\boldsymbol{u}(\tau)\mathrm{d}\tau = \int_{t_0}^{t} \boldsymbol{G}(\tau)\boldsymbol{u}(t-\tau)\mathrm{d}\tau \text{ , } t \geqslant t_0 \tag{1.1.50}$$

where t_0 is initial moment of the system.

1.1.4 Transfer-function matrix of LTI systems

For a MIMO LTI system

$$\boldsymbol{y} = L(\boldsymbol{u}) \tag{1.1.51}$$

where $\boldsymbol{u}(t)$ is a p dimensional input vector, $\boldsymbol{y}(t)$ is q dimensional output vector, we will discus the transfer function matrix derived from the impulse response matrix of the system.

(1) Laplace transform is defined as

$$\boldsymbol{y}(s) = L(\boldsymbol{y}(t)) = \int_{0}^{+\infty} \boldsymbol{y}(t)\mathrm{e}^{-st}\mathrm{d}t \tag{1.1.52}$$

where $\boldsymbol{y}(s)$ is the Laplace transform of $\boldsymbol{y}(t)$.

(2) Definition. Transfer function matrix of a LTI system. For the system eq. (1.1.51), its transfer function matrix is

$$\boldsymbol{G}(s) = \begin{bmatrix} g_{11}(s) & g_{12}(s) & \cdots & g_{1p}(s) \\ g_{21}(s) & g_{22}(s) & \cdots & g_{2p}(s) \\ \vdots & \vdots & & \vdots \\ g_{q1}(s) & g_{q2}(s) & \cdots & g_{qp}(s) \end{bmatrix} \tag{1.1.53}$$

where,

$$g_{ij}(s) = \int_{0}^{+\infty} g_{ij}(t)\mathrm{e}^{-st}\mathrm{d}t = \frac{L(y_i(t))}{L(u_j(t))}, i = 1,2,\cdots q; j = 1,2,\cdots,p \tag{1.1.54}$$

(3) The relationship between the unit impulse response matrix and the transfer function matrix of a LTI system.

Conclusion 1.1.7 The relationship between the unit impulse response matrix and the transfer function matrix of a LTI system is

$$\boldsymbol{G}(s) = L(\boldsymbol{G}(t)) = \int_{0}^{+\infty} \boldsymbol{G}(t)\mathrm{e}^{-st}\mathrm{d}t$$

$$\boldsymbol{G}(t) = L^{-1}(\boldsymbol{G}(s)) \tag{1.1.55}$$

where $\boldsymbol{G}(s)$ is the transfer function matrix of the LTI system, $\boldsymbol{u}(s)$ is the Laplace transform of input vector $\boldsymbol{u}(t)$, $\boldsymbol{G}(t)$ is the unit impulse response matrix of the system.

Thus, there is

$$\boldsymbol{y}(s) = \boldsymbol{G}(s)\boldsymbol{u}(s) \tag{1.1.56}$$

Proof Assume $t_0 = 0$ and initial conditions are zero, from eq. (1.1.45) and eq. (1.1.52), we can get

$$\begin{aligned} y(s) &= \int_0^{+\infty} \left(\int_0^{+\infty} G(t-\tau)u(\tau)\mathrm{d}\tau \right) \mathrm{e}^{-st}\mathrm{d}t \\ &= \int_0^{+\infty} \left(\int_0^{+\infty} G(t-\tau)\mathrm{e}^{-s(t-\tau)}\mathrm{d}t \right) u(\tau)\mathrm{e}^{-s\tau}\mathrm{d}\tau \\ &= \int_0^{+\infty} G(v)\mathrm{e}^{-sv}\mathrm{d}v \cdot \int_0^{+\infty} u(\tau)\mathrm{e}^{-s\tau}\mathrm{d}\tau \end{aligned}$$

The $\boldsymbol{G}(s) = \int_0^{+\infty} \boldsymbol{G}(v)\mathrm{e}^{-sv}\mathrm{d}v$ is the Laplace transform of $\boldsymbol{G}(t)$ and $\boldsymbol{u}(s) = \int_0^{+\infty} \boldsymbol{u}(t)\mathrm{e}^{-s\tau}\mathrm{d}\tau$ is the Laplace transform of $\boldsymbol{u}(t)$. Thus we can get

$$\boldsymbol{y}(s) = \boldsymbol{G}(s)\boldsymbol{u}(s)$$

(4) Proper rational fraction matrix, strictly proper rational fraction matrix and improper rational fraction matrix.

In general, each entry $g_{ij}(s)$ $(i = 1,2,\cdots,q; j = 1,2,\cdots,p)$ of $\boldsymbol{G}(s)$ is rational fraction, which can be expressed as $n_{ij}(s)/d_{ij}(s)$, where $n_{ij}(s)$ and $d_{ij}(s)$ are polynomials of s. Thus $\boldsymbol{G}(s)$ is called rational fraction matrix. When $g_{ij}(s)$ satisfies

$$\lim_{s\to+\infty} g_{ij}(s) = 0 \quad \text{or} \quad \deg n_{ij}(s) < \deg d_{ij}(s)$$

the $g_{ij}(s)$ is called strictly proper fraction, where deg denotes the degree of a polynomial. When $g_{ij}(s)$ satisfies

$$\lim_{s\to\infty} g_{ij}(s) = \text{non-zero constant} \quad \text{or} \quad \deg n_{ij}(s) = \deg d_{ij}(s)$$

the $g_{ij}(s)$ is called proper fraction. When $g_{ij}(s)$ satisfies

$$\lim_{s\to\infty} g_{ij}(s) = \infty \quad \text{or} \quad \deg n_{ij}(s) > \deg d_{ij}(s)$$

the $g_{ij}(s)$ is called improper fraction.

A rational fraction matrix $\boldsymbol{G}(s)$ is said to be proper if its every entry is proper or strictly proper or if $\boldsymbol{G}(\infty)$ is a nonzero constant matrix; it is strictly proper if its every entry is strictly proper or if $\boldsymbol{G}(\infty)$ is a zero matrix. Otherwise it is improper.

A system which transfer function is improper is difficult to use in practice, e. g.

$$y(s) = g(s)u(s) = \frac{s^2}{s-1}u(s) = \left(s + 1 + \frac{1}{s-1}\right)u(s)$$

If we take the inverse Laplace transform of both sides of the above equation, we get

$$y(t) = L^{-1}(y(s)) = \dot{u}(t) + u(t) + L^{-1}\left(\frac{1}{s-1}u(s)\right)$$

where the first item $\dot{u}(t)$ causes the high frequency noise contained in $u(t)$ to be greatly amplified and reflected in $y(t)$.

1.2 State space descriptions of linear systems

1.2.1 Restriction of input-output descriptions

The input-output description only represents the mathematical relationship between the input and the output vector under zero initial conditions. The input-output description cannot describe the behavior excited by non-zero initial conditions. In addition, the input-output description cannot reveal all behaviors of systems, that is, it cannot give a comprehensive description of systems.

Example 1.2.1 Compare the following two systems.

The two systems in Figure 1.2.1 have the same external characteristics, that is, their transfer functions are:

$$g(s) = \frac{1}{s-1} \cdot \frac{s-1}{s+1} = \frac{s-1}{s+1} \cdot \frac{1}{s-1} = \frac{1}{s+1}$$

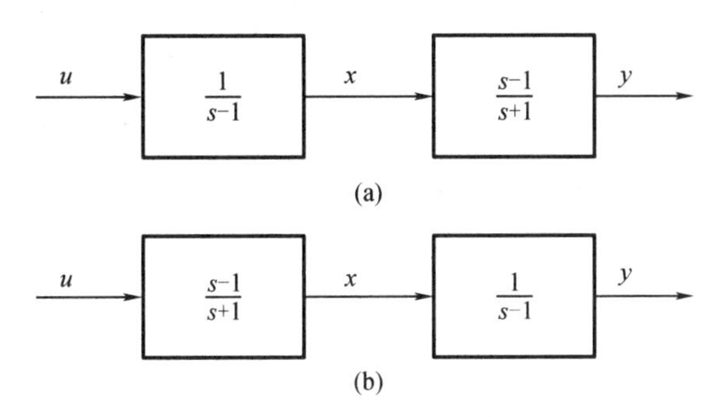

Figure 1.2.1 Limitations of input-output descriptions

However, these two systems have completely different internal structures. The state space of system (a) in Figure 1.2.1 can be described as follows:

$$x_1 = x, \quad x_2 = y$$

$$\dot{\boldsymbol{x}} = \begin{bmatrix} 1 & 0 \\ 0 & -1 \end{bmatrix} \boldsymbol{x} + \begin{bmatrix} 1 \\ 1 \end{bmatrix} u$$

$$y = \begin{bmatrix} 0 & 1 \end{bmatrix} \boldsymbol{x}$$

It can be seen that the state of the systemis completely controllable but not completely observable. Controllability and observability of system's state will be discussed in Chapter 3.

The system (b) in Figure 1.2.1 is transformed into Figure 1.2.2, and its state space description is as follows:

$$x_1 = z, \quad x_2 = y$$

$$\dot{\boldsymbol{x}} = \begin{bmatrix} -1 & 0 \\ -2 & 1 \end{bmatrix} \boldsymbol{x} + \begin{bmatrix} 1 \\ 1 \end{bmatrix} u$$

$$y = [0 \quad 1] \boldsymbol{x}$$

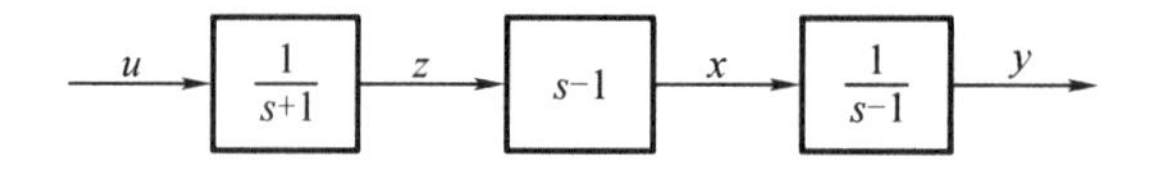

Figure 1.2.2 The equivalent system of Figure 1.2.1(b)

It can be seen that the state of the system is not completely controllable but completely observable.

Through the above analysis, it can be seen that the system input-output description can not fully describe all the characteristics of systems, and it has insufficient ability in system modeling. Now, we introduce another mathematical description of systems, the state space description.

1.2.2 State and state space

1. State

Definition 1.2.1 (State) The state of systems refers to the variables that describe the past, present and future behavior of systems. It is the smallest possible subset of system variables that can fully describe systems and its response to any given set of inputs at any given time.

The number of state variables used to describe systems is called the order of systems. For the following system

$$y^{(n)}(t) + a_{n-1}y^{(n-1)}(t) + \cdots + a_1\dot{y} + a_0 y = b_n u^{(n)}(t) + \cdots + b_1\dot{u}(t) + b_0 u(t)$$

where $y^{(i)}(t) = \dfrac{\mathrm{d}^i y}{\mathrm{d}t^i}$, y is the output variable of the system, u is the input variable of the system, the order of the system is n, i. e. the order of the system is the highest order of the derivative of the output y with respect to time t.

State is an important concept in the description of state space. The state variables that constitute the state are discussed in detail below.

2. State variables

Definition 1.2.2 (State variables) State variables are each of the variables that constitute the state of the systems. State variables, the input and output vector are shown in Figure 1.2.3.

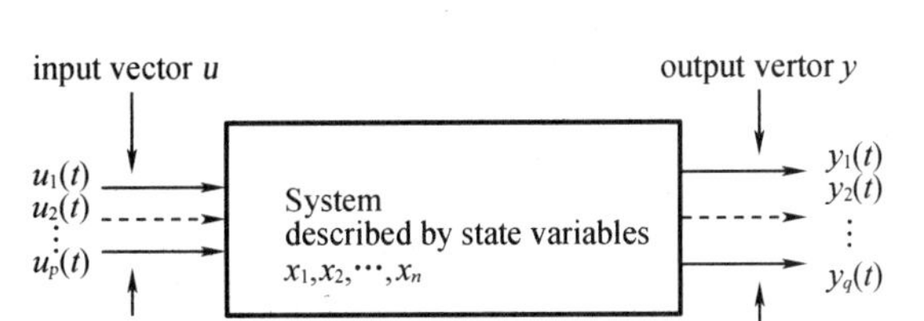

Figure 1.2.3 State variables, input and output vector

The notes about state variables of systems are stated as follows:

(1) State variables can fully describe the system's behavior. A mathematical description of systems in terms of a minimum set of variables $x_i(t)$, $i=1,2,\cdots,n$, together with knowledge of those variables at an initial time t_0 and the system inputs for time $t \geqslant t_0$, are sufficient to predict the state and outputs for all time $t > t_0$, i. e.

$$\begin{cases} x_1(0), x_2(0), \cdots, x_n(0) \\ u_1(t), u_2(t), \cdots, u_p(t) \end{cases} \rightarrow \begin{cases} x_1(t), x_2(t), \cdots, x_n(t) \\ y_1(t), y_2(t), \cdots, y_q(t) \end{cases}$$

(2) Minimum set of variables. The number of state variables, n is the smallest number to fully describe system's behavior. Using $n-1$ state variables can not fully describe system's behavior. It's not necessary to use $n+1$ variables to describe a system. Where the number n is defined to be the order of the system.

(3) State is not unique. There is no unique set of state variables that describe any given system. Many different sets of variables may be selected to yield a complete system's description. However, for a given system its order n is unique, and is independent of the particular set of state variables chosen.

(4) Output variable can be chosen as state variable.

(5) Input variable can not be chosen as state variable.

(6) Sometimes state variable is not measurable. In a practical system, there are some state variables that cannot be measured by sensors. For example, for the angular servo system, the angle, angular velocity and angular acceleration can be measured by sensors, but if the system has a state variable of the fourth derivative of the angle, then there is no corresponding sensor to measure. In addition, due to the cost limit in system development, some state variables are not considered to be measured by sensors. However, when designing the control law, it is often necessary to know the value of each state variable. This problem can be solved by the state observer introduced in the following chapters.

(7) State variable is defined in time-domain, not in frequency-domain.

4. State vector

Definition 1.2.3 (State vector) State vector $\boldsymbol{x}(t)$ is a column vector composed of state variables, i. e.

$$x(t) = \begin{bmatrix} x_1(t) \\ x_2(t) \\ \vdots \\ x_n(t) \end{bmatrix}, \quad t \geqslant t_0$$

4. State space

Definition 1.2.4(State space) State space is a set consists of all possible state values of a dynamical system.

Considering that the state variables of a practical system can only take real numbers, the state space is the vector space defined in the real number field. Let its dimension be n, then the state space is denoted as a $\mathbf{R}^n$. For a certain moment, the state vector is a point in the state space, and the change of the state vector with time constitutes a trajectory in the state space.

1.2.3 State space description of linear systems

The mathematical description of systems includes the external description and the internal description. The external description is also called input-output description. As mentioned above, the external description regards the system as a black box and only describes the relationship between the input vector and the output vector of the system, which is usually expressed in the form of differential equations, transfer function matrix, or difference equations. The internal description is also called state-space description, which is a kind of mathematical model based on internal structure analysis of the system. The state-space description consists of a state equation and an output equation. The state equation describes the relationship between the system state vector and the input vector. The output equation describes the relationship between the output vector, the state vector and the input vector. The specific form of state-space description of a linear system is as follows:

$$\begin{cases} \dot{\boldsymbol{x}} = \boldsymbol{A}(t)\boldsymbol{x} + \boldsymbol{B}(t)\boldsymbol{u}, \boldsymbol{x}(t_0) = \boldsymbol{x}_0, t \in [t_0, t_\alpha] & (1.2.1a) \\ \boldsymbol{y} = \boldsymbol{C}(t)\boldsymbol{x} + \boldsymbol{D}(t)\boldsymbol{u} & (1.2.1b) \end{cases}$$

where, $\boldsymbol{x}$ is an n-dimensional state vector, $\boldsymbol{u}$ is a p-dimensional input vector, $\boldsymbol{y}$ is a q-dimensional output vector, $\boldsymbol{A}(t)$、$\boldsymbol{B}(t)$, $\boldsymbol{C}(t)$ and $\boldsymbol{D}(t)$ are the $n \times n$, $n \times p$, $q \times n$ and $q \times p$ time-varying matrices respectively , i. e.

$$\boldsymbol{x} = \begin{bmatrix} x_1 \\ x_2 \\ \vdots \\ x_n \end{bmatrix} \in \mathbf{R}^n, \ \boldsymbol{y} = \begin{bmatrix} y_1 \\ y_2 \\ \vdots \\ y_q \end{bmatrix} \in \mathscr{Y}^q, \ \boldsymbol{u} = \begin{bmatrix} u_1 \\ u_2 \\ \vdots \\ u_p \end{bmatrix} \in \mathscr{U}^p$$

$$
\boldsymbol{A}(t)=\begin{bmatrix} a_{11}(t) & a_{12}(t) & \cdots & a_{1n}(t) \\ a_{21}(t) & a_{22}(t) & \cdots & a_{2n}(t) \\ \vdots & \vdots & & \vdots \\ a_{n1}(t) & a_{n2}(t) & \cdots & a_{nn}(t) \end{bmatrix} \quad \boldsymbol{B}(t)=\begin{bmatrix} b_{11}(t) & b_{12}(t) & \cdots & b_{1p}(t) \\ b_{21}(t) & b_{22}(t) & \cdots & b_{2p}(t) \\ \vdots & \vdots & & \vdots \\ b_{n1}(t) & b_{n2}(t) & \cdots & b_{np}(t) \end{bmatrix}
$$

$$
\boldsymbol{C}(t)=\begin{bmatrix} c_{11}(t) & c_{12}(t) & \cdots & c_{1n}(t) \\ c_{21} & c_{22}(t) & \cdots & c_{2n}(t) \\ \vdots & \vdots & & \vdots \\ c_{q1}(t) & c_{q2}(t) & \cdots & c_{qn}(t) \end{bmatrix} \quad \boldsymbol{D}(t)=\begin{bmatrix} d_{11}(t) & d_{12}(t) & \cdots & d_{1p}(t) \\ d_{21}(t) & d_{22}(t) & \cdots & d_{2p}(t) \\ \vdots & \vdots & & \vdots \\ d_{q1}(t) & d_{q2}(t) & \cdots & d_{qp}(t) \end{bmatrix}
$$

The $\boldsymbol{A}(t)$, $\boldsymbol{B}(t)$, $\boldsymbol{C}(t)$ and $\boldsymbol{D}(t)$ are time-continues matrices on $(-\infty, +\infty)$ and are called parameter matrices of the state-space description of systems or parameter matrices for short. The matrix $\boldsymbol{A}(t)$ reflects many important properties of systems, such as stability. We often call the matrix $\boldsymbol{A}(t)$ the system matrix. $\boldsymbol{B}(t)$ is an $n \times p$ input matrix, $\boldsymbol{C}(t)$ is a $q \times n$ output matrix, $\boldsymbol{D}(t)$ is a $q \times p$ direct transmission matrix. The system eq. (1.2.1) can be abbreviated as $\{\boldsymbol{A}(t), \boldsymbol{B}(t), \boldsymbol{C}(t), \boldsymbol{D}(t)\}$. Eq. (1.2.1a) is called state equation of the system, eq. (1.2.1b) is called output equation of the system. For a linear time-invariant system, four parameter matrices $\{\boldsymbol{A}, \boldsymbol{B}, \boldsymbol{C}, \boldsymbol{D}\}$ are constant. The block diagram of the system is shown in Figure 1.2.4.

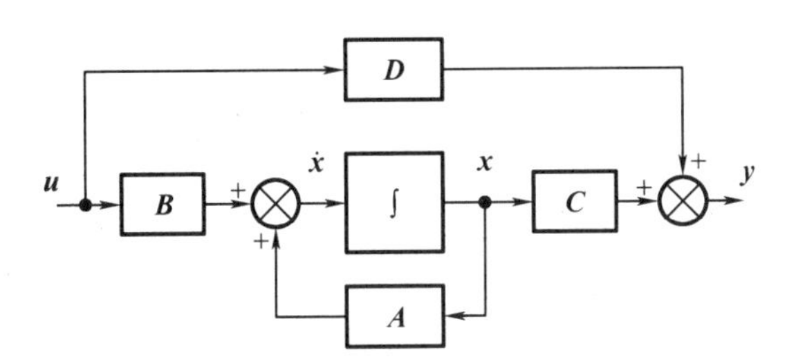

Figure 1.2.4 The block diagram of the system

Next, we give some examples to show how to establish the state-space description of systems.

Example 1.2.2 Consider the system shown in Figure 1.2.5. The input variable is the voltage at both terminals of the voltage source, the output variable is the voltage at both terminals of the capacitor. Establish the state space description of the system.

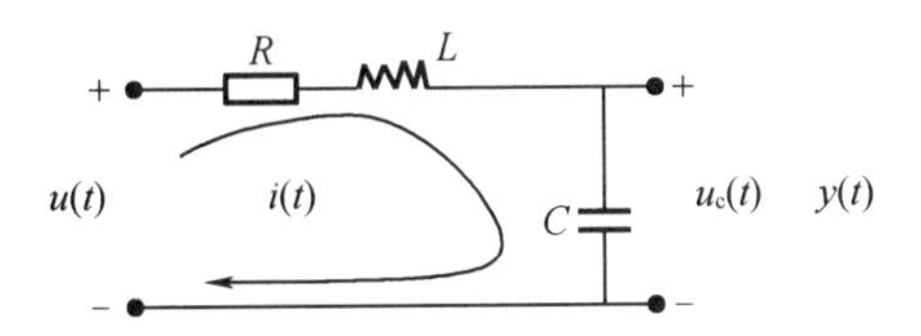

Figure 1.2.5 The figure for example 1.2.2

Answer **Method 1** Chose the state variables as

$$x_1(t)=i(t),\ x_2(t)=u_c(t)$$

Because

$$i(t)=C\frac{\mathrm{d}u_c(t)}{\mathrm{d}t}$$

So,

$$\dot{x}_2=\frac{1}{C}x_1$$

And according to circuit laws, we get

$$Ri(t)+L\frac{\mathrm{d}i(t)}{\mathrm{d}t}+u_c(t)=u(t) \tag{1.2.2}$$

Thus

$$\dot{x}_1=\frac{\mathrm{d}i(t)}{\mathrm{d}t}=-\frac{R}{L}x_1-\frac{1}{L}x_2+\frac{1}{L}u$$

Rewrite above results as matrix equation form, we get the state equation

$$\dot{\boldsymbol{x}}=\begin{bmatrix}-\frac{R}{L} & -\frac{1}{L}\\ \frac{1}{C} & 0\end{bmatrix}\begin{bmatrix}x_1\\ x_2\end{bmatrix}+\begin{bmatrix}\frac{1}{L}\\ 0\end{bmatrix}\boldsymbol{u}$$

The output equation of the system is

$$y=\begin{bmatrix}0 & 1\end{bmatrix}\begin{bmatrix}x_1\\ x_2\end{bmatrix}$$

Method 2 Chose the state variables as

$$\bar{x}_1(t)=u_c(t),\bar{x}_2(t)=\dot{u}_c(t)=\dot{\bar{x}}_1(t)$$

From $i(t)=C\frac{\mathrm{d}u_c(t)}{\mathrm{d}t}$ and eq. (1.2.2), we get

$$RC\frac{\mathrm{d}u_c(t)}{\mathrm{d}t}+LC\frac{\mathrm{d}^2u_c(t)}{\mathrm{d}t^2}+u_c(t)=u(t)$$

$$\dot{\bar{x}}_2=\frac{\mathrm{d}^2u_c(t)}{\mathrm{d}t^2}=-\frac{1}{LC}\bar{x}_1-\frac{R}{L}\bar{x}_2+\frac{1}{LC}u$$

Rewrite above results as matrix equation form, we get the state equation

$$\dot{\bar{\boldsymbol{x}}}=\begin{bmatrix}0 & 1\\ -\frac{1}{LC} & -\frac{R}{L}\end{bmatrix}\begin{bmatrix}\bar{x}_1\\ \bar{x}_2\end{bmatrix}+\begin{bmatrix}0\\ \frac{1}{LC}\end{bmatrix}\boldsymbol{u}$$

The output equation of the system is

$$y=\begin{bmatrix}1 & 0\end{bmatrix}\begin{bmatrix}\bar{x}_1\\ \bar{x}_2\end{bmatrix}$$

Note the relationship between the two sets of state variables is as follow:

$$\bar{x} = \begin{bmatrix} 0 & 1 \\ \frac{1}{C} & 0 \end{bmatrix} x = Px$$

where, the matrix P is invertible.

Example 1.2.3 Figure 1.2.6 shows a DC permanent magnet torque motor circuit. The current flowing through the rotor windings of the motor is $i(t)$, the inductance and resistance of the rotor windings are L and R respectively, and the control voltage applied to both ends of the rotor windings of the motor is $u_a(t)$. Establish the state-space description of the system.

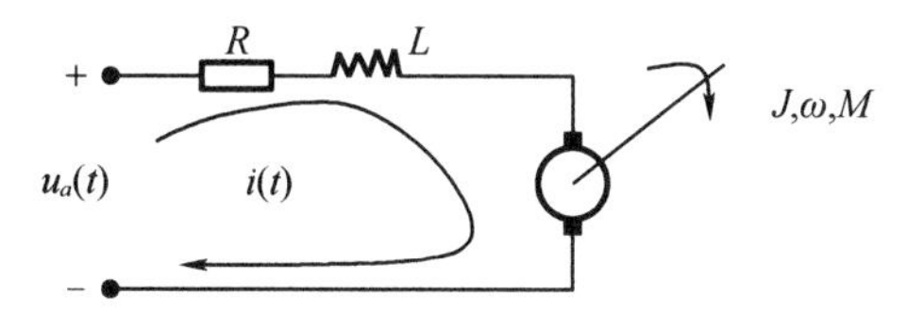

Figure 1.2.6 The figure for example 1.2.3

Answer There are the following equations

$$u_a(t) = i(t)R + L\frac{\mathrm{d}i(t)}{\mathrm{d}t} + C_e\varphi\omega(t) \tag{1.2.3}$$

$$M(t) = k_t i(t) \tag{1.2.4}$$

Where, C_e is the potential coefficient of the motor, φ is main magnetic flux, ω is angular speed of motor rotation, k_t is moment coefficient , $M(t)$ is torque generated by the motor. Assume the moment of inertia of the rotor is J, the load torque is $M_L(t)$, the back-EMF is $c\omega(t)$, we get

$$J\frac{\mathrm{d}\omega(t)}{\mathrm{d}t} = M(t) - M_L(t) - c\omega(t) = k_t i(t) - M_L(t) - c\omega(t) \tag{1.2.5}$$

From eq. (1.2.3), we get

$$\frac{\mathrm{d}i(t)}{\mathrm{d}t} = -\frac{R}{L}i(t) - \frac{C_e\varphi}{L}\omega(t) + \frac{1}{L}u_a(t) \tag{1.2.6}$$

Chose the state vector as

$$x(t) = \begin{bmatrix} x_1(t) \\ x_2(t) \end{bmatrix} = \begin{bmatrix} i(t) \\ \omega(t) \end{bmatrix}$$

Chose input vector as

$$u(t) = \begin{bmatrix} u_1(t) \\ u_2(t) \end{bmatrix} = \begin{bmatrix} u_a(t) \\ M_L(t) \end{bmatrix}$$

Chose output as $y(t) = \omega(t)$. From eq. (1.2.5) and eq. (1.2.6), we get the state equation and the output equation of the system as:

$$\begin{bmatrix} \dot{x}_1(t) \\ \dot{x}_2(t) \end{bmatrix} = \begin{bmatrix} -\frac{R}{L} & -\frac{C_e\varphi}{L} \\ \frac{k_t}{J} & -\frac{c}{J} \end{bmatrix} \begin{bmatrix} x_1(t) \\ x_2(t) \end{bmatrix} + \begin{bmatrix} \frac{1}{L} & 0 \\ 0 & -\frac{1}{J} \end{bmatrix} \begin{bmatrix} u_a(t) \\ M_L(t) \end{bmatrix} \tag{1.2.7}$$

$$y(t)=\begin{bmatrix}0 & 1\end{bmatrix}\begin{bmatrix}x_1(t)\\x_2(t)\end{bmatrix} \tag{1.2.8}$$

Example 1.2.4 Determine the state-space description of the inverted pendulum system as shown in Figure 1.2.7. The car mass is M, the length of the stick is l, the mass of the pendant is m, the angle between the pole and the plumb line is θ. For simplicity, the car and the pendulum are assumed to move in only one plane, and the friction and the mass of the stick, and gust of wind are disregarded. The problem is to maintain the pendulum at the vertical position.

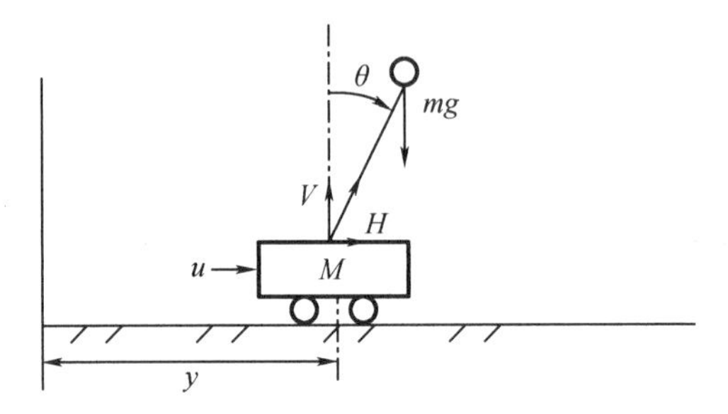

Figure 1.2.7 Car with Inverted pendulum

Answer Let H and V be respectively, the horizontal and vertical forces exerted by the car on the pendulum as shown in the Figure 1.2.7. The application of Newton's second law to the car yields

$$M\frac{\mathrm{d}^2y}{\mathrm{d}t^2}=u-H \tag{1.2.9}$$

The application of Newton's second law to the pendant in horizontal direction yields

$$H=m\frac{\mathrm{d}^2}{\mathrm{d}t^2}(y+l\sin\theta)=m\ddot{y}+ml\cos\theta\cdot\ddot{\theta}-ml\sin\theta\cdot(\dot{\theta})^2 \tag{1.2.10}$$

The application of Newton's second law to the pendant in vertical direction yields

$$mg-V=m\frac{\mathrm{d}^2}{\mathrm{d}t^2}(l\cos\theta)=ml[-\sin\theta\cdot\ddot{\theta}-\cos\theta\cdot(\dot{\theta})^2] \tag{1.2.11}$$

The relationship between horizontal and vertical forces of the car on the pendant is

$$\tan\theta=\frac{H}{V} \tag{1.2.12}$$

From eq. (1.2.12), we get

$$V\sin\theta=H\cos\theta \tag{1.2.13}$$

and

$$Vl\sin\theta=Hl\cos\theta \tag{1.2.14}$$

Assume θ and $\dot{\theta}$ to be small. Under this assumption, we can use the approximation $\sin\theta=\theta$ and $\cos\theta=1$. Dropping the terms with θ^2, $\dot{\theta}^2$ and $\theta\cdot\ddot{\theta}$, from eq. (1.2.11) and eq. (1.2.14), we obtain

$$mg - V = 0 \tag{1.2.15}$$

From eq. (1.2.15), we get

$$mg\sin\theta - V\sin\theta = 0 \tag{1.2.16}$$

From eq. (1.2.16) and eq. (1.2.13), we get

$$mg\sin\theta - H\cos\theta = 0 \tag{1.2.17}$$

Considering $\cos\theta = 1$ and $\sin\theta = \theta$, from eq. (1.2.17), we get

$$H = mg\theta \tag{1.2.18}$$

Substituting eq. (1.2.18) into eq. (1.2.9), yields

$$M\ddot{y} = u - mg\theta \tag{1.2.19}$$

Substituting eq. (1.2.18) into eq. (1.2.10), yields

$$mg\theta = m\ddot{y} + ml\ddot{\theta} \tag{1.2.20}$$

Substituting eq. (1.2.19) into eq. (1.2.20), yields

$$\ddot{\theta} = -\frac{\ddot{y}}{l} + \frac{mg\theta}{ml} = -\frac{1}{l}\left(-\frac{mg}{M}\theta + \frac{u}{M}\right) + \frac{g}{l}\theta = \frac{m+M}{lM}g\theta - \frac{1}{Ml}u \tag{1.2.21}$$

We select state variables as $x_1 = \theta, x_2 = \dot{x}_1 = \dot{\theta}, x_3 = y, x_4 = \dot{x}_3 = \dot{y}$. Then from eq. (1.2.21) and eq. (1.2.19), we obtain the state-space description of the inverted pendulum as

$$\begin{bmatrix} \dot{x}_1 \\ \dot{x}_2 \\ \dot{x}_3 \\ \dot{x}_4 \end{bmatrix} = \begin{bmatrix} 0 & 1 & 0 & 0 \\ \frac{M+m}{Ml}g & 0 & 0 & 0 \\ 0 & 0 & 0 & 1 \\ -\frac{mg}{M} & 0 & 0 & 0 \end{bmatrix} \begin{bmatrix} x_1 \\ x_2 \\ x_3 \\ x_4 \end{bmatrix} + \begin{bmatrix} 0 \\ -\frac{1}{Ml} \\ 0 \\ \frac{1}{M} \end{bmatrix} u \tag{1.2.22}$$

$$y = [0 \quad 0 \quad 1 \quad 0]x$$

Example 1.2.5 Establish the state-space description of mechanical displacement system composed of spring-mass-damper.

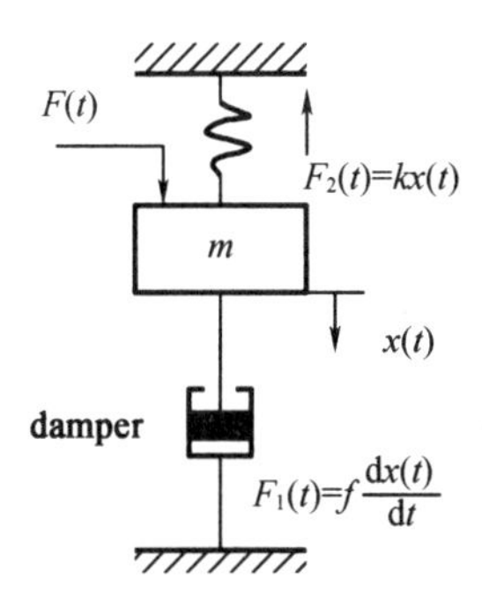

Figure 1.2.8 spring-mass-damper system

Answer Assume the spring is in a natural state of extension at initial moment. In Figure 1.2.8, k is spring constant, $F(t)$ is applied force which direction is downward, $F_1(t)$ is damping force of the damper which direction is upwards, $F_2(t)$ is spring tension which direction is upwards. The displacement of the mass is $x(t)$. By using Newton's law, we get

$$F(t)+mg-F_1(t)-F_2(t)=m\frac{\mathrm{d}^2x(t)}{\mathrm{d}t^2} \tag{1.2.23}$$

$$F(t)+mg-f\frac{\mathrm{d}x(t)}{\mathrm{d}t}-kx(t)=m\frac{\mathrm{d}^2x(t)}{\mathrm{d}t^2} \tag{1.2.24}$$

We select state variables as $x_1=x, x_2=\dot{x}$, and obtain

$$\begin{cases}\dot{x}_1=x_2\\ \dot{x}_2=-\dfrac{k}{m}x_1-\dfrac{f}{m}x_2+\dfrac{F(t)}{m}+g\end{cases} \tag{1.2.25}$$

1.2.4 Existence and uniqueness conditions for solutions to state equations

For any initial state, the analysis of systems is meaningful only if the solution to the state equation of the linear system eq. (1.2.1) exists and is unique. Mathematically, this requires that the coefficient matrix and input action in the equation of state satisfy certain assumptions that are necessary to ensure the existence and uniqueness of the solution to the state equation.

For the LTV system eq. (1.2.1), if all the elements of the system matrices $\boldsymbol{A}(t)$ and $\boldsymbol{B}(t)$ are real-valued continuous functions of t on the time interval $t\in[t_0,t_\alpha]$, and the elements of the input $\boldsymbol{u}(t)$ are real-valued continuous functions on the time interval $t\in[t_0,t_\alpha]$, then the solution $\boldsymbol{x}(t)$ of the state equation exists and is unique.

In general, these conditions are always satisfied for real physical systems. However, these conditionsare sufficient and the constraint is too strong, so there are often reduced to:

(1) The elements of $\boldsymbol{A}(t)$ are absolutely integrable on $[t_0,t_\alpha]$, i. e.

$$\int_{t_0}^{t_\alpha}|a_{ij}(t)|\mathrm{d}t<\infty,\quad i,j=1,2,\cdots,n \tag{1.2.26}$$

(2) For a given input $\boldsymbol{u}(t)$, the elements of $\boldsymbol{B}(t)\boldsymbol{u}(t)$ are absolutely integrable on $[t_0,t_\alpha]$, i. e.

$$\int_{t_0}^{t_\alpha}|\boldsymbol{B}(t)\boldsymbol{u}(t)|\mathrm{d}t<\infty$$

1.2.5 State parameter matrix representation of transfer function matrix

For a LTI system with multiple inputs and multiple outputs, the transfer function matrix is the most basic form to characterize the input-output characteristics of the system. Next, we derive the expression of $G(s)$ by the parameter matrices of the state-space description.

Conclusion For the state space description

$$\begin{cases}\dot{\boldsymbol{x}}=\boldsymbol{A}\boldsymbol{x}+\boldsymbol{B}\boldsymbol{u}, & \boldsymbol{x}(0)=\boldsymbol{x}_0,\ t\geqslant 0\\ \boldsymbol{y}=\boldsymbol{C}\boldsymbol{x}+\boldsymbol{D}\boldsymbol{u}\end{cases} \tag{1.2.27}$$

corresponding transfer matrix is

$$G(s) = C(sI - A)^{-1}B + D \tag{1.2.28}$$

When $D \neq O$, *the rational fraction matrix* $G(s)$ is proper. When $D = 0$, the rational fraction matrix $G(s)$ is strictly proper and

$$\lim_{s \to \infty} G(s) = 0 \tag{1.2.29}$$

Proof Applying the Laplace transform to eq. (1.2.25) yields

$$\begin{cases} sx(s) - x(0) = Ax(s) + Nu(s) \\ y(s) = Cx(s) + Du(s) \end{cases} \tag{1.2.30}$$

Then, from the first equation in eq. (1.2.30), we obtain

$$(sI - A)x(s) = x(0) + Bu(s) \tag{1.2.31}$$

And considering that $(sI - A)$ is always non-singular as a polynomial matrix, eq. (1.2.31) can be rewritten as

$$x(s) = (sI - A)^{-1}x(0) + (sI - A)^{-1}Bu(s) \tag{1.2.32}$$

Substitute eq. (1.2.32) into the second equation of (1.2.30), yields

$$y(s) = C(sI - A)^{-1}x(0) + C(sI - A)^{-1}Bu(s) + Du(s) \tag{1.2.33}$$

If the initial state x_0 is zero, the eq. (1.2.33) is reduced to

$$y(s) = C(sI - A)^{-1}Bu(s) + Du(s)$$

Thus, the transfer matrix of the system is

$$G(s) = C(sI - A)^{-1}B + D \tag{1.2.34}$$

Because

$$(sI - A)^{-1} = \frac{\operatorname{adj}(sI - A)}{\det(sI - A)} \tag{1.2.35}$$

And the degree of each element of $\operatorname{adj}(sI - A)$ is less than the degree of $\det(sI - A)$, so it has to be

$$\lim_{s \to \infty}(sI - A)^{-1} = 0 \tag{1.2.36}$$

Therefore, it can be seen from eq. (1.2.36) and (1.2.34) that when $D \neq 0$, $G(\infty)$ is a non-zero constant matrix, so $G(s)$ given by eq. (1.2.28) is proper. When $D = 0$, $G(\infty)$ is a zero matrix, so $G(s)$ is strictly proper.

1.2.6 Practical calculation method of transfer function matrix $G(s)$

When the order of a system is high, it is inconvenient to directly calculate the transfer function matrix of the system by eq. (1.2.28). The practical algorithm for computing $G(s)$ from $\{A, B, C, D\}$ suitable for computer is given below.

Theorem 1.2.1 Given the parameter matrices $\{A, B, C, D\}$ of the state-space description of a system, computer the characteristic polynomial of the system

$$\alpha(s) \triangleq \det(sI - A) = s^n + a_{n-1}s^{n-1} + \cdots + a_1 s + a_0 \tag{1.2.37}$$

and

$$\begin{cases} E_{n-1} = CB \\ E_{n-2} = CAB + a_{n-1}CB \\ \quad\vdots \\ E_1 = CA^{n-2}B + a_{n-1}CA^{n-3}B + \cdots + a_2CB \\ E_0 = CA^{n-1}B + a_{n-1}CA^{n-2}B + \cdots + a_1CB \end{cases} \tag{1.2.38}$$

Then , the calculation formula of the corresponding transfer function matrix is as follows:

$$G(s) = \frac{1}{\alpha(s)}(E_{n-1}s^{n-1} + E_{n-2}s^{n-2} + \cdots + E_1s + E_0) + D \tag{1.2.39}$$

Proof First compute $(sI - A)^{-1}$ by using

$$(sI - A)^{-1} = \frac{\text{adj}(sI - A)}{\det(sI - A)} = \frac{R(s)}{\alpha(s)} = \frac{1}{\alpha(s)}(R_{n-1}s^{n-1} + R_{n-2}s^{n-2} + \cdots + R_1s + R_0) \tag{1.2.40}$$

where $R_{n-1}, R_{n-2}, \cdots, R_0$ are $n \times n$ matrices. Multiply both sides of eq. (1.2.40) by $\alpha(s)(sI - A)$ form the right side, and get

$$\alpha(s)I = (R_{n-1}s^{n-1} + R_{n-2}s^{n-2} + \cdots + R_1s + R_0)(sI - A) \tag{1.2.41}$$

Substituting eq. (1.2.37) into above equation, yields

$$Is^n + a_{n-1}Is^{n-1} + \cdots + a_1Is + a_0I = R_{n-1}s^n + (R_{n-2} - R_{n-1}A)s^{n-1} + \cdots + (R_0 - R_1A)s - R_0A \tag{1.2.42}$$

The coefficient matrix of $s^i (i = 1, 2, \cdots, n)$ on both sides of the above equation are equal, so

$$\begin{cases} R_{n-1} = I \\ R_{n-2} = R_{n-1}A + a_{n-1}I \\ \quad\vdots \\ R_1 = R_2A + a_2I \\ R_0 = R_1A + a_0I \end{cases} \tag{1.2.43}$$

In eq. (1.2.43), the previous equation is substituted into the next equation in turn, obtain

$$\begin{cases} R_{n-1} = I \\ R_{n-2} = A + a_{n-1}I \\ R_{n-3} = R_{n-2}A + a_{n-2}I = (A + a_{n-1}I)A + a_{n-2}I = A^2 + a_{n-1}A + a_{n-2}I \\ \quad\vdots \\ R_1 = A^{n-2} + a_{n-1}A^{n-3} + \cdots + a_2I \\ R_0 = A^{n-1} + a_{n-1}A^{n-2} + \cdots + a_1I \end{cases} \tag{1.2.44}$$

Therefore, using eq. (1.2.28), eq. (1.2.40) and eq. (1.2.44), we get

$$\begin{aligned} G(s) &= C(sI - A)^{-1}B + D \\ &= \frac{1}{\alpha(s)}(CR_{n-1}Bs^{n-1} + CR_{n-2}Bs^{n-2} + \cdots + CR_1Bs + CR_0B) + D \\ &= \frac{1}{\alpha(s)}[CBs^{n-1} + C(A + a_{n-1}I)Bs^{n-2} + \cdots + C(A^{n-2} + \cdots + a_2I)Bs + \end{aligned}$$

$$
\begin{aligned}
& C(A^{n-1}+\cdots+a_1 I)B]+D \\
&=\frac{1}{\alpha(s)}[CBs^{n-1}+(CA+a_{n-1}CB)s^{n-2}+\cdots+(CA^{n-2}B+\cdots+a_2CB)s+ \\
& (CA^{n-1}B+\cdots+a_1CB)]+D \\
&=\frac{1}{\alpha(s)}(E_{n-1}s^{n-1}+E_{n-2}s^{n-2}+\cdots+E_1 s+E_0)+D
\end{aligned}
$$

In the above discussion, when calculating the transfer matrix by theorem 1. 2. 1, it is assumed that the coefficients $a_0, a_1, \cdots, a_{n-1}$ in the characteristic polynomial (1. 2. 37) of the matrix $\boldsymbol{A}$ have already been calculated, but when the system order is higher, it is often difficult to calculate the characteristic polynomial of the matrix $\boldsymbol{A}$, manually. Below, the algorithm for computing characteristic polynomial is given without proof.

Theorem 1.2.2 An algorithm for computing characteristic polynomials

Givenan $n \times n$ constant matrix $\boldsymbol{A}$, its characteristic polynomial is expressed as

$$\alpha(s)=\det(s\boldsymbol{I}-\boldsymbol{A})=s^n+a_{n-1}s^{n-1}+\cdots+a_1 s+a_0 \tag{1.2.45}$$

Then its coefficient $a_i (i=0,1,\cdots,n-1)$ can be calculated recursively in the following order:

$$
\begin{cases}
\boldsymbol{R}_{n-1}=\boldsymbol{I}, a_{n-1}=-\dfrac{\operatorname{tr}(\boldsymbol{R}_{n-1}\boldsymbol{A})}{1} \\
\boldsymbol{R}_{n-2}=\boldsymbol{R}_{n-1}\boldsymbol{A}+a_{n-1}\boldsymbol{I},\ a_{n-2}=-\dfrac{\operatorname{tr}(\boldsymbol{R}_{n-2}\boldsymbol{A})}{2} \\
\qquad \vdots \\
\boldsymbol{R}_1=\boldsymbol{R}_2\boldsymbol{A}+a_2\boldsymbol{I}, a_1=-\dfrac{\operatorname{tr}(\boldsymbol{R}_1\boldsymbol{A})}{n-1} \\
\boldsymbol{R}_0=\boldsymbol{R}_1\boldsymbol{A}+a_1\boldsymbol{I}, a_0=-\dfrac{\operatorname{tr}(\boldsymbol{R}_0\boldsymbol{A})}{n}
\end{cases}
\tag{1.2.46}
$$

where tr $\boldsymbol{A}$ is trace of the matric $\boldsymbol{A}$, i. e. the sum of the diagonal elements of $\boldsymbol{A}$.

1.3 Conversion from input-output description to state-space description

The problem of determining state-space description from the input-output description is called realization, that is, the input-output description is converted to the state-space description. In this section, we only discuss the realization for single input single output LTI systems.

1.3.1 The state-space description derived from a differential equation or a transfer function

Problem Consider a single input single output LTI system, and let y and u be the output and input variable of the system, respectively. The differential equation of the input and output variable of the system is described as follows:

$$y^{(n)}+a_{n-1}y^{(n-1)}+\cdots+a_1y^{(1)}+a_0y=b_mu^{(m)}+b_{m-1}u^{(m-1)}+\cdots+b_2u^{(2)}+b_1u^{(1)}+b_0u \tag{1.3.1}$$

where $y^{(i)}\triangleq\frac{\mathrm{d}y^i}{\mathrm{d}t^i}$, $u^{(i)}\triangleq\frac{\mathrm{d}u^i}{\mathrm{d}t^i}$, $m\leqslant n$.

On the other hand, the state space description of the single input-single output LTI system has the following form:

$$\begin{cases}\dot{\boldsymbol{x}}=\boldsymbol{A}\boldsymbol{x}+\boldsymbol{b}u\\ y=\boldsymbol{c}\boldsymbol{x}+du\end{cases} \tag{1.3.2}$$

where $\boldsymbol{A}$ is an $n\times n$ matrix, $\boldsymbol{b}$ is an $n\times 1$ vector, $\boldsymbol{c}$ is a $1\times n$ vector and d is a scale. The problem of determining the state-space description eq. (1.3.2) from the input-output description eq. (1.3.1) is summed up as how to select appropriate state variables and determine coefficient matrices $(\boldsymbol{A},\boldsymbol{b},\boldsymbol{c},d)$. Wherein, eq. (1.3.1) and eq. (1.3.2) satisfy $g(s)=\boldsymbol{c}(s\boldsymbol{I}-\boldsymbol{A})^{-1}\boldsymbol{b}+d$. In addition, with the selection of different state vectors, the coefficient matrices in the description of state space will be correspondingly different. Below, we will discuss three typical realization methods.

Method 1 There are two cases are discussed

Case a The simplest situation. When $m=0$, that is, when the right side of eq. (1.3.1) does not contain the derivative term of the input variable, eq. (1.3.1) can be rewritten as

$$y^{(n)}+a_{n-1}y^{(n-1)}+\cdots+a_1y^{(1)}+a_0y=b_0u \tag{1.3.3}$$

The highest derivative of the output variabley with respect to time t on the left-hand side of eq. (1.3.3) is n, so the order of the system is n. The n state variables of the system are selected as:

$$\begin{cases}x_1=y\\ x_2=\dot{x}_1=\dot{y}\\ x_3=\dot{x}_2=y^{(2)}\\ \quad\vdots\\ x_n=\dot{x}_{n-1}=y^{(n-1)}\end{cases} \tag{1.3.4}$$

Substituting eq. (1.3.4) into eq. (1.3.3), yields

$$\dot{x}_n=y^{(n)}=b_0u-a_{n-1}y^{(n-1)}-\cdots-a_1y^{(1)}-a_0y=b_0u-a_{n-1}x_n-\cdots-a_1x_2-a_0x_1 \tag{1.3.5}$$

Denote the state vectorx as $[x_1\quad x_2\quad\cdots\quad x_n]^{\mathrm{T}}$, and write the above equations in the form of matrix equation, namely, the state space description, as

$$\dot{x}=\begin{bmatrix}\dot{x}_1\\ \vdots\\ \dot{x}_{n-1}\\ \dot{x}_n\end{bmatrix}=\left[\begin{array}{c|ccc}0 & 1 & \cdots & 0\\ \vdots & & \ddots & \\ 0 & & & 1\\ \hline -a_0 & -a_1 & \cdots & -a_{n-1}\end{array}\right]\begin{bmatrix}x_1\\ \vdots\\ x_{n-1}\\ x_n\end{bmatrix}+\begin{bmatrix}0\\ \vdots\\ 0\\ b_0\end{bmatrix}u$$

$$y=x_1=[1\quad 0\quad \cdots\quad 0]\begin{bmatrix}x_1\\ x_2\\ \vdots\\ x_n\end{bmatrix}=[1\quad 0\quad \cdots\quad 0]\,\boldsymbol{x} \tag{1.3.6}$$

Case b When $m=n$, or when the right side of eq. (1.3.1) contains the derivative terms of the input variable, eq. (1.3.1) can be rewritten as:

$$y^{(n)}+a_{n-1}y^{(n-1)}+\cdots+a_1y^{(1)}+a_0y=b_nu^{(n)}+b_{n-1}u^{(n-1)}+\cdots+b_2u^{(2)}+b_1u^{(1)}+b_0u \tag{1.3.7}$$

If the state variables are still selected according to the above method, namely

$$\begin{cases}x_1=y\\ x_2=\dot{x}_1=\dot{y}\\ x_3=\dot{x}_2=y^{(2)}\\ \vdots\\ x_n=\dot{x}_{n-1}=y^{(n-1)}\end{cases}$$

We have

$$\begin{cases}\dot{x}_1=x_2\\ \dot{x}_2=x_3\\ \vdots\\ \dot{x}_{n-1}=x_n\\ \dot{x}_n=y^{(n)}=-a_{n-1}x_n-\cdots-a_1x_2-a_0x_1+b_nu^{(n)}+\cdots+b_0u\end{cases}$$

Obviously, the right-hand side of above last equation contains the derivative terms of the input variable which doesn't satisfy the definition of state equations. Therefore, it is not feasible to select the state variables according to the above method.

In order to eliminate the derivative terms of the input variable of state equations, we chose the state variables as follows:

$$\begin{cases}x_1=y-\beta_0u\\ x_2=\dot{y}-\beta_0u^{(1)}-\beta_1u\\ x_3=y^{(2)}-\beta_0u^{(2)}-\beta_1u^{(1)}-\beta_2u\\ \vdots\\ x_n=y^{(n-1)}-\beta_0u^{(n-1)}-\beta_1u^{(n-1)}-\cdots-\beta_{n-2}u^{(1)}-\beta_{n-1}u\end{cases} \tag{1.3.8}$$

From eq. (1.3.8), we obtain

$$\begin{cases} y = x_1 + \beta_0 u \\ \dot{y} = x_2 + \beta_0 u^{(1)} + \beta_1 u \\ y^{(2)} = x_3 + \beta_0 u^{(2)} + \beta_1 u^{(1)} + \beta_2 u \\ \vdots \\ y^{(n-1)} = x_n + \beta_0 u^{(n-1)} + \beta_1 u^{(n-2)} + \cdots + \beta_{n-2} u^{(1)} + \beta_{n-1} u \end{cases} \tag{1.3.9}$$

Take the first-order derivative to time t of both sides of the last equation above, and we get

$$y^{(n)} = \dot{x}_n + \beta_0 u^{(n)} + \beta_1 u^{(n-1)} + \cdots + \beta_{n-2} u^{(2)} + \beta_{n-1} u^{(1)} \tag{1.3.10}$$

Substituting the above equation into eq. (1.3.7), yields

$$\begin{aligned} & \dot{x}_n + \beta_0 u^{(n)} + \beta_1 u^{(n-1)} + \cdots + \beta_{n-2} u^{(2)} + \beta_{n-1} u^{(1)} + \\ & a_{n-1}(x_n + \beta_0 u^{(n-1)} + \beta_1 u^{(n-2)} + \cdots + \beta_{n-2} u^{(1)} + \beta_{n-1} u) + \cdots + \\ & a_1(x_2 + \beta_0 u^{(1)} + \beta_1 u) + a_0(x_1 + \beta_0 u) \\ = {} & b_n u^{(n)} + b_{n-1} u^{(n-1)} + \cdots + b_2 u^{(2)} + b_1 u^{(1)} + b_0 u \end{aligned} \tag{1.3.11}$$

Collating the similar terms in the above equation, yields

$$\begin{aligned} \dot{x}_n + a_{n-1}x_n + \cdots + a_1 x_2 + a_0 x_1 = {} & (b_n - \beta_0) u^{(n)} + (b_{n-1} - \beta_1 - a_{n-1}\beta_0) u^{(n-1)} + \\ & (b_{n-2} - \beta_2 - a_{n-1}\beta_1 - a_{n-2}\beta_0) u^{(n-2)} + \cdots + \\ & (b_1 - \beta_{n-1} - a_{n-1}\beta_{n-2} - a_{n-2}\beta_{n-1} - \cdots - a_1\beta_0) u^{(1)} + \\ & (b_0 - a_{n-1}\beta_{n-1} - a_{n-2}\beta_{n-2} - \cdots - a_1\beta_1 - a_0\beta_0) u \end{aligned} \tag{1.3.12}$$

Let

$$\begin{cases} b_n - \beta_0 = 0 \\ b_{n-1} - \beta_1 - a_{n-1}\beta_0 = 0 \\ b_{n-2} - \beta_2 - a_{n-1}\beta_1 - a_{n-2}\beta_0 = 0 \\ \qquad \vdots \\ b_1 - \beta_{n-1} - a_{n-1}\beta_{n-2} - a_{n-2}\beta_{n-1} - \cdots - a_1\beta_0 = 0 \\ b_0 - a_{n-1}\beta_{n-1} - a_{n-2}\beta_{n-2} - \cdots - a_1\beta_1 - a_0\beta_0 = \beta_n \end{cases} \tag{1.3.13}$$

We get

$$\begin{cases} \beta_0 = b_n \\ \beta_1 = b_{n-1} - a_{n-1}\beta_0 \\ \beta_2 = b_{n-2} - a_{n-1}\beta_1 - a_{n-2}\beta_0 \\ \qquad \vdots \\ \beta_{n-1} = b_1 - a_{n-1}\beta_{n-2} - a_{n-2}\beta_{n-1} - \cdots - a_1\beta_0 \\ \beta_n = b_0 - a_{n-1}\beta_{n-1} - a_{n-2}\beta_{n-2} - \cdots - a_1\beta_1 - a_0\beta_0 \end{cases} \tag{1.3.14}$$

Substituting the above equation into eq. (1.3.12), yields

$$\begin{aligned} & \dot{x}_n + a_{n-1}x_n + \cdots + a_1 x_2 + a_0 x_1 = \beta_n u \\ & \dot{x}_n = -a_{n-1}x_n - \cdots - a_1 x_2 - a_0 x_1 + \beta_n u \end{aligned} \tag{1.3.15}$$

Take the first-order derivative of time t on both sides of each equation in (1.3.8) in turn, and then through simple transformation, we get

$$\begin{cases} \dot{x}_1 = y^{(1)} - \beta_0 u^{(1)} = x_2 + \beta_1 u \\ \dot{x}_2 = y^{(2)} - \beta_0 u^{(2)} - \beta_1 u^{(1)} = x_3 + \beta_2 u \\ \vdots \\ \dot{x}_{n-1} = x_n + \beta_{n-1} u \end{cases} \tag{1.3.16}$$

According to eq. (1.3.15) and eq. (1.3.16), the state equation of the system is

$$\begin{bmatrix} \dot{x}_1 \\ \vdots \\ \dot{x}_{n-1} \\ \dot{x}_n \end{bmatrix} = \left[\begin{array}{c:cccc} 0 & 1 & \cdots & 0 \\ \vdots & & \ddots & \\ 0 & & & 1 \\ \hdashline -a_0 & -a_1 & \cdots & -a_{n-1} \end{array}\right] \begin{bmatrix} x_1 \\ \vdots \\ x_{n-1} \\ x_n \end{bmatrix} + \begin{bmatrix} \beta_1 \\ \vdots \\ \beta_{n-1} \\ \beta_n \end{bmatrix} u \tag{1.3.17}$$

From the first equation of eq. (1.3.9), the output equation of the system is

$$y = x_1 + \beta_0 u = [1 \quad 0 \quad \cdots \quad 0] \begin{bmatrix} x_1 \\ x_2 \\ \vdots \\ x_n \end{bmatrix} + \beta_0 u \tag{1.3.18}$$

Example 1.3.1 The external description of a system is given as

$$y^{(2)} + 3y^{(1)} + 2y = \dot{u} + 3u$$

Find the state space description of the system.

Answer The order of the system n is 2 based on its external description.

According to eq. (1.3.14), we get

$$\begin{cases} \beta_0 = b_2 = 0 \\ \beta_1 = b_1 - a_1\beta_0 = 1 - 3 \times 0 = 1 \\ \beta_2 = b_0 - a_1\beta_1 - a_0\beta_0 = 3 - 3 \times 1 - 2 \times 0 = 0 \end{cases}$$

From eq. (1.3.17) and eq. (1.3.18), the state space of the system is described as

$$\begin{cases} \begin{bmatrix} \dot{x}_1 \\ \dot{x}_2 \end{bmatrix} = \begin{bmatrix} 0 & 1 \\ -2 & -3 \end{bmatrix} \begin{bmatrix} x_1 \\ x_2 \end{bmatrix} + \begin{bmatrix} 1 \\ 0 \end{bmatrix} u \\ y = [1 \quad 0] \begin{bmatrix} x_1 \\ x_2 \end{bmatrix} \end{cases}$$

Method 2 Intermediate variable method

Without loss of generality, let the differential equation between the output variable and the input variable of a system be:

$$y^{(n)} + a_{n-1}y^{(n-1)} + \cdots + a_1 y^{(1)} + a_0 y = b_n u^{(n)} + b_{n-1}u^{(n-1)} + \cdots + b_2 u^{(2)} + b_1 u^{(1)} + b_0 u \tag{1.3.19}$$

Let the initial conditions of the system be 0 and $u(t_0)=0$. Take the Laplace transform on both sides of the above equation, and get

$$s^n y(s)+a_{n-1}s^{n-1}y(s)+\cdots+a_1 sy(s)+a_0 y(s)=b_n s^n u(s)+\cdots+b_1 su(s)+b_0 u(s)$$

Then, the transfer function of the system is

$$g(s)=\frac{y(s)}{u(s)}=\frac{b_n s^n+\cdots+b_1 s+b_0}{s^n+a_{n-1}s^{n-1}+\cdots+a_1 s+a_0} \tag{1.3.20}$$

The above formula can be expressed as

$$u(s)\rightarrow\boxed{\frac{b_n s^n+\cdots+b_1 s+b_0}{s^n+a_{n-1}s^{n-1}+\cdots+a_1 s+a_0}}\rightarrow y(s)$$

Introduce the intermediate variable $z(t)$ whose Laplace transform is $z(s)$. The above formula can be expressed as

$$u(s)\rightarrow\boxed{\frac{1}{s^n+a_{n-1}s^{n-1}+\cdots+a_1 s+a_0}}\xrightarrow{z(s)}\boxed{b_n s^n+\cdots+b_1 s+b_0}\rightarrow y(s)$$

So, there are

$$\frac{z(s)}{u(s)}=\frac{1}{s^n+a_{n-1}s^{n-1}+\cdots+a_1 s+a_0} \tag{1.3.21}$$

$$\frac{y(s)}{z(s)}=b_n s^n+\cdots+b_1 s+b_0 \tag{1.3.22}$$

Then take the inverse Laplace transform on the above two equations, we get

$$z^{(n)}(t)+a_{n-1}z^{(n-1)}(t)+\cdots+a_1 z^{(1)}(t)+a_0 z(t)=u(t) \tag{1.3.23}$$

$$y(t)=b_n z^{(n)}(t)+b_{n-1}z^{(n-1)}(t)+\cdots+b_2 z^{(2)}(t)+b_1 z^{(1)}(t)+b_0 z(t) \tag{1.3.24}$$

For the subsystem described in eq. (1.3.23), state variables are selected according to method 1, i. e.

$$\begin{cases} x_1=z \\ x_2=\dot{x}_1=\dot{z} \\ x_3=\dot{x}_2=z^{(2)} \\ \vdots \\ x_n=\dot{x}_{n-1}=z^{(n-1)} \end{cases} \tag{1.3.25}$$

From eq. (1.3.23) and eq. (1.3.25), we get

$$\dot{x}_n=z^{(n)}(t)=-a_{n-1}x_n-\cdots-a_1 x_2-a_0 x_1+u \tag{1.3.26}$$

Thus, the state equation of the system is

$$\begin{bmatrix} \dot{x}_1 \\ \vdots \\ \dot{x}_{n-1} \\ \dot{x}_n \end{bmatrix}=\left[\begin{array}{c|ccc} 0 & 1 & \cdots & 0 \\ \vdots & & \ddots & \\ 0 & & & 1 \\ \hline -a_0 & -a_1 & \cdots & -a_{n-1} \end{array}\right]\begin{bmatrix} x_1 \\ \vdots \\ x_{n-1} \\ x_n \end{bmatrix}+\begin{bmatrix} 0 \\ \vdots \\ 0 \\ 1 \end{bmatrix}u \tag{1.3.27}$$

from eq. (1.3.24) and eq. (1.3.25), the output equation of the system is

$$y = b_n(-a_0x_1 - a_1x_2 - \cdots - a_{n-1}x_n + u) + [b_0 \quad b_1 \quad \cdots \quad b_{n-1}]\begin{bmatrix} x_1 \\ x_2 \\ \vdots \\ x_n \end{bmatrix}$$

$$= b_n[-a_0 \quad -a_1 \quad \cdots \quad -a_{n-1}]\begin{bmatrix} x_1 \\ x_2 \\ \vdots \\ x_n \end{bmatrix} + [b_0 \quad b_1 \quad \cdots \quad b_{n-1}]\begin{bmatrix} x_1 \\ x_2 \\ \vdots \\ x_n \end{bmatrix} + b_n u \quad (1.3.28)$$

Example 1.3.2 The external description of a system is given as

$$y^{(3)} + 3y^{(2)} + 2y = u^{(3)} + 2\dot{u} + 3u$$

Find the state space description of the system.

Answer According to the intermediate variable method, we have

$$\begin{cases} \begin{bmatrix} \dot{x}_1 \\ \dot{x}_2 \\ \dot{x}_3 \end{bmatrix} = \begin{bmatrix} 0 & 1 & 0 \\ 0 & 0 & 1 \\ -2 & 0 & -3 \end{bmatrix}\begin{bmatrix} x_1 \\ x_2 \\ x_3 \end{bmatrix} + \begin{bmatrix} 0 \\ 0 \\ 1 \end{bmatrix}u \\ y = [-2 \quad 0 \quad -3]\begin{bmatrix} x_1 \\ x_2 \\ x_3 \end{bmatrix} + [3 \quad 2 \quad 0]\begin{bmatrix} x_1 \\ x_2 \\ x_3 \end{bmatrix} + u = [1 \quad 2 \quad -3]\begin{bmatrix} x_1 \\ x_2 \\ x_3 \end{bmatrix} + u \end{cases}$$

Example 1.3.3 The external description of a system is given as

$$y^{(2)} + 3y^{(1)} + 2y = \dot{u} + 3u$$

Find the state space description of the system.

Answer According to the intermediate variable method, we have

$$\begin{cases} \begin{bmatrix} \dot{x}_1 \\ \dot{x}_2 \end{bmatrix} = \begin{bmatrix} 0 & 1 \\ -2 & -3 \end{bmatrix}\begin{bmatrix} x_1 \\ x_2 \end{bmatrix} + \begin{bmatrix} 0 \\ 1 \end{bmatrix}u \\ y = [3 \quad 1]\begin{bmatrix} x_1 \\ x_2 \end{bmatrix} \end{cases}$$

Method 3 Observable canonical form method

The differential equation betweenthe output variable y and the input variable u of a system is given as follows:

$$y^{(n)} + a_{n-1}y^{(n-1)} + \cdots + a_1y^{(1)} + a_0y = b_nu^{(n)} + b_{n-1}u^{(n-1)} + \cdots + b_2u^{(2)} + b_1u^{(1)} + b_0u \quad (1.3.29)$$

Let

$$\begin{cases} x_n(t) = y(t) - b_n u(t) \\ x_{n-1}(t) = y^{(1)}(t) - b_n \dot{u}(t) + a_{n-1}y(t) - b_{n-1}u(t) \\ \vdots \\ x_1(t) = y^{(n-1)}(t) - b_n u^{(n-1)}(t) + a_{n-1}y^{(n-2)}(t) - b_{n-1}u^{(n-2)}(t) + \cdots + a_1 y(t) - b_1 u(t) \end{cases} \tag{1.3.30}$$

Taking simple transformation on the above equation, we get

$$y(t) = x_n(t) + b_n u(t) \tag{1.3.31}$$

and

$$\begin{cases} x_{n-1}(t) = \dot{x}_n(t) + a_{n-1}(x_n(t) + b_n u(t)) - b_{n-1}u(t) \\ x_{n-2}(t) = \dot{x}_{n-1}(t) + a_{n-2}(x_n(t) + b_n u(t)) - b_{n-2}u(t) \\ \vdots \\ x_1(t) = \dot{x}_2(t) + a_1(x_n(t) + b_n u(t)) - b_1 u(t) \end{cases} \tag{1.3.32}$$

Taking simple transformation on eq. (1.3.32), we get

$$\begin{cases} \dot{x}_n(t) = x_{n-1}(t) - a_{n-1}(x_n(t) + b_n u(t)) + b_{n-1}u(t) \\ \dot{x}_{n-1}(t) = x_{n-2}(t) - a_{n-2}(x_n(t) + b_n u(t)) + b_{n-2}u(t) \\ \vdots \\ \dot{x}_2(t) = x_1(t) - a_1(x_n(t) + b_n u(t)) + b_1 u(t) \end{cases} \tag{1.3.33}$$

Taking the first-order derivative of time t on both sides of the last equation of eq. (1.3.30), yields

$$\dot{x}_1(t) = y^{(n)}(t) - b_n u^{(n)}(t) + a_{n-1}y^{(n-1)}(t) - b_{n-1}u^{(n-1)}(t) + \cdots + a_1 y^{(1)}(t) - b_1 u^{(1)}(t) \tag{1.3.34}$$

From eq. (1.3.29), eq. (1.3.30) and eq. (1.3.34), we get

$$\dot{x}_1(t) = -a_0 y(t) + b_0 u(t) = -a_0(x_n(t) + b_n u(t)) + b_0 u(t) \tag{1.3.35}$$

According to eq. (1.3.33), eq. (1.3.35) and eq. (1.3.31), the state-space description of the system can be described as

$$\begin{cases} \begin{bmatrix} \dot{x}_1 \\ \vdots \\ \dot{x}_{n-1} \\ \dot{x}_n \end{bmatrix} = \begin{bmatrix} 0 & 0 & \cdots & 0 & -a_0 \\ 1 & 0 & \cdots & 0 & -a_1 \\ \vdots & \vdots & & \vdots & \vdots \\ 0 & 0 & \cdots & 1 & -a_{n-1} \end{bmatrix} \begin{bmatrix} x_1 \\ \vdots \\ x_{n-1} \\ x_n \end{bmatrix} + \begin{bmatrix} -a_0 b_n + b_0 \\ -a_1 b_n + b_1 \\ \vdots \\ -a_{n-1}b_n + b_{n-1} \end{bmatrix} u \\ y = [0 \quad 0 \quad \cdots \quad 1] \begin{bmatrix} x_1 \\ x_2 \\ \vdots \\ x_n \end{bmatrix} + b_n u \end{cases} \tag{1.3.36}$$

1.3.2 Deriving state space description from block diagram of systems

Block diagrams based on transfer functions are widely used to describe LTI systems. Now we discuss how to derive state space descriptions from block diagrams of systems. By using the block diagram of a LTI system shown in Figure 1.3.1(a), we illustrates the method and steps of deriving state space descriptions from block diagrams.

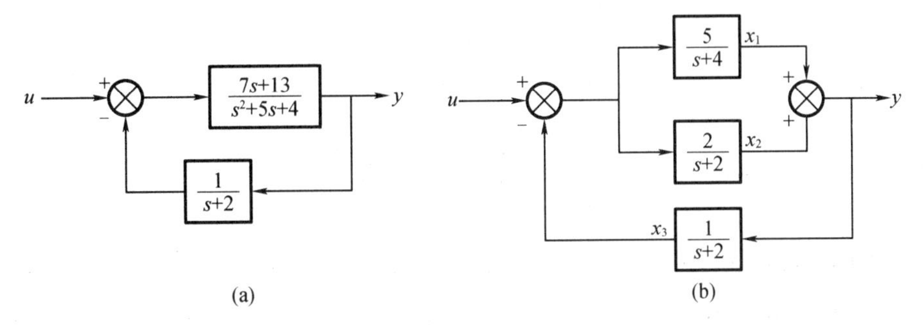

Figure 1.3.1 A block diagram of a SISO LTI system

(1) Transform the given block diagram into the canonical form. A block graph is said to be a canonical form if the transfer functions of its components are only first-order inertia element $(k_i/(s+s_i))$ or proportional component k_{0j}. For the block diagram shown in Figure 1.3.1(a), the transfer function of the second-order inertia element is transformed into the sum of the two first-order inertia elements, i. e.

$$\frac{7s+13}{s^2+5s+4}=\frac{5}{s+4}+\frac{2}{s+1}$$

So we get the canonical form shown in Figure 1.3.1(b).

(2) Specify state variables for canonical block diagrams. The basic principle is that only the output of the first-order inertia element can be chosen as a state variable. For the canonical block diagram shown in Figure 1.3.1(b), there are three first-order inertia element, based on which three state variables x_1, x_2 and x_3 are chosen.

(3) Write the relationship equations between variables. Based on the canonical block diagram, write the corresponding relation equations according to the output and input relation of the first order inertia elements and summation operations. For the canonical block diagram shown in Figure 1.3.1(b), from the output and input relations of three first-order inertia elements and one summation operation, the relational equation set can be easily listed as

$$x_1=\frac{5}{s+4}(u-x_3)$$

$$x_2=\frac{2}{s+1}(u-x_3)$$

$$x_3=\frac{1}{s+2}(x_1+x_2)$$

$$y = x_1 + x_2$$

Where, for simplicity of writing, the Laplace transform of the variable is represented by the the variable itself.

(4) Driving the state variable equation and output variable equation in frequency domain. For the block diagram shown in Figure 1.3.1(b), the state variable equation in the transform domain can be determined as follows by simple deduction of the relation equations derived in step (3).

$$sx_1 = -4x_1 - 5x_3 + 5u$$

$$sx_2 = -x_2 - 2x_3 + 2u$$

$$sx_3 = x_1 + x_2 - 2x_3$$

And output variable equation in transform domain is

$$y = x_1 + x_2$$

(5) Determining the state space description. Firstly, using inverse Laplace transform, sx_i which is in the frequency domain equations is replaced by dx_i/dt, and Laplace transforms of variables are replaced by time domain variables. We get

$$\dot{x}_1 = -4x_1 - 5x_3 + 5u$$

$$\dot{x}_2 = -x_2 - 2x_3 + 2u$$

$$\dot{x}_3 = x_1 + x_2 - 2x_3$$

and

$$y = x_1 + x_2$$

Then rewrite the equations of time domain as matrix form and the state space description corresponding to the block diagram shown in Figure 1.3.1(a) is obtained as

$$\begin{bmatrix} \dot{x}_1 \\ \dot{x}_2 \\ \dot{x}_3 \end{bmatrix} = \begin{bmatrix} -4 & 0 & -5 \\ 0 & -1 & -2 \\ 1 & 1 & -2 \end{bmatrix} \begin{bmatrix} x_1 \\ x_2 \\ x_3 \end{bmatrix} + \begin{bmatrix} 5 \\ 2 \\ 0 \end{bmatrix} u$$

$$y = [1 \quad 1 \quad 0] \begin{bmatrix} x_1 \\ x_2 \\ x_3 \end{bmatrix}$$

1.4 Diagonal canonical form and Jordan canonical form of state equations

We only discuss canonical form of LTI systems in this section. In the analysis and synthesis of linear systems, if the system matrix $\boldsymbol{A}$ in the equation of state has the form of diagonal or block diagonal, it will bring a lot of convenience to solve the problems. The state equation can be transformed into a canonical form represented by eigenvalues by appropriate linear nonsingular transformation. The state equation of systems can be transformed to the diagonal canonical form if the matrix $\boldsymbol{A}$ satisfies a certain condition, otherwise, the equation of state can only be transformed to the Jordan canonical form.

1.4.1 Diagonal canonical form

The state equation of an n-order LTI system is given as

$$\dot{\boldsymbol{x}} = \boldsymbol{A}\boldsymbol{x} + \boldsymbol{B}\boldsymbol{u} \tag{1.4.1}$$

The eigenvalues of the system matrix $\boldsymbol{A}$ are solved from the its characteristic equation

$$\det(\lambda \boldsymbol{I} - \boldsymbol{A}) = 0 \tag{1.4.2}$$

An n-order system has and only has n eigenvalues. For the ith eigenvalue λ_i of the matrix $\boldsymbol{A}$, if there exists a nonzero vector $\boldsymbol{v}_i$ such that

$$(\lambda_i \boldsymbol{I} - \boldsymbol{A})\boldsymbol{v}_i = \boldsymbol{0} \tag{1.4.3}$$

then the $\boldsymbol{v}_i$ is called the eigenvector of $\boldsymbol{A}$ associated with the eigenvalue λ_i. Notice that the eigenvectors are not unique. However, when the n eigenvalues $\lambda_1, \lambda_2, \cdots, \lambda_n$ are distinct, any a group of eigenvectors $\boldsymbol{v}_1, \boldsymbol{v}_2, \cdots, \boldsymbol{v}_n$, must be linearly independent.

1. Conditions for transforming a matrix into the diagonal form

Conclusion 1.4.1 When the n eigenvectors $\boldsymbol{v}_1, \boldsymbol{v}_2, \cdots, \boldsymbol{v}_n$ of the system matrix $\boldsymbol{A}$ are linearly independent, the state eq. (1.4.1) can be transformed into a diagonal canonical form by a linear nonsingular transformation. Obviously, when the n eigenvalues $\lambda_1, \lambda_2, \cdots, \lambda_n$ are distinct, the corresponding n eigenvectors are linearly independent, satisfying the condition that matrix $\boldsymbol{A}$ being transformed into the diagonal form.

2. The method transforming matrix $\boldsymbol{A}$ into the diagonal form

(1) When the matrix $\boldsymbol{A}$ is in general form, the following conclusion is made.

Conclusion 1.4.2 Suppose that the system eq. (1.4.1) satisfies the condition of being converted into the diagonal form, then the state equation must be converted into the following diagonal canonical form under transformation $\bar{\boldsymbol{x}} = \boldsymbol{P}^{-1}\boldsymbol{x}$:

$$\dot{\bar{\boldsymbol{x}}} = \begin{bmatrix} \lambda_1 & & & \\ & \lambda_2 & & \\ & & \ddots & \\ & & & \lambda_n \end{bmatrix} \bar{\boldsymbol{x}} + \bar{\boldsymbol{B}}\boldsymbol{u},\ \bar{\boldsymbol{B}} \triangleq \boldsymbol{P}^{-1}\boldsymbol{B} \tag{1.4.4}$$

where $\lambda_1, \lambda_2, \cdots, \lambda_n$ are n eigenvalues of the system matrix $\boldsymbol{A}$, $\boldsymbol{v}_1, \boldsymbol{v}_2, \cdots, \boldsymbol{v}_n$ are n corresponding eigenvectors, $\boldsymbol{P} = [\boldsymbol{v}_1 \quad \boldsymbol{v}_2 \quad \cdots \quad \boldsymbol{v}_n]$.

Proof The inverse of the matrix $\boldsymbol{P}$ exists because the n eigenvectors $\boldsymbol{v}_1, \boldsymbol{v}_2, \cdots, \boldsymbol{v}_n$ of the matrix $\boldsymbol{A}$ are linearly independent. From $\bar{\boldsymbol{x}} = \boldsymbol{P}^{-1}\boldsymbol{x}$ and eq. (1.4.1), we get

$$\dot{\bar{\boldsymbol{x}}} = \boldsymbol{P}^{-1}\dot{\boldsymbol{x}} = \boldsymbol{P}^{-1}\boldsymbol{A}\boldsymbol{P}\boldsymbol{x} + \boldsymbol{P}^{-1}\boldsymbol{B}\boldsymbol{u} = \bar{\boldsymbol{A}}\,\bar{\boldsymbol{x}} + \bar{\boldsymbol{B}}\boldsymbol{u} \tag{1.4.5}$$

where $\bar{A}=P^{-1}AP$. From $P=[v_1 \quad v_2 \quad \cdots \quad v_n]$ and $Av_i=\lambda_i v_i$, we get

$$
\begin{aligned}
AP &= [Av_1 \quad Av_2 \quad \cdots \quad Av_n] \\
&= [\lambda_1 v_1 \quad \lambda_2 v_2 \quad \cdots \quad \lambda_n v_n] \\
&= [v_1 \quad v_2 \quad \cdots \quad v_n]\begin{bmatrix}\lambda_1 & & & \\ & \lambda_2 & & \\ & & \ddots & \\ & & & \lambda_n\end{bmatrix} \\
&= P\begin{bmatrix}\lambda_1 & & & \\ & \lambda_2 & & \\ & & \ddots & \\ & & & \lambda_n\end{bmatrix}
\end{aligned}
\tag{1.4.6}
$$

Multiplying both sides of eq. (1.4.6) by P^{-1} from left side, yields

$$
\bar{A}=P^{-1}AP=\begin{bmatrix}\lambda_1 & & & \\ & \lambda_2 & & \\ & & \ddots & \\ & & & \lambda_n\end{bmatrix}
\tag{1.4.7}
$$

Substituting eq. (1.4.7) into eq. (1.4.5), we get eq. (1.4.4).

It can be seen from conclusion 1.4.2 that the transformation matrix P, which converts the state equation into the diagonal canonical form, is a square matrix composed of n linearly independent eigenvectors of A.

Meanwhile, it can be seen from eq. (1.4.4) that, in the diagonal canonical form, the state variables are fully decoupled and the state equation is expressed as n independent state variable equations.

Example 1.4.1 The state equation of a LTI system is given as

$$
\dot{x}=\begin{bmatrix}0 & 1 & -1 \\ -6 & -11 & 6 \\ -6 & -11 & 5\end{bmatrix}x+\begin{bmatrix}1 \\ 2 \\ 3\end{bmatrix}u
$$

Determine the diagonal canonical form of the system.

Answer The characteristic equation of the system is $\det(sI-A)=0$, i. e.

$$
\begin{vmatrix}s & -1 & 1 \\ 6 & s+11 & -6 \\ 6 & 11 & s-5\end{vmatrix}=s^3+6s^2+11s+6=(s+1)(s+2)(s+3)=0
$$

Therefore, the three eigenvalues of A are

$$
s_1=-1,\ s_1=-2,\ s_1=-3
$$

The eigenvector p_1 associated with the eigenvalue s_1 is determined by the following formula

$$
(s_1 I-A)p_1=\mathbf{0}
$$

$$\begin{bmatrix} -1 & -1 & 1 \\ 6 & 10 & -6 \\ 6 & 11 & -6 \end{bmatrix} \begin{bmatrix} p_{11} \\ p_{21} \\ p_{31} \end{bmatrix} = \mathbf{0}$$

There are infinite solutions to the above equation, let $p_{11} = 1$, we have

$$\begin{cases} 6 + 10p_{21} - 6p_{31} = 0 \\ 6 + 11p_{21} - 6p_{31} = 0 \end{cases}, \quad \text{get} \quad \begin{cases} p_{21} = 0 \\ p_{31} = 1 \end{cases}$$

So

$$\boldsymbol{p}_1 = \begin{bmatrix} 1 \\ 0 \\ 1 \end{bmatrix}$$

Similarly, using the same method, we can obtain the eigenvectors $\boldsymbol{p}_2$ and $\boldsymbol{p}_3$ associated with the eigenvalues s_2 and s_3, respectively.

$$\boldsymbol{p}_2 = \begin{bmatrix} 1 \\ 2 \\ 4 \end{bmatrix}, \quad \boldsymbol{p}_3 = \begin{bmatrix} 1 \\ 6 \\ 9 \end{bmatrix}$$

So

$$\boldsymbol{P} = [p_1 \quad p_2 \quad p_3] = \begin{bmatrix} 1 & 1 & 1 \\ 0 & 2 & 6 \\ 1 & 4 & 9 \end{bmatrix}, \quad \boldsymbol{P}^{-1} = \begin{bmatrix} 3 & \frac{5}{2} & -2 \\ -3 & -4 & 3 \\ 1 & \frac{3}{2} & -1 \end{bmatrix}$$

$$\boldsymbol{P}^{-1}\boldsymbol{A}\boldsymbol{P} = \begin{bmatrix} -1 & 0 & 0 \\ 0 & -2 & 0 \\ 0 & 0 & -3 \end{bmatrix}, \boldsymbol{P}^{-1}\boldsymbol{B} = \begin{bmatrix} 2 \\ -2 \\ 1 \end{bmatrix}$$

The diagonal form of the system is

$$\dot{\bar{\boldsymbol{x}}} = \begin{bmatrix} -1 & 0 & 0 \\ 0 & -2 & 0 \\ 0 & 0 & -3 \end{bmatrix} \bar{\boldsymbol{x}} + \begin{bmatrix} 2 \\ -2 \\ 1 \end{bmatrix} u$$

(2) When $\boldsymbol{A}$ is a companion matrix, i. e.

$$\boldsymbol{A} = \left[\begin{array}{c|cccc} 0 & 1 & 0 & \cdots & 0 \\ 0 & 0 & 1 & \cdots & 0 \\ \vdots & \vdots & \vdots & & \vdots \\ 0 & 0 & 0 & \cdots & 1 \\ \hline -a_0 & -a_1 & -a_2 & \cdots & -a_{n-1} \end{array}\right] \tag{1.4.8}$$

The characteristic polynomial of the companion matrix $\boldsymbol{A}$ is

$$\det(s\boldsymbol{I} - \boldsymbol{A}) = (s - s_1)(s - s_2)\cdots(s - s_n) = s^n + a_{n-1}s^{n-1} + \cdots + a_1 s + a_0 \tag{1.4.9}$$

When n eigenvalues of the companion matrix $\boldsymbol{A}$ are distinct, it can be transformed to the diagonal canonical form, i. e.

$$P^{-1}AP=\begin{bmatrix} s_1 & 0 & \cdots & 0 \\ 0 & s_2 & \cdots & 0 \\ \vdots & \vdots & & \vdots \\ 0 & 0 & \cdots & s_n \end{bmatrix} \tag{1.4.10}$$

where matrix P is computed by

$$P=\begin{bmatrix} 1 & 1 & \cdots & 1 \\ s_1 & s_2 & \cdots & s_n \\ s_1^2 & s_2^2 & \cdots & s_n^2 \\ \vdots & \vdots & & \vdots \\ s_1^{n-1} & s_2^{n-1} & \cdots & s_n^{n-1} \end{bmatrix} \tag{1.4.11}$$

The matrix P in the eq. (1.4.11) is called the Vandermonde matrix.

Proof Let $P=[p_1 \quad p_2 \quad \cdots \quad p_n]$, where p_i, $i=1, 2, \cdots, n$ is i^{th} column of P. From eq. (1.4.10), we get

$$AP=P\begin{bmatrix} s_1 & 0 & \cdots & 0 \\ 0 & s_2 & \cdots & 0 \\ \vdots & \vdots & & \vdots \\ 0 & 0 & \cdots & s_n \end{bmatrix}$$

$$[Ap_1 \quad Ap_2 \quad \cdots \quad Ap_n]=[s_1 p_1 \quad s_2 p_2 \quad \cdots \quad s_n p_n]$$

So

$$s_i p_i = Ap_i, (s_i I - A)p_i = \mathbf{0}, i=1,2,\cdots,n$$

$$\begin{bmatrix} s_i & -1 & 0 & \cdots & 0 & 0 \\ 0 & s_i & -1 & \cdots & 0 & 0 \\ \vdots & \vdots & \vdots & & \vdots & \vdots \\ 0 & 0 & 0 & \cdots & s_i & -1 \\ a_0 & a_1 & a_2 & \cdots & a_{n-2} & s_i + a_{n-1} \end{bmatrix}\begin{bmatrix} P_{1i} \\ P_{2i} \\ \vdots \\ P_{ni} \end{bmatrix}=\mathbf{0}$$

and

$$\begin{cases} s_i P_{1i} - P_{2i} = 0 \\ s_i P_{2i} - P_{3i} = 0 \\ \quad \vdots \\ s_i P_{n-1,i} - P_{ni} = 0 \end{cases} \tag{1.4.12}$$

Because $\det(s_i I - A) = 0$ and $\operatorname{rank}(s_i I - A) = n-1$, there are infinite solutions $p_i = \begin{bmatrix} P_{1i} \\ P_{2i} \\ \vdots \\ P_{ni} \end{bmatrix}$

witch satisfy eq. (1.4.12). One of the solutions can be obtained as follows:

Let $P_{1i}=1$, from eq. (1.4.12), we get

$$\begin{cases} P_{2i}=s_i \\ P_{3i}=s_i^2 \\ \vdots \\ P_{ni}=s_i^{n-1} \end{cases}$$

So

$$\boldsymbol{p}_i=\begin{bmatrix} 1 \\ s_i \\ \vdots \\ s_i^{n-1} \end{bmatrix}, i=1,2,\cdots,n$$

Example 1.4.2 The state equation of a LTI system is given as

$$\dot{\boldsymbol{x}}=\begin{bmatrix} 0 & 1 & 0 \\ 0 & 0 & 1 \\ -6 & -11 & -6 \end{bmatrix}\boldsymbol{x}+\begin{bmatrix} 1 \\ 2 \\ 3 \end{bmatrix}u$$

Determine the diagonal canonical form of the system.

Answer The matrix $\boldsymbol{A}$ is a companion matrix. The characteristic equation of the system is

$$\det(s\boldsymbol{I}-\boldsymbol{A})=s^3+6s^2+11s+6=(s+1)(s+2)(s+3)$$

So the eigenvalues of $\boldsymbol{A}$ are $s_1=-1$, $s_2=-2$, $s_3=-3$. According to eq. (1.4.11), we get

$$\boldsymbol{P}=\begin{bmatrix} 1 & 1 & 1 \\ -1 & -2 & -3 \\ 1 & 4 & 9 \end{bmatrix}, \boldsymbol{P}^{-1}=\frac{1}{2}\begin{bmatrix} 6 & 5 & 1 \\ -6 & -8 & -2 \\ 2 & 3 & 1 \end{bmatrix}$$

$$\boldsymbol{P}^{-1}\boldsymbol{A}\boldsymbol{P}=\begin{bmatrix} -1 & 0 & 0 \\ 0 & -2 & 0 \\ 0 & 0 & -3 \end{bmatrix}, \boldsymbol{P}^{-1}\boldsymbol{B}=\begin{bmatrix} 2 \\ -2 \\ 1 \end{bmatrix}$$

The diagonal canonical form of the system is

$$\dot{\bar{\boldsymbol{x}}}=\begin{bmatrix} -1 & 0 & 0 \\ 0 & -2 & 0 \\ 0 & 0 & -3 \end{bmatrix}\bar{\boldsymbol{x}}+\begin{bmatrix} 2 \\ -2 \\ 1 \end{bmatrix}u$$

1.4.2 Jordan canonical form

For a matrix $\boldsymbol{A}$ which has repeated eigenvalues, generally speaking, it is not be transformed into the diagonal canonical form by using similarity transformation, but can be transformed to Jordan canonical form. In the following part, the definition of Jordan canonical form and how to transform $\boldsymbol{A}$ into the Jordan canonical form are discussed in detail.

1. Conditions for transforming matrix $\boldsymbol{A}$ to Jordan form

An eigenvalue with multiplicity 2 or higher is called repeated eigenvalues. When an $n \times n$ matrix $\boldsymbol{A}$ has repeated eigenvalues, there are two cases:

Case a Although the matrix $\boldsymbol{A}$ has repeated eigenvalues, but it has n linearly independent eigenvectors, then the matrix $\boldsymbol{A}$ also can be transformed to a diagonal form.

Case b If the number of linearly independent eigenvectors of $\boldsymbol{A}$ is less than system order n, the matrix $\boldsymbol{A}$ only be transformed to a Jordan form. In general, If the matrix $\boldsymbol{A}$ has repeated eigenvalues, the number of linearly independent eigenvectors of $\boldsymbol{A}$ is often less than system order n, the matrix $\boldsymbol{A}$ only be transformed to Jordan form.

2. Jordan canonical form

Here we assume the system eq. (1.4.1) satisfies the condition of being transformed to a Jordan canonical form. The eigenvalues of the system matrix $\boldsymbol{A}$ are λ_1 (multiplicity σ_1), λ_2 (multiplicity σ_2), $\cdots \lambda_l$ (multiplicity σ_l) and $\sigma_1 + \sigma_2 + \cdots + \sigma_l = n$. Then there exists a invertible transformation matrix $\boldsymbol{Q}$. By introducing the transformation $\hat{\boldsymbol{x}} = \boldsymbol{Q}^{-1}\boldsymbol{x}$, the equation of state (1.4.1) is transformed into the Jordan canonical form as follows:

$$\dot{\hat{\boldsymbol{x}}} = \boldsymbol{Q}^{-1}\hat{\boldsymbol{x}} + \boldsymbol{Q}^{-1}\boldsymbol{B}\boldsymbol{u} = \begin{bmatrix} \boldsymbol{J}_1 & & & & \\ & \ddots & & & \\ & & \boldsymbol{J}_i & & \\ & & & \ddots & \\ & & & & \boldsymbol{J}_l \end{bmatrix} \hat{\boldsymbol{x}} + \hat{\boldsymbol{B}}\boldsymbol{u} \tag{1.4.13}$$

where the $\sigma_i \times \sigma_i$ matrix $\boldsymbol{J}_i$ is called the Jordan block corresponding to λ_i. It has following form:

$$\boldsymbol{J}_i = \begin{bmatrix} \boldsymbol{J}_{i1} & & & \\ & \boldsymbol{J}_{i1} & & \\ & & \ddots & \\ & & & \boldsymbol{J}_{i\alpha_i} \end{bmatrix}_{\sigma_i \times \sigma_i} \tag{1.4.14}$$

where the $\sigma_{ik} \times \sigma_{ik}$ matrix $\boldsymbol{J}_{ik}$ is the kth block of $\boldsymbol{J}_i$ corresponding to $\lambda_i (k = 1, 2, \cdots, \alpha_i)$. $\boldsymbol{J}_{ik}$ is also called the kth small Jordan block corresponding to λ_i which has following form:

$$\boldsymbol{J}_{ik} = \begin{bmatrix} \lambda_i & 1 & & & \\ & \lambda_i & 1 & & \\ & & \ddots & \ddots & \\ & & & \lambda_i & 1 \\ & & & & \lambda_i \end{bmatrix}_{\sigma_{ik} \times \sigma_{ik}} \tag{1.4.15}$$

and

$$\sigma_{i1} + \sigma_{i2} + \cdots + \sigma_{i\alpha_i} = \sigma_i \tag{1.4.16}$$

3. Algebraic multiplicity and geometric multiplicity of eigenvalues

(1) Algebraic multiplicity

λ_i is an eigenvalue of $\boldsymbol{A}$ and it satisfies

$$\begin{cases}\det(\lambda \boldsymbol{I}-\boldsymbol{A})=(\lambda-\lambda_i)^{\sigma_i}\beta_i(\lambda)\\ \beta_i(\lambda_i)\neq 0\end{cases} \tag{1.4.17}$$

Then σ_i is called algebraic multiplicity of λ_i. Obviously, the algebraic multiplicity of λ_i is the number of λ_i and is also the sum of the orders of all small Jordan blocks belonging to λ_i.

(2) Geometric multiplicity

The geometric multiplicity α_i of λ_i can be computed as

$$\alpha_i = n-\operatorname{rank}(\lambda_i \boldsymbol{I}-\boldsymbol{A}) \tag{1.4.18}$$

It is the number of small Jordan blocks corresponding to eigenvalue λ_i and also the number of linearly independent eigenvectors of λ_i. If and only if the algebraic multiplicity is equal to its geometric multiplicity for all eigenvalues, the matrix $\boldsymbol{A}$ can be transformed into a diagonal form.

4. Generalized eigenvector

A non-zero vector $\boldsymbol{v}_i$, if it satisfy

$$\begin{cases}(\boldsymbol{A}-\lambda_i \boldsymbol{I})^k \boldsymbol{v}_i=\boldsymbol{0}\\ (\boldsymbol{A}-\lambda_i \boldsymbol{I})^{k-1}\boldsymbol{v}_i\neq\boldsymbol{0}\end{cases} \tag{1.4.19}$$

$\boldsymbol{v}_i$ is called a generalized vector of grade k with λ_i. The generalized vectors have the three basic properties.

Property 1 $\boldsymbol{v}_i$ is a generalized vector of grade k associated with λ_i of $\boldsymbol{A}$. We define

$$\begin{cases}\boldsymbol{v}_i^{(k)} \triangleq \boldsymbol{v}_i\\ \boldsymbol{v}_i^{(k-1)} \triangleq (\boldsymbol{A}-\lambda_i \boldsymbol{I})\boldsymbol{v}_i\\ \vdots\\ \boldsymbol{v}\boldsymbol{v}_i^{(1)} \triangleq (\boldsymbol{A}-\lambda_i \boldsymbol{I})^{k-1}\boldsymbol{v}_i\end{cases} \tag{1.4.20}$$

$\boldsymbol{v}_i^{(1)},\boldsymbol{v}_i^{(2)},\cdots,\boldsymbol{v}_i^{(k)}$ are called a chain of generalized eigenvectors of length k and they are linearly independent.

Proof We just need to prove that all the constants $\beta_1,\beta_2,\cdots,\beta_k$ that make

$$\beta_1\boldsymbol{v}_i^{(1)}+\beta_2\boldsymbol{v}_i^{(2)}+\cdots+\beta_{k-1}\boldsymbol{v}_i^{(k-1)}+\beta_k\boldsymbol{v}_i^{(k)}=\boldsymbol{0} \tag{1.4.21}$$

true are zero. In order do it, from eq. (1.4.20), we have

$$\begin{cases}(\lambda_i \boldsymbol{I}-\boldsymbol{A})^{k-1}\boldsymbol{v}_i^{(k-1)}=(\lambda_i \boldsymbol{I}-\boldsymbol{A})^k\boldsymbol{v}_i=\boldsymbol{0}\\ (\lambda_i \boldsymbol{I}-\boldsymbol{A})^{k-1}\boldsymbol{v}_i^{(k-2)}=(\lambda_i \boldsymbol{I}-\boldsymbol{A})(\lambda_i \boldsymbol{I}-\boldsymbol{A})^k\boldsymbol{v}_i=\boldsymbol{0}\\ \vdots\\ (\lambda_i \boldsymbol{I}-\boldsymbol{A})^{k-1}\boldsymbol{v}_i^{(1)}=(\lambda_i \boldsymbol{I}-\boldsymbol{A})^{k-2}(\lambda_i \boldsymbol{I}-\boldsymbol{A})^k\boldsymbol{v}_i=\boldsymbol{0}\end{cases} \tag{1.4.22}$$

Multiply both sides of eq. (1.4.21) by $(\lambda_i \boldsymbol{I}-\boldsymbol{A})^{k-1}$, and from eq. (1.4.22), we get

$$\beta_k(\lambda_i \boldsymbol{I}-\boldsymbol{A})^{k-1}\boldsymbol{v}_i^{(k)}=\boldsymbol{0}$$

From eq. (1.4.20), we get $\beta_k(\lambda_i \boldsymbol{I}-\boldsymbol{A})^{k-1}\boldsymbol{v}_i=\beta_k(\lambda_i \boldsymbol{I}-\boldsymbol{A})^{k-1}\boldsymbol{v}_i^{(k)}=\boldsymbol{0}$. Because $(\lambda_i \boldsymbol{I}-$

$\boldsymbol{A})^{k-1}\boldsymbol{v}_i \neq \boldsymbol{0}$, so we get

$$\beta_k = 0$$

In the similar way, multiply both sides of eq. (1.4.21) by $(\lambda_i \boldsymbol{I} - \boldsymbol{A})^{k-2}$, we get $\beta_{k-1} = 0$. Like this, we can get $\beta_1 = \beta_2 = \cdots = \beta_{k-2} = 0$.

Property 2 We use the following example to show property 2. Assume λ_i is an eigenvalue of $\boldsymbol{A}_{n\times n}$ with multiplicity σ_i. Compute

$$\gamma_m = n - \mathrm{rank}(\lambda_i \boldsymbol{I} - \boldsymbol{A})^m, \quad m = 0, 1, 2, \cdots \tag{1.4.23}$$

until $m = m_0$ and $\gamma_{m0} = \sigma_i$. Generate generalized chains as following method. Where we assume $n = 10$, $\sigma_i = 8$, $m_0 = 4$, $\gamma_0 = 0$, $\gamma_1 = 3$, $\gamma_2 = 6$, $\gamma_3 = 7$, $\gamma_4 = 8$. Establish Table 1.4.1 as following:

Table 1.4.1 Relevant data

$\gamma_4 - \gamma_3 = 1$	$\gamma_3 - \gamma_2 = 1$	$\gamma_2 - \gamma_1 = 3$	$\gamma_1 - \gamma_0 = 3$
		$\boldsymbol{v}_{i3}^{(2)} \triangleq \boldsymbol{v}_{i3}$	$\boldsymbol{v}_{i3}^{(1)} \triangleq (\boldsymbol{A} - \lambda_i \boldsymbol{I})\boldsymbol{v}_{i3}$
		$\boldsymbol{v}_{i2}^{(2)} \triangleq \boldsymbol{v}_{i2}$	$\boldsymbol{v}_{i2}^{(1)} \triangleq (\boldsymbol{A} - \lambda_i \boldsymbol{I})\boldsymbol{v}_{i2}$
$\boldsymbol{v}_{i1}^{(4)} \triangleq \boldsymbol{v}_{i1}$	$\boldsymbol{v}_{i1}^{(3)} \triangleq (\boldsymbol{A} - \lambda_i \boldsymbol{I})\boldsymbol{v}_{i1}$	$\boldsymbol{v}_{i1}^{(2)} \triangleq (\boldsymbol{A} - \lambda_i \boldsymbol{I})^2\boldsymbol{v}_{i1}$	$\boldsymbol{v}_{i1}^{(1)} \triangleq (\boldsymbol{A} - \lambda_i \boldsymbol{I})^3\boldsymbol{v}_{i1}$

Where, $\boldsymbol{v}_{i1}$ is a non-zero column vector which satisfies

$$(\boldsymbol{A} - \lambda_i \boldsymbol{I})^4 \boldsymbol{v}_{i1} = \boldsymbol{0} \quad \text{and} \quad (\boldsymbol{A} - \lambda_i \boldsymbol{I})^3 \boldsymbol{v}_{i1} \neq \boldsymbol{0}$$

$\boldsymbol{v}_{i2}, \boldsymbol{v}_{i3}$ are non-zero column vector which satisfy

$$\begin{cases} \{\boldsymbol{v}_{i1}^{(2)}, \boldsymbol{v}_{i2}, \boldsymbol{v}_{i3}\} \quad \text{linearly independent} \\ (\boldsymbol{A} - \lambda_i \boldsymbol{I})^2 \boldsymbol{v}_{i2} = \boldsymbol{0} \quad \text{and} \quad (\boldsymbol{A} - \lambda_i \boldsymbol{I}) \boldsymbol{v}_{i2} \neq \boldsymbol{0} \\ (\boldsymbol{A} - \lambda_i \boldsymbol{I})^2 \boldsymbol{v}_{i3} = \boldsymbol{0} \quad \text{and} \quad (\boldsymbol{A} - \lambda_i \boldsymbol{I}) \boldsymbol{v}_{i3} \neq \boldsymbol{0} \end{cases}$$

So, we get σ_i generalized eigenvectors

$$\{\boldsymbol{v}_{i1}^{(k)}, k = 1, 2, 3, 4, \ \boldsymbol{v}_{i2}^{(j)}, \boldsymbol{v}_{i3}^{(j)}, j = 1, 2\}$$

shown in Table 1.4.1, which must be linearly independent. And generalize to a more general case, we get σ_i generalized eigenvectors

$$\{\boldsymbol{v}_{i1}^{(k)}, k = 1, 2, \cdots, r_{i1};\ \boldsymbol{v}_{i2}^{(j)}, j = 1, 2, \cdots, r_{i2};\boldsymbol{v}_{i\alpha_i}^{(\beta)}, \beta = 1, 2, \cdots, r_{\alpha_i}\}$$

which must be linearly independent, where $r_{i1} \geqslant r_{i2} \geqslant \cdots \geqslant r_{i\alpha_i}$ and $r_{i1} + r_{i2} + \cdots + r_{i\alpha_i} = \sigma_i$.

Property 3 The generalized eigenvectors of $\boldsymbol{A}$ belonging to different eigenvalues must be linearly independent.

5. The construction of the transformation matrix

The transformation matrix $\boldsymbol{Q}$, which transforms the state equation into the Jordan canonical form, can be computed as follows:

$$\boldsymbol{Q} = [\boldsymbol{Q}_1 \vdots \cdots \vdots \boldsymbol{Q}_l]_{n\times n} \tag{1.4.24}$$

where

$$\boldsymbol{Q}_i = [\boldsymbol{Q}_{i1} \vdots \boldsymbol{Q}_{i2} \vdots \cdots \vdots \boldsymbol{Q}_{i\alpha_i}]_{n\times\sigma_i} \tag{1.4.25}$$

and

$$\boldsymbol{Q}_{ik} = [\boldsymbol{v}_{ik}^{(1)} \quad \boldsymbol{v}_{ik}^{(2)} \quad \cdots \quad \boldsymbol{v}_{ik}^{(r_{ik})}]_{n\times r_{ik}} \tag{1.4.26}$$

$$i = 1,2,\cdots,l; \quad k = 1,2,\cdots,\alpha_i$$

Using transformation $\hat{\boldsymbol{x}} = \boldsymbol{Q}^{-1}\boldsymbol{x}$, we can obtain the Jordan canonical form eq. (1.4.13)—eq. (1.4.16).

We can use the MATLAB function $[q, d]$ = jordan($\boldsymbol{A}$) to get the Jordan form of $\boldsymbol{A}$, where q is the transformation matrix and d is the result.

Example 1.4.3 Given a state equation of a LTI system

$$\dot{\boldsymbol{x}} = \boldsymbol{A}\boldsymbol{x} + \boldsymbol{B}\boldsymbol{u}$$

where

$$\boldsymbol{A} = \begin{bmatrix} 3 & -1 & 1 & 1 & 0 & 0 \\ 1 & 1 & -1 & -1 & 0 & 0 \\ 0 & 0 & 2 & 0 & 1 & 1 \\ 0 & 0 & 0 & 2 & -1 & -1 \\ 0 & 0 & 0 & 0 & 1 & 1 \\ 0 & 0 & 0 & 0 & 1 & 1 \end{bmatrix}, \boldsymbol{B} = \begin{bmatrix} 1 & 0 \\ -1 & 1 \\ 2 & 1 \\ 0 & -1 \\ 0 & 2 \\ 1 & 0 \end{bmatrix}$$

Compute the Jordan canonical form of the system.

Answer

(1) Compute the eigenvalues of $\boldsymbol{A}$:

$$\det(\lambda \boldsymbol{I} - \boldsymbol{A}) = (\lambda - 2)^5\lambda, \lambda_1 = 2(\sigma_1 = 5) \text{ and } \lambda_2 = 0(\sigma_2 = 1)$$

(2) For $\lambda_1 = 2$, compute $\gamma_m = n\text{-rank}(\lambda_1 \boldsymbol{I} - \boldsymbol{A})^m$, $m = 0,1,2,\cdots$

$$(2\boldsymbol{I} - \boldsymbol{A})^0 = \boldsymbol{I}, \gamma_0 = 0$$

$$(2\boldsymbol{I} - \boldsymbol{A})^1 = \begin{bmatrix} -1 & 1 & -1 & -1 & 0 & 0 \\ -1 & 1 & 1 & 1 & 0 & 0 \\ 0 & 0 & 0 & 0 & -1 & -1 \\ 0 & 0 & 0 & 0 & 1 & 1 \\ 0 & 0 & 0 & 0 & 1 & -1 \\ 0 & 0 & 0 & 0 & -1 & 1 \end{bmatrix}, \gamma_1 = 2$$

$$(2\boldsymbol{I} - \boldsymbol{A})^2 = \begin{bmatrix} 0 & 0 & 2 & 2 & 0 & 0 \\ 0 & 0 & 2 & 2 & 0 & 0 \\ 0 & 0 & 0 & 0 & 0 & 0 \\ 0 & 0 & 0 & 0 & 0 & 0 \\ 0 & 0 & 0 & 0 & 2 & -2 \\ 0 & 0 & 0 & 0 & -2 & 2 \end{bmatrix}, \gamma_2 = 4$$

$$(2\boldsymbol{I}-\boldsymbol{A})^3=\begin{bmatrix}0&0&0&0&0&0\\0&0&0&0&0&0\\0&0&0&0&0&0\\0&0&0&0&0&0\\0&0&0&0&4&-4\\0&0&0&0&-4&4\end{bmatrix},\gamma_3=5$$

Until $\gamma_3=\sigma_3=5$.

(3) Determine the 5 generalized eigenvectors of belonging to $\lambda_1=2$.

First construct Table 1.4.2.

Table 1.4.2 Relevant data

$\gamma_3-\gamma_2=1$	$\gamma_2-\gamma_1=2$	$\gamma_1-\gamma_0=2$
	$\boldsymbol{v}_{12}^{(2)}\triangleq\boldsymbol{v}_{12}$	$\boldsymbol{v}_{12}^{(1)}\triangleq(\boldsymbol{A}-2\boldsymbol{I})\boldsymbol{v}_{12}$
$\boldsymbol{v}_{11}^{(3)}\triangleq\boldsymbol{v}_{11}$	$\boldsymbol{v}_{11}^{(2)}\triangleq(\boldsymbol{A}-2\boldsymbol{I})\boldsymbol{v}_{11}$	$\boldsymbol{v}_{11}^{(1)}\triangleq(\boldsymbol{A}-2\boldsymbol{I})^2\boldsymbol{v}_{11}$

Because

$$(\boldsymbol{A}-2\boldsymbol{I})^3\boldsymbol{v}_{11}=\text{ and }(\boldsymbol{A}-2\boldsymbol{I})^2\boldsymbol{v}_{11}\neq\boldsymbol{0}$$

We get $\boldsymbol{v}_{11}=[0\ \ 0\ \ 1\ \ 0\ \ 0\ \ 0]^{\mathrm{T}}$, and

$$\boldsymbol{v}_{11}^{(1)}=(\boldsymbol{A}-2\boldsymbol{I})^2\boldsymbol{v}_{11}=\begin{bmatrix}2\\2\\0\\0\\0\\0\end{bmatrix},\boldsymbol{v}_{11}^{(2)}=(\boldsymbol{A}-2\boldsymbol{I})\boldsymbol{v}_{11}=\begin{bmatrix}1\\-1\\0\\0\\0\\0\end{bmatrix},\boldsymbol{v}_{11}^{(3)}=\boldsymbol{v}_{11}=\begin{bmatrix}0\\0\\1\\0\\0\\0\end{bmatrix}$$

Then from

$$\begin{cases}\{\boldsymbol{v}_{11}^{(2)},\ \boldsymbol{v}_{12}\}\quad\text{linearly independent}\\(\boldsymbol{A}-2\boldsymbol{I})^2\boldsymbol{v}_{12}=\boldsymbol{0}\quad\text{and}\quad(\boldsymbol{A}-2\boldsymbol{I})\boldsymbol{v}_{12}\neq\boldsymbol{0}\end{cases}$$

we get $\boldsymbol{v}_{12}=[0\ \ 0\ \ 1\ \ -1\ \ 1\ \ 1]^{\mathrm{T}}$, and

$$\boldsymbol{v}_{12}^{(1)}=(\boldsymbol{A}-2\boldsymbol{I})\boldsymbol{v}_{12}=\begin{bmatrix}0\\0\\2\\-2\\0\\0\end{bmatrix},\boldsymbol{v}_{12}^{(2)}=\boldsymbol{v}_{12}=\begin{bmatrix}0\\0\\1\\-1\\1\\1\end{bmatrix}$$

(4) Determine the eigenvector $\boldsymbol{v}_2$ of $\boldsymbol{A}$ belonging to $\lambda_2=0$.

From $(\boldsymbol{A}-\lambda_2\boldsymbol{I})\boldsymbol{v}_2=\boldsymbol{0}$ we get $\boldsymbol{v}_2=[0\ \ 0\ \ 0\ \ 0\ \ 1\ \ -1]^{\mathrm{T}}$。

(5) Compose the transformation matrix $\boldsymbol{Q}$:

$$\boldsymbol{Q}=[\boldsymbol{v}_{11}^{(1)}\quad \boldsymbol{v}_{11}^{(2)}\quad \boldsymbol{v}_{11}^{(3)}\quad \boldsymbol{v}_{12}^{(1)}\quad \boldsymbol{v}_{12}^{(2)}\quad \boldsymbol{v}_{2}]$$

$$=\begin{bmatrix} 2 & -1 & 0 & 0 & 0 & 0 \\ 2 & 1 & 0 & 0 & 0 & 0 \\ 0 & 0 & 1 & 2 & 1 & 0 \\ 0 & 0 & 0 & -2 & -1 & 0 \\ 0 & 0 & 0 & 0 & 1 & 1 \\ 0 & 0 & 0 & 0 & 1 & -1 \end{bmatrix}$$

The inverse of $\boldsymbol{Q}$ is

$$\boldsymbol{Q}^{-1}=\begin{bmatrix} \frac{1}{4} & \frac{1}{4} & 0 & 0 & 0 & 0 \\ \frac{1}{2} & -\frac{1}{2} & 0 & 0 & 0 & 0 \\ 0 & 0 & 1 & 1 & 0 & 0 \\ 0 & 0 & 0 & -\frac{1}{2} & -\frac{1}{4} & -\frac{1}{4} \\ 0 & 0 & 0 & 0 & \frac{1}{2} & \frac{1}{2} \\ 0 & 0 & 0 & 0 & \frac{1}{2} & -\frac{1}{2} \end{bmatrix}$$

(6) The Jordan canonical form of the state equation is

$$\dot{\hat{\boldsymbol{x}}}=\boldsymbol{Q}^{-1}\boldsymbol{AQx}+\boldsymbol{Q}^{-1}\boldsymbol{Bu}$$

$$\dot{\hat{\boldsymbol{x}}}=\left[\begin{array}{ccc:cc:c} 2 & 1 & 0 & 0 & 0 & 0 \\ 0 & 2 & 1 & 0 & 0 & 0 \\ 0 & 0 & 2 & 0 & 0 & 0 \\ \hdashline 0 & 0 & 0 & 2 & 1 & 0 \\ 0 & 0 & 0 & 0 & 2 & 0 \\ \hdashline 0 & 0 & 0 & 0 & 0 & 0 \end{array}\right]\hat{\boldsymbol{x}}+\begin{bmatrix} 0 & \frac{1}{4} \\ 1 & -\frac{1}{2} \\ 2 & 0 \\ -\frac{1}{4} & 0 \\ \frac{1}{2} & 1 \\ -\frac{1}{2} & 1 \end{bmatrix}\boldsymbol{u}$$

Example 1.4.4 Compute the eigenvalues and the determinant of $\boldsymbol{A}$ in eq. (1.4.13).

Answer Because $\boldsymbol{Q}^{-1}\boldsymbol{AQ}=\begin{bmatrix} \boldsymbol{J}_1 & & & & \\ & \ddots & & & \\ & & \boldsymbol{J}_i & & \\ & & & \ddots & \\ & & & & \boldsymbol{J}_l \end{bmatrix}$, so

$$\det \boldsymbol{A} = \det \boldsymbol{Q} \det \begin{bmatrix} \boldsymbol{J}_1 & & & & \\ & \ddots & & & \\ & & \boldsymbol{J}_i & & \\ & & & \ddots & \\ & & & & \boldsymbol{J}_l \end{bmatrix} \det \boldsymbol{Q}^{-1} = \det \begin{bmatrix} \boldsymbol{J}_1 & & & & \\ & \ddots & & & \\ & & \boldsymbol{J}_i & & \\ & & & \ddots & \\ & & & & \boldsymbol{J}_l \end{bmatrix}$$

Jordan-form matrices are triangular and block diagonal, so the determinant of $\boldsymbol{A}$ is the production of all diagonal entries of $\boldsymbol{J}$, and we have

$$\det \boldsymbol{A} = \det \begin{bmatrix} \boldsymbol{J}_1 & & & & \\ & \ddots & & & \\ & & \boldsymbol{J}_i & & \\ & & & \ddots & \\ & & & & \boldsymbol{J}_l \end{bmatrix} = \prod_{i=1}^{l} \lambda_i^{\sigma_i} \tag{1.4.27}$$

where $\lambda_i, i = 1, 2, \cdots, l$, is the element on the diagonal line of Jordan form of $\boldsymbol{A}$, and is also eigenvalue of $\boldsymbol{A}$. eq. (1.4.27) indicates the determinant of $\boldsymbol{A}$ is the production of all eigenvalues of $\boldsymbol{A}$.

1.5 Similarity transformation of linear systems

The similarity transformation is to transform the representation of a system in one coordinate system in state space into a representation in another coordinate system. It is also called coordinate transformation. As we know from the previous discussion, the selection of state variables is not unique. And in mathematics, state vectors has different representations for different coordinate systems in the state space. The relation between these different representations can be established by coordinate transformation. These different representations are equivalent to describing the dynamic behavior of systems. Transforming a state equation into a diagonal canonical form or a Jordan canonical form is a coordinate transformation. Using coordinate transformation can highlight the system's characteristics in certain aspects, and simplify the analysis and calculation of problems. Therefore, coordinate transformation is widely used in the analysis and design of systems.

1.5.1 Similarity transformation

1. Basis

The set $\boldsymbol{B}$ of vectors in a linear space $\boldsymbol{V}$ is called a basis if every element of $\boldsymbol{V}$ may be written as a linear combination of the elements in $\boldsymbol{B}$. Equivalently, a set $\boldsymbol{B}$ is a basis if its elements are linearly independent and every element of $\boldsymbol{V}$ is a linear combination of the elements in $\boldsymbol{B}$. The coefficients of this linear combination are referred to as components or coordinates of the vector with respect to $\boldsymbol{B}$. The elements of a basis are called basis vectors. A basis determines a coordinate system in a linear space, and different bases determine different coordinate systems.

2. Coordinate transformation

Coordinate transformation is taking a representation of the system on one basis of the state space and turning it into a representation on another basis. The essence of coordinate transformation is to change basis, that is, to change one basis for another in the state space.

Theorem 1.5.1 In an n-dimensional vector space, any n linearly independent vectors can serve as a basis.

Theorem 1.5.2 Coordinate transformation is a linear nonsingular transformation.

Proof In an n-dimensional state space, the representation of a state on the basis $\{\boldsymbol{e}_1, \boldsymbol{e}_2, \cdots, \boldsymbol{e}_n\}$ is

$$\boldsymbol{x} = [x_1 \quad x_2 \quad \cdots \quad x_n]^{\mathrm{T}} \tag{1.5.1}$$

And the representation of the state on the basis $\{\bar{\boldsymbol{e}}_1 \quad \bar{\boldsymbol{e}}_2 \quad \cdots \quad \bar{\boldsymbol{e}}_n\}$ is

$$\bar{\boldsymbol{x}} = [\bar{x}_1 \quad \bar{x}_2 \quad \cdots \quad \bar{x}_n]^{\mathrm{T}} \tag{1.5.2}$$

Now we determine the relationship between $\boldsymbol{x}$ and $\bar{\boldsymbol{x}}$.

Because $\{\bar{\boldsymbol{e}}_1 \quad \bar{\boldsymbol{e}}_2 \quad \cdots \quad \bar{\boldsymbol{e}}_n\}$ are the basis vectors, they are linearly independent and

$$\begin{cases} \boldsymbol{e}_1 = p_{11}\bar{\boldsymbol{e}}_1 + p_{21}\bar{\boldsymbol{e}}_2 + \cdots + p_{n1}\bar{\boldsymbol{e}}_n \\ \boldsymbol{e}_2 = p_{12}\bar{\boldsymbol{e}}_1 + p_{22}\bar{\boldsymbol{e}}_2 + \cdots + p_{n2}\bar{\boldsymbol{e}}_n \\ \quad\vdots \\ \boldsymbol{e}_n = p_{1n}\bar{\boldsymbol{e}}_1 + p_{2n}\bar{\boldsymbol{e}}_2 + \cdots + p_{nn}\bar{\boldsymbol{e}}_n \end{cases} \tag{1.5.3}$$

Let

$$\boldsymbol{P} = \begin{bmatrix} p_{11} & p_{12} & \cdots & p_{1n} \\ p_{21} & p_{22} & \cdots & p_{2n} \\ \vdots & \vdots & & \vdots \\ p_{n1} & p_{n2} & \cdots & p_{nn} \end{bmatrix} \tag{1.5.4}$$

Then, eq. (1.5.3) can be expressed as

$$[\boldsymbol{e}_1 \quad \boldsymbol{e}_2 \quad \cdots \quad \boldsymbol{e}_n] = [\bar{\boldsymbol{e}}_1 \quad \bar{\boldsymbol{e}}_2 \quad \cdots \quad \bar{\boldsymbol{e}}_n]\boldsymbol{P} \tag{1.5.5}$$

Further more, the coordinates $\boldsymbol{x}$ and $\bar{\boldsymbol{x}}$ represent the same state vector under the two basses, so we have

$$[\bar{\boldsymbol{e}}_1 \quad \bar{\boldsymbol{e}}_2 \quad \cdots \quad \bar{\boldsymbol{e}}_n]\begin{bmatrix} \bar{x}_1 \\ \bar{x}_2 \\ \vdots \\ \bar{x}_n \end{bmatrix} = [\boldsymbol{e}_1 \quad \boldsymbol{e}_2 \quad \cdots \quad \boldsymbol{e}_n]\begin{bmatrix} x_1 \\ x_2 \\ \vdots \\ x_n \end{bmatrix} \tag{1.5.6}$$

Substituting eq. (1.5.5) into eq. (1.5.6), yields

$$[\bar{\boldsymbol{e}}_1 \quad \bar{\boldsymbol{e}}_2 \quad \cdots \quad \bar{\boldsymbol{e}}_n]\begin{bmatrix}\bar{x}_1\\ \bar{x}_2\\ \vdots\\ \bar{x}_n\end{bmatrix} = [\bar{\boldsymbol{e}}_1 \quad \bar{\boldsymbol{e}}_2 \quad \cdots \quad \bar{\boldsymbol{e}}_n]\boldsymbol{P}\begin{bmatrix}x_1\\ x_2\\ \vdots\\ x_n\end{bmatrix} \tag{1.5.7}$$

So, there is

$$\bar{\boldsymbol{x}} = \boldsymbol{P}\boldsymbol{x} \tag{1.5.8}$$

Similarly, according to the above methodwe can get

$$\boldsymbol{x} = \boldsymbol{Q}\bar{\boldsymbol{x}} \tag{1.5.9}$$

Substituting eq. (1.5.8) into eq. (1.5.9), yields

$$\boldsymbol{x} = \boldsymbol{Q}\boldsymbol{P}\boldsymbol{x} \tag{1.5.10}$$

Substituting eq. (1.5.9) into eq. (1.5.8), yields

$$\bar{\boldsymbol{x}} = \boldsymbol{Q}\boldsymbol{P}\bar{\boldsymbol{x}} \tag{1.5.11}$$

So, we have

$$\boldsymbol{Q}\boldsymbol{P} = \boldsymbol{P}\boldsymbol{Q} = \boldsymbol{I} \tag{1.5.12}$$

This indicates $\boldsymbol{P}^{-1} = \boldsymbol{Q}$, i. e. the transformation between $\boldsymbol{x}$ and $\bar{\boldsymbol{x}}$ is linear and non-singular.

1.5.2 Properties of algebraically equivalent systems

Conclusion 1.5.1 Consider an n-order LTI system

$$\begin{cases}\dot{\boldsymbol{x}} = \boldsymbol{A}\boldsymbol{x} + \boldsymbol{B}\boldsymbol{u}\\ \boldsymbol{y} = \boldsymbol{C}\boldsymbol{x} + \boldsymbol{D}\boldsymbol{u}\end{cases} \tag{1.5.13}$$

Take a linear non-singular transformation $\bar{\boldsymbol{x}} = \boldsymbol{P}^{-1}\boldsymbol{x}$, and the obtained state-space description is as follows:

$$\begin{cases}\dot{\bar{\boldsymbol{x}}} = \bar{\boldsymbol{A}}\bar{\boldsymbol{x}} + \bar{\boldsymbol{B}}\boldsymbol{u}\\ \boldsymbol{y} = \bar{\boldsymbol{C}}\bar{\boldsymbol{x}} + \bar{\boldsymbol{D}}\boldsymbol{u}\end{cases} \tag{1.5.14}$$

The four parameter matrices in eq. (1.5.13) and eq. (1.5.14) have the following relations

$$\bar{\boldsymbol{A}} = \boldsymbol{P}^{-1}\boldsymbol{A}\boldsymbol{P},\quad \bar{\boldsymbol{B}} = \boldsymbol{P}^{-1}\boldsymbol{B},\quad \bar{\boldsymbol{C}} = \boldsymbol{C}\boldsymbol{P},\quad \bar{\boldsymbol{D}} = \boldsymbol{D} \tag{1.5.15}$$

Proof From $\bar{\boldsymbol{x}} = \boldsymbol{P}^{-1}\boldsymbol{x}$, we have

$$\dot{\bar{\boldsymbol{x}}} = \boldsymbol{P}^{-1}\dot{\boldsymbol{x}} = \boldsymbol{P}^{-1}(\boldsymbol{A}\boldsymbol{x} + \boldsymbol{B}\boldsymbol{u}) = \boldsymbol{P}^{-1}\boldsymbol{A}\boldsymbol{P}\bar{\boldsymbol{x}} + \boldsymbol{P}^{-1}\boldsymbol{B}\boldsymbol{u}$$

$$\boldsymbol{y} = \boldsymbol{C}\boldsymbol{x} + \boldsymbol{D}\boldsymbol{u} = \boldsymbol{C}\boldsymbol{P}\bar{\boldsymbol{x}} + \boldsymbol{D}\boldsymbol{u}$$

So, we get $\bar{\boldsymbol{A}} = \boldsymbol{P}^{-1}\boldsymbol{A}\boldsymbol{P}$, $\bar{\boldsymbol{B}} = \boldsymbol{P}^{-1}\boldsymbol{B}$, $\bar{\boldsymbol{C}} = \boldsymbol{C}\boldsymbol{P}$, $\bar{\boldsymbol{D}} = \boldsymbol{D}$.

Conclusion 1.5.2 For state space description eq. (1.5.13) and eq. (1.5.14), there are

(1) They have the same eigenvalues, i. e.

$$\lambda_i(A)=\lambda_i(\overline{A}),i=1,2,\cdots,n \tag{1.5.16}$$

(2) Their transfer matrices are the same.

Proof (1) From $\overline{A}=P^{-1}AP$, we have

$$\begin{aligned}0&=\det(\lambda_i I-\overline{A})\\&=\det(\lambda_i I-P^{-1}AP)\\&=\det(P^{-1}\lambda_i P-P^{-1}AP)\\&=\det[P^{-1}(\lambda_i I-A)P)]\\&=\det(\lambda_i I-A)\det(P^{-1})\det(P)\\&=\det(\lambda_i I-A)\det(P^{-1}P)\\&=\det(\lambda_i I-A)\end{aligned}$$

The above equation indicates that λ_i is the eigenvalues of $\overline{A}$ and A at the same time, that is, eq. (1.5.16) is hold.

(2) The transfer matrix of state space description eq. (1.5.13) is

$$G(s)=C(sI-A)^{-1}B+D$$

The transfer matrix of state space description eq. (1.5.14) is

$$\overline{G}(s)=\overline{C}(sI-\overline{A})^{-1}\overline{B}+\overline{D}$$

and

$$\overline{A}=P^{-1}AP,\quad \overline{B}=P^{-1}B,\quad \overline{C}=CP,\quad \overline{D}=D$$

Thus, we get

$$\begin{aligned}\overline{G}(s)&=CP(sI-P^{-1}AP)^{-1}P^{-1}B+D=CP(P^{-1}sP-P^{-1}AP)^{-1}P^{-1}B+D\\&=CP[P^{-1}(sI-A)P]^{-1}P^{-1}B+D=CP[P^{-1}(sI-A)^{-1}P]P^{-1}B+D\\&=C(sI-A)^{-1}B+D\end{aligned}$$

Conclusion 1.5.2 indicates that different state-space descriptions of the same system have the same input-output characteristics $G(s)$, which are determined not by the selection of state variables, but by the inherent characteristic of the system.

Conclusion 1.5.3 Consider a LTV system

$$\begin{cases}\dot{x}=A(t)x+B(t)u\\y=C(t)x+D(t)u\end{cases} \tag{1.5.17}$$

Takea linear nonsingular transformation $\overline{x}=P^{-1}(t)x$, where $P(t)$ is invertible and continuously differentiable. Let obtained state-space description is

$$\begin{cases}\dot{\overline{x}}=\overline{A}(t)\overline{x}+\overline{B}(t)u\\y=\overline{C}(t)\overline{x}+\overline{D}(t)u\end{cases} \tag{1.5.18}$$

The four parameter matrices in eq. (1.5.17) and eq. (1.5.18) have the following relationships:

$$\begin{cases} \bar{A}(t) = -P^{-1}(t)\dot{P}(t)P^{-1}(t)A(t)P(t) \\ \bar{B}(t) = P^{-1}(t)B(t),\ \bar{C} = C(t)P(t),\ \bar{D} = D \end{cases} \tag{1.5.19}$$

where $\dot{P}(t) \triangleq \frac{dP(t)}{dt}$.

Proof Taking the derivative of both sides of $P(t)P^{-1}(t) = I$ with respect to time t, yields

$$(P^{-1}(t))'P(t) + P^{-1}(t)\dot{P}(t) = 0$$

And then, we have

$$(P^{-1}(t))' = -P^{-1}(t)\dot{P}(t)P^{-1}(t) \tag{1.5.20}$$

Where $(P^{-1}(t))' = \frac{d P^{-1}(t)}{dt}$.

From $\bar{x} = P^{-1}(t)x$ and $x = P(t)\bar{x}$, we have

$$\begin{aligned} \dot{\bar{x}} &= (P^{-1}(t))'x + P^{-1}(t)\dot{x} \\ &= -P^{-1}(t)\dot{P}(t)P^{-1}(t)x + P^{-1}(t)(A(t)x + B(t)u) \\ &= -P^{-1}(t)\dot{P}(t)P^{-1}(t)P(t)\bar{x} + P^{-1}(t)A(t)P(t)\bar{x} + P^{-1}(t)B(t)u \\ &= (-P^{-1}(t)\dot{P}(t) + P^{-1}(t)A(t)P(t))\bar{x} + P^{-1}(t)B(t)u \end{aligned}$$

$$y = C(t)P(t)\bar{x} + D(t)u$$

Algebraically equivalence The relationship between the two state-space descriptions is expressed in eq. (1.5.15), i. e. $\bar{x} = P^{-1}x$, the two state-space descriptions are said to be algebraically equivalent. They share some algebraic properties, such as:

Property 1 The two state-space descriptions derived from the same system using different state variable groups are algebraic equivalent.

Forthe system in example 1.2.2, state vectors $x = [i(t) \quad u_c(t)]^T$ and $\bar{x} = [u_c(t) \quad \dot{u}_c(t)]^T$ are selected respectively, and the relationship between states x and $\bar{x}$ is as follows:

$$\bar{x}_1 = x_2$$

$$\bar{x}_2 = \frac{du_c}{dt} = \frac{1}{c}i(t) = \frac{1}{c}x_1(t)$$

So, there is

$$\bar{x} = \begin{bmatrix} 0 & 1 \\ \frac{1}{c} & 0 \end{bmatrix} x$$

This shows that the two state space descriptions derived from the system using different state variable groups are algebraically equivalent.

Property 2 For LTI systems, two algebraically equivalent state space descriptions can be transformed to the same diagonal canonical form or Jordan canonical form.

The inherent characteristics of systems are determined by the internal structure and parameters, and have nothing to do with the selection of state. For example, the eigenvalues and transfer function matrices of the same system remain unchanged in different coordinate systems.

1.6 State space description of composite systems

A system composed of two or more subsystems connected in a certain way is called a composite system. The basic connect ways of subsystems include parallel, cascade and feedback connections. A more complex practical system is usually combined by the several connection ways. In this section, only the three basic connection ways mentioned above are discussed respectively to establish the corresponding state space description and input-output description of the composite systems.

1.6.1 Parallel connection

Considera composite system connected by the following two subsystems

$$S_i:\begin{cases}\dot{\boldsymbol{x}}_i=\boldsymbol{A}_i\boldsymbol{x}_i+\boldsymbol{B}_i\boldsymbol{u}_i\\ \boldsymbol{y}_i=\boldsymbol{C}_i\boldsymbol{x}_i+\boldsymbol{D}_i\boldsymbol{u}_i\end{cases},i=1,2 \tag{1.6.1}$$

in parallel which is shown in Figure 1.6.1. Where $\boldsymbol{u}$ and $\boldsymbol{y}$ are the input and output vector of the composite system, respectively.

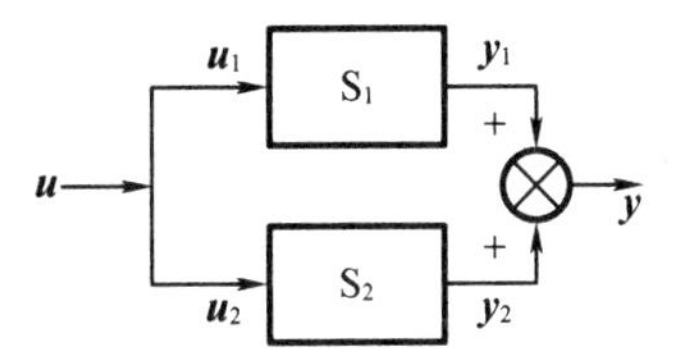

Figure 1.6.1 Parallel connection

The two subsystems can be connected in parallel under the following conditions:

$$\begin{aligned}\dim(\boldsymbol{u}_1)&=\dim(\boldsymbol{u}_2)\\ \dim(\boldsymbol{y}_1)&=\dim(\boldsymbol{y}_2)\end{aligned} \tag{1.6.2}$$

Where dim(·) denotes the dimension of a vector.

The characteristic of the parallel system is

$$\boldsymbol{u}_1=\boldsymbol{u}_2=\boldsymbol{u} \text{ and } \boldsymbol{y}=\boldsymbol{y}_1+\boldsymbol{y}_2 \tag{1.6.3}$$

So, from eq. (1.6.1) and eq. (1.6.3), the state equation and output equation of the parallel system are

$$\begin{cases} \dot{\boldsymbol{x}}_1 = \boldsymbol{A}_1\boldsymbol{x}_1 + \boldsymbol{B}_1\boldsymbol{u} \\ \dot{\boldsymbol{x}}_2 = \boldsymbol{A}_2\boldsymbol{x}_2 + \boldsymbol{B}_2\boldsymbol{u} \\ \boldsymbol{y} = \boldsymbol{y}_1 + \boldsymbol{y}_2 = \boldsymbol{C}_1\boldsymbol{x}_1 + \boldsymbol{C}_2\boldsymbol{x}_2 + \boldsymbol{D}_1\boldsymbol{u} + \boldsymbol{D}_2\boldsymbol{u} \end{cases} \tag{1.6.4}$$

They are written in matrix form as

$$\begin{aligned} \begin{bmatrix} \dot{\boldsymbol{x}}_1 \\ \dot{\boldsymbol{x}}_2 \end{bmatrix} &= \begin{bmatrix} \boldsymbol{A}_1 & \\ & \boldsymbol{A}_2 \end{bmatrix} \begin{bmatrix} \boldsymbol{x}_1 \\ \boldsymbol{x}_2 \end{bmatrix} + \begin{bmatrix} \boldsymbol{B}_1 \\ \boldsymbol{B}_2 \end{bmatrix} \boldsymbol{u} \\ \boldsymbol{y} &= [\boldsymbol{C}_1 \quad \boldsymbol{C}_2] \begin{bmatrix} \boldsymbol{x}_1 \\ \boldsymbol{x}_2 \end{bmatrix} + (\boldsymbol{D}_1 + \boldsymbol{D}_2)\boldsymbol{u} \end{aligned} \tag{1.6.5}$$

According to the above method, the parallel system of N subsystems can be described as

$$\begin{cases} \begin{bmatrix} \dot{x}_1 \\ \dot{x}_2 \\ \vdots \\ \dot{x}_N \end{bmatrix} = \begin{bmatrix} \boldsymbol{A}_1 & & & \\ & \boldsymbol{A}_2 & & \\ & & \ddots & \\ & & & \boldsymbol{A}_N \end{bmatrix} \begin{bmatrix} x_1 \\ x_2 \\ \vdots \\ x_N \end{bmatrix} + \begin{bmatrix} \boldsymbol{B}_1 \\ \boldsymbol{B}_2 \\ \vdots \\ \boldsymbol{B}_N \end{bmatrix} \boldsymbol{u} \\ \boldsymbol{y} = [\boldsymbol{C}_1 \quad \boldsymbol{C}_2 \quad \cdots \quad \boldsymbol{C}_N] \begin{bmatrix} x_1 \\ x_2 \\ \vdots \\ x_N \end{bmatrix} + (\boldsymbol{D}_1 + \boldsymbol{D}_2 + \cdots + \boldsymbol{D}_N) u \end{cases} \tag{1.6.6}$$

Next, we compute the transfer matrix of the system described by eq. (1.6.6). Assume $\boldsymbol{G}_i(s)$ is the transfer matrix of ith subsystem. The Laplace transform of the output of the parallel system is

$$\boldsymbol{y}(s) = \boldsymbol{y}_1(s) + \boldsymbol{y}_2(s) + \cdots + \boldsymbol{y}_N(s) = \sum_{i=1}^{N} (\boldsymbol{G}_i(s)\boldsymbol{u}_i(s)) = \left(\sum_{i=1}^{N} \boldsymbol{G}_i(s)\right)\boldsymbol{u}(s)$$

So the transfer matrix of the parallel system is

$$\boldsymbol{G}(s) = \sum_{i=1}^{N} \boldsymbol{G}_i(s) \tag{1.6.7}$$

1.6.2 Cascade connection

Consider a composite system connected by two subsystems in cascade which is shown in Figure 1.6.2. Where the subsystem are described by eq. (1.6.1).

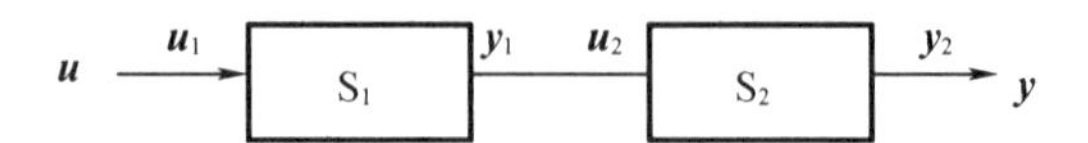

Figure 1.6.2 Cascade connection

The two subsystems can be connected in cascade under the following conditions:

$$\dim(\boldsymbol{y}_1) = \dim(\boldsymbol{u}_2) \tag{1.6.8}$$

The characteristic of the cascade system is

$$\boldsymbol{u}_1 = \boldsymbol{u}, \boldsymbol{u}_2 = \boldsymbol{y}_1, \boldsymbol{y} = \boldsymbol{y}_2 \tag{1.6.9}$$

From eq. (1.6.1) and eq. (1.6.9), the state space description of the cascade system is established as

$$\dot{\boldsymbol{x}}_1 = \boldsymbol{A}_1\boldsymbol{x}_1 + \boldsymbol{B}_1\boldsymbol{u}$$

$$\dot{\boldsymbol{x}}_2 = \boldsymbol{A}_2\boldsymbol{x}_2 + \boldsymbol{B}_2\boldsymbol{u}_2 = \boldsymbol{A}_2\boldsymbol{x}_2 + \boldsymbol{B}_2(\boldsymbol{C}_1\boldsymbol{x}_1 + \boldsymbol{D}_1\boldsymbol{u}_1) = \boldsymbol{A}_2\boldsymbol{x}_2 + \boldsymbol{B}_2\boldsymbol{C}_1\boldsymbol{x}_1 + \boldsymbol{B}_2\boldsymbol{D}_1\boldsymbol{u}$$

$$\boldsymbol{y} = \boldsymbol{y}_2 = \boldsymbol{C}_2\boldsymbol{x}_2 + \boldsymbol{D}_2\boldsymbol{u}_2 = \boldsymbol{C}_2\boldsymbol{x}_2 + \boldsymbol{D}_2(\boldsymbol{C}_1\boldsymbol{x}_1 + \boldsymbol{D}_1\boldsymbol{u}_1) = \boldsymbol{C}_2\boldsymbol{x}_2 + \boldsymbol{D}_2\boldsymbol{C}_1\boldsymbol{x}_1 + \boldsymbol{D}_2\boldsymbol{D}_1\boldsymbol{u}$$

It can be written in matrix form as

$$\begin{cases} \begin{bmatrix} \dot{\boldsymbol{x}}_1 \\ \dot{\boldsymbol{x}}_2 \end{bmatrix} = \begin{bmatrix} \boldsymbol{A}_1 & \boldsymbol{O} \\ \boldsymbol{B}_2\boldsymbol{C}_1 & \boldsymbol{A}_2 \end{bmatrix} \begin{bmatrix} \boldsymbol{x}_1 \\ \boldsymbol{x}_2 \end{bmatrix} + \begin{bmatrix} \boldsymbol{B}_1 \\ \boldsymbol{B}_2\boldsymbol{D}_1 \end{bmatrix} \boldsymbol{u} \\ \boldsymbol{y} = [\boldsymbol{D}_2\boldsymbol{C}_1 \quad \boldsymbol{C}_2] \begin{bmatrix} \boldsymbol{x}_1 \\ \boldsymbol{x}_2 \end{bmatrix} + (\boldsymbol{D}_2\boldsymbol{D}_1)\boldsymbol{u} \end{cases} \tag{1.6.10}$$

Furthermore, compute the transfer function matrix of N subsystems in series. The Laplace transform of the output of the cascade system is

$$\begin{aligned} \boldsymbol{y}(s) &= \boldsymbol{y}_N(s) \\ &= \boldsymbol{G}_N(s)\boldsymbol{u}_N(s) \\ &= \boldsymbol{G}_N(s)\boldsymbol{y}_{N-1}(s) \\ &= \boldsymbol{G}_N(s)\boldsymbol{G}_{N-1}(s)\boldsymbol{u}_{N-1}(s) \\ &\ \ \vdots \\ &= \boldsymbol{G}_N(s)\boldsymbol{G}_{N-1}(s)\cdots\boldsymbol{G}_1(s)\boldsymbol{u}_1(s) \\ &= \boldsymbol{G}_N(s)\boldsymbol{G}_{N-1}(s)\cdots\boldsymbol{G}_1(s)\boldsymbol{u}(s) \end{aligned}$$

So the transfer matrix of the cascade system is

$$\boldsymbol{G}(s) = \boldsymbol{G}_N(s)\boldsymbol{G}_{N-1}(s)\cdots\boldsymbol{G}_1(s) \tag{1.6.11}$$

1.6.3 Feedback connection

Consider a combined system composed of two subsystems, S_1 and S_2, which are connected by feedback, as shown in Figure 1.6.3. The state space description of the subsystems is eq. (1.6.1). In order to simplify computation, let $\boldsymbol{D}_i = \boldsymbol{0}$.

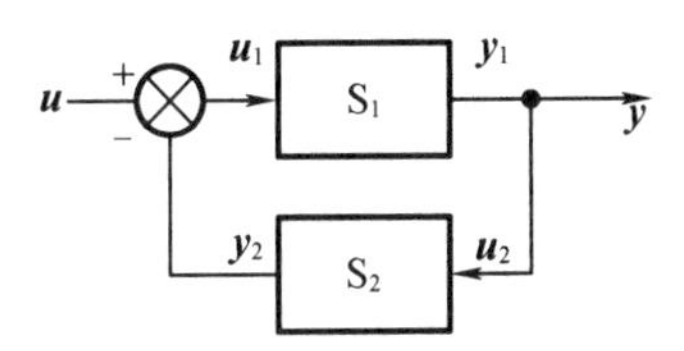

Figure 1.6.3 Feedback connection

The two subsystems can be connected by feedback under the following conditions:

$$\dim(\boldsymbol{u}_1) = \dim(\boldsymbol{y}_2) \text{ and } \dim(\boldsymbol{u}_2) = \dim(\boldsymbol{y}_1) \tag{1.6.12}$$

The characteristic of the feedback system is

$$\boldsymbol{u}_1 = \boldsymbol{u} - \boldsymbol{y}_2 \quad \boldsymbol{y}_1 = \boldsymbol{y} = \boldsymbol{u}_2 \tag{1.6.13}$$

From eq. (1.6.1) with $\boldsymbol{D}_i = 0$ and eq. (1.6.12), the state space description of the feedback system is established as

$$\begin{aligned}\dot{\boldsymbol{x}}_1 &= \boldsymbol{A}_1\boldsymbol{x}_1 + \boldsymbol{B}_1\boldsymbol{u}_1\\ &= \boldsymbol{A}_1\boldsymbol{x}_1 + \boldsymbol{B}_1(\boldsymbol{u} - \boldsymbol{y}_2)\\ &= \boldsymbol{A}_1\boldsymbol{x}_1 + \boldsymbol{B}_1(\boldsymbol{u} - \boldsymbol{C}_2\boldsymbol{x}_2)\\ &= \boldsymbol{A}_1\boldsymbol{x}_1 + \boldsymbol{B}_1\boldsymbol{u} - \boldsymbol{B}_1\boldsymbol{C}_2\boldsymbol{x}_2\end{aligned}$$

$$\dot{\boldsymbol{x}}_2 = \boldsymbol{A}_2\boldsymbol{x}_2 + \boldsymbol{B}_2\boldsymbol{u}_2 = \boldsymbol{A}_2\boldsymbol{x}_2 + \boldsymbol{B}_2\boldsymbol{y}_1 = \boldsymbol{A}_2\boldsymbol{x}_2 + \boldsymbol{B}_2\boldsymbol{C}_1\boldsymbol{x}_1$$

$$\boldsymbol{y} = \boldsymbol{y}_1 = \boldsymbol{C}_1\boldsymbol{x}_1$$

It can be written in matrix form as

$$\begin{cases}\begin{bmatrix}\dot{\boldsymbol{x}}_1\\ \dot{\boldsymbol{x}}_2\end{bmatrix} = \begin{bmatrix}\boldsymbol{A}_1 & -\boldsymbol{B}_1\boldsymbol{C}_2\\ \boldsymbol{B}_2\boldsymbol{C}_1 & \boldsymbol{A}_2\end{bmatrix}\begin{bmatrix}\boldsymbol{x}_1\\ \boldsymbol{x}_2\end{bmatrix} + \begin{bmatrix}\boldsymbol{B}_1\\ \boldsymbol{0}\end{bmatrix}\boldsymbol{u}\\ \boldsymbol{y} = \begin{bmatrix}\boldsymbol{C}_1 & \boldsymbol{0}\end{bmatrix}\begin{bmatrix}\boldsymbol{x}_1\\ \boldsymbol{x}_2\end{bmatrix}\end{cases} \tag{1.6.14}$$

Next we determine the transfer matrix of the feedback system. The Laplace transform of the output of the feedback system is

$$\begin{aligned}\boldsymbol{y}(s) &= \boldsymbol{y}_1(s)\\ &= \boldsymbol{G}_1(s)\boldsymbol{u}_1(s)\\ &= \boldsymbol{G}_1(s)(\boldsymbol{u}(s) - \boldsymbol{y}_2(s))\\ &= \boldsymbol{G}_1(s)(\boldsymbol{u}(s) - \boldsymbol{G}_2(s)\boldsymbol{u}_2(s))\\ &= \boldsymbol{G}_1(s)(\boldsymbol{u}(s) - \boldsymbol{G}_2(s)\boldsymbol{y}(s))\\ &= \boldsymbol{G}_1(s)\boldsymbol{u}(s) - \boldsymbol{G}_1(s)\boldsymbol{G}_2(s)\boldsymbol{y}(s)\end{aligned}$$

And then,

$$(\boldsymbol{I} + \boldsymbol{G}_1(s)\boldsymbol{G}_2(s))\boldsymbol{y}(s) = \boldsymbol{G}_1(s)\boldsymbol{u}(s)$$

If $\det(\boldsymbol{I} + \boldsymbol{G}_1(s)\boldsymbol{G}_2(s)) \neq 0$, multiply both sides of above equation by $(\boldsymbol{I} + \boldsymbol{G}_1(s)\boldsymbol{G}_2(s))^{-1}$, and we get

$$\boldsymbol{y}(s) = (\boldsymbol{I} + \boldsymbol{G}_1(s)\boldsymbol{G}_2(s))^{-1}\boldsymbol{G}_1(s)\boldsymbol{u}(s)$$

Therefore, the transfer matrix of thefeedback system is

$$\boldsymbol{G}(s) = (\boldsymbol{I} + \boldsymbol{G}_1(s)\boldsymbol{G}_2(s))^{-1}\boldsymbol{G}_1(s) \tag{1.6.15}$$

Similarly, according to eq. (1.6.13), we have

$$\begin{aligned}\boldsymbol{u}_1(s) &= \boldsymbol{u}(s) - \boldsymbol{y}(s) = \boldsymbol{u}(s) - \boldsymbol{G}_2(s)\boldsymbol{u}_2(s) = \boldsymbol{u}(s) - \boldsymbol{G}_2(s)\boldsymbol{y}_1(s)\\ &= \boldsymbol{u}(s) - \boldsymbol{G}_2(s)\boldsymbol{G}_1(s)\boldsymbol{u}_1(s)\end{aligned}$$

Then

$$(\boldsymbol{I}+\boldsymbol{G}_2(s)\boldsymbol{G}_1(s))\boldsymbol{u}_1(s)=\boldsymbol{u}(s)$$

If $\det(\boldsymbol{I}+\boldsymbol{G}_2(s)\boldsymbol{G}_1(s))\neq 0$, multiply both sides of above equation by $(\boldsymbol{I}+\boldsymbol{G}_2(s)\boldsymbol{G}_1(s))^{-1}$, we get

$$\boldsymbol{u}_1(s)=(\boldsymbol{I}+\boldsymbol{G}_2(s)\boldsymbol{G}_1(s))^{-1}\boldsymbol{u}(s)$$

And the Laplace transform of the output of the feedback system is

$$\begin{aligned}\boldsymbol{y}(s)&=\boldsymbol{y}_1(s)\\&=\boldsymbol{G}_1(s)\boldsymbol{u}_1(s)\\&=\boldsymbol{G}_1(s)\boldsymbol{u}_1(s)\\&=\boldsymbol{G}_1(s)(\boldsymbol{I}+\boldsymbol{G}_2(s)\boldsymbol{G}_1(s))^{-1}\boldsymbol{u}(s)\end{aligned}$$

So , another expression of the transfer matrix of the feedback system is

$$\boldsymbol{G}(s)=\boldsymbol{G}_1(s)(\boldsymbol{I}+\boldsymbol{G}_2(s)\boldsymbol{G}_1(s))^{-1} \tag{1.6.16}$$

Problems

1.1 Find state equations and output equations to describe the circuits shown in Figure P1.1(a) and P1.1(b). State variables, input variable and output variable are chosen as:

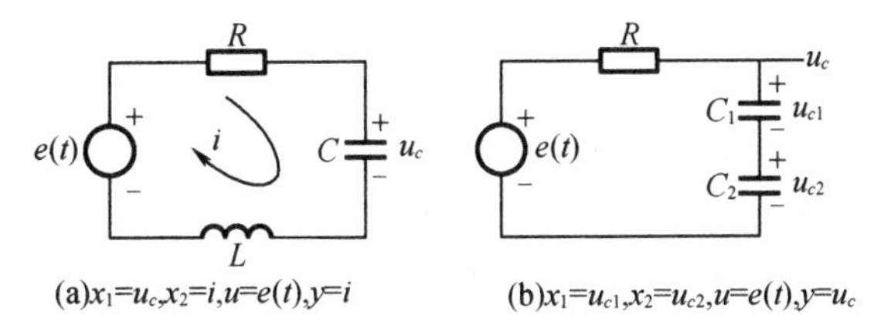

Figure P1.1

1.2 Consider a separately excited DC motor system. The magnetic flow Φ is constant. The back EMF is $C_e\omega$. The electromagnetic torque is $C_M i_a$. Where C_e and C_M are constant. Now chose the state variables $x_1=i_a$ and $x_2=\omega$, the input variable $u=e(t)$. Find the state equations of the DC motor system.

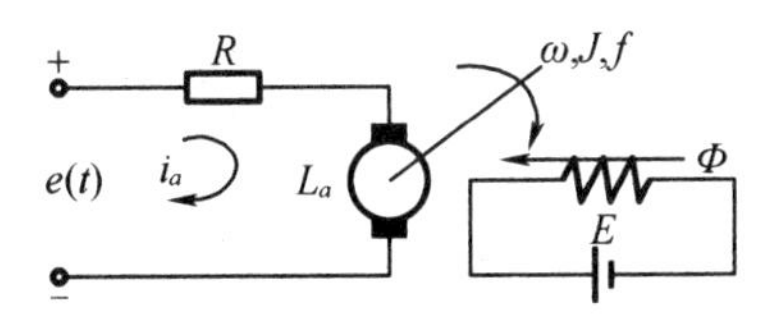

Figure P1.2

1.3 Figure P1.3 shows the schematic diagram of the lunar module's soft landing on the moon, and its motion equation can be expressed as $m\ddot{y}=k\dot{m}-mg$. Where, m is the mass of the lunar module, g is the gravitational constant of the lunar surface, k is a constant, $k\dot{m}$ represents the reverse thrust, y is the distance of the lunar module to the lunar surface, and upward is positive. Choose the state variable as $x_1 = y, x_2 = \dot{y}$ and $x_3 = m$, the input variable $u = \dot{m}$. Find the state equation of the system.

1.4 Figure P1.4 is a block diagram of a system, in which u and y are its input variable and output variable, respectively. Choose state variables as $x_1 = y$ and $x_2 = \dot{y}$. Find the state equation and output of the system.

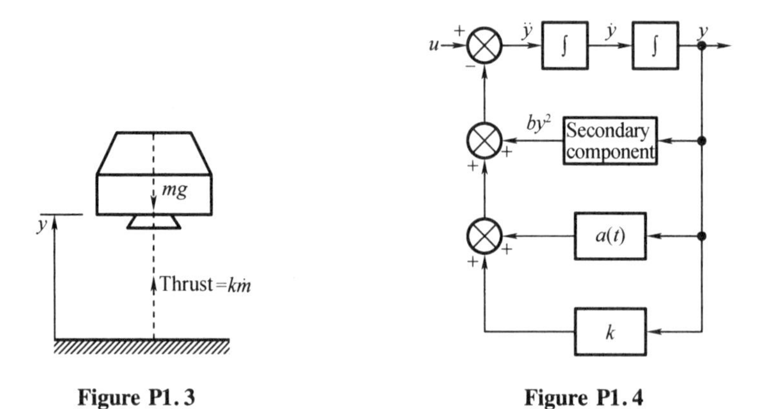

Figure P1.3 **Figure P1.4**

1.5 Use two different methods to find the state-space descriptions of the following input-output descriptions.

(1) $y^{(3)} + 2\ddot{y} + 6\dot{y} + 3y = 5u$

(2) $y^{(3)} + 2\ddot{y} + 6\dot{y} + 3y = 7\dot{u} + 5u$

(3) $3y^{(3)} + 6\ddot{y} + 12\dot{y} + 9y = 3u^{(3)} + 6\dot{u} + 3u$

1.6 Use two different methods to find the state-space descriptions of the following transfer functions.

(1) $\dfrac{\hat{y}(s)}{\hat{u}(s)} = \dfrac{2s^2 + 18s + 40}{s^3 + 6s^2 + 11s + 6}$

(2) $\dfrac{\hat{y}(s)}{\hat{u}(s)} = \dfrac{3(s+5)}{(s+3)^2(s+1)}$

1.7 Figure P1.5 is a block diagram of a system, in which y and u are its output variable and input variable, respectively. Find a state space description of the system.

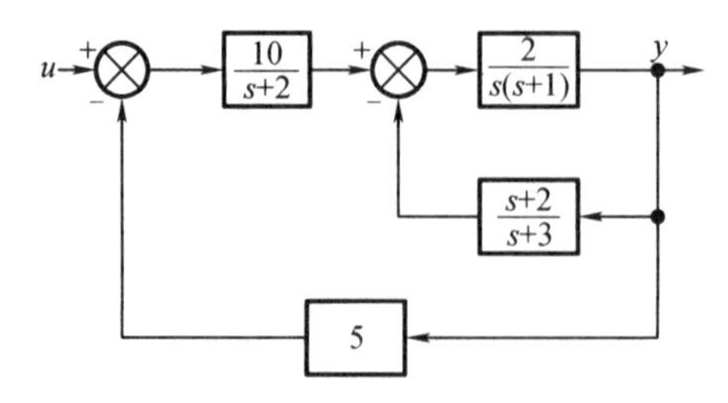

Figure P1.5

1.8 Find the characteristic equation and eigenvalues of the following square matrix $\boldsymbol{A}$.

(1) $\boldsymbol{A}=\begin{bmatrix} 2 & 5 \\ -2 & -3 \end{bmatrix}$

(2) $\boldsymbol{A}=\begin{bmatrix} 0 & 1 & 0 \\ 0 & 0 & 1 \\ 0 & -1 & -1 \end{bmatrix}$

1.9 Given the nonsingular matrices of the same dimension, $\boldsymbol{A}$ and $\boldsymbol{B}$, prove that the eigenvalues of $\boldsymbol{AB}$ must be equal to the eigenvalues of $\boldsymbol{BA}$.

1.10 Given an n-dimensional non-singular constant matrix $\boldsymbol{A}$ whose eigenvalues $\lambda_1,\lambda_2,\cdots,\lambda_n$ are distinct, prove that the eigenvalues of $\boldsymbol{A}^{-1}$ are $\{\lambda_1^{-1},\lambda_2^{-1},\cdots,\lambda_n^{-1}\}$.

1.11 Convert the followingstate equations to the diagonal canonical forms or Jordan canonical forms.

(1) $\dot{\boldsymbol{x}}=\begin{bmatrix} 8 & -8 & -2 \\ 4 & -3 & -2 \\ 3 & -4 & 1 \end{bmatrix}\boldsymbol{x}+\begin{bmatrix} 2 & 3 \\ 1 & 5 \\ 7 & 1 \end{bmatrix}\boldsymbol{u}$

(2) $\dot{\boldsymbol{x}}=\begin{bmatrix} 0 & 1 \\ -9 & -6 \end{bmatrix}\boldsymbol{x}+\begin{bmatrix} 4 \\ 2 \end{bmatrix}\boldsymbol{u}$

1.12 Compute the transfer function $g(s)$ for the following state space description:

$$\dot{\boldsymbol{x}}=\begin{bmatrix} -5 & -1 \\ 3 & -1 \end{bmatrix}\boldsymbol{x}+\begin{bmatrix} 2 \\ 5 \end{bmatrix}\boldsymbol{u}$$

$$y=[1 \quad 2]\boldsymbol{x}+4u$$

1.13 Thestate equation of the given system is

$$\begin{bmatrix} \dot{x}_1 \\ \dot{x}_2 \\ \dot{x}_3 \end{bmatrix}=\begin{bmatrix} 0 & 2 & 0 \\ 0 & 0 & 2 \\ 1 & -3 & 5 \end{bmatrix}\begin{bmatrix} x_1 \\ x_2 \\ x_3 \end{bmatrix}+\begin{bmatrix} 2 \\ 3 \\ 5 \end{bmatrix}u$$

Now, if $y=2x_2+3x_3$, write the scalar differential equation between input y and output u.

1.14 Compute the transfer matrix $\boldsymbol{G}(s)$ for the following state space description.

$$\dot{\boldsymbol{x}}=\begin{bmatrix} 0 & 1 & 0 \\ 0 & 0 & 1 \\ -3 & -1 & -2 \end{bmatrix}\boldsymbol{x}+\begin{bmatrix} 1 & 0 \\ 0 & 1 \\ 1 & 1 \end{bmatrix}\boldsymbol{u}$$

$$y=[1 \quad 0 \quad 1]\boldsymbol{x}$$

1.15 Given that the square matrices $\boldsymbol{A}$ and $\widetilde{\boldsymbol{A}}$ with the same dimension are

$$\boldsymbol{A}=\left[\begin{array}{c|ccc} 0 & 1 & & \\ \vdots & & \ddots & \\ 0 & & & 1 \\ \hline a_0 & a_1 & \cdots & a_{n-1} \end{array}\right],\widetilde{\boldsymbol{A}}=\left[\begin{array}{ccc|c} 0 & \cdots & 0 & a_0 \\ \hline 1 & & & a_1 \\ & \ddots & & \vdots \\ & & 1 & a_{n-1} \end{array}\right]$$

Try to find a transformation matrix $\boldsymbol{P}$ such that $\widetilde{\boldsymbol{A}} = \boldsymbol{P}^{-1}\boldsymbol{AP}$.

1.16 Given the square constant matrix $\boldsymbol{A}$ as

$$\boldsymbol{A} = \begin{bmatrix} 0 & 1 & 0 \\ 0 & 0 & 1 \\ -6 & -1 & 4 \end{bmatrix}$$

Try to calculate $\boldsymbol{A}^{100}$.

1.17 Let $\boldsymbol{A}$ be the square constant matrix, and define

$$\mathrm{e}^{\boldsymbol{A}} \triangleq \boldsymbol{I} + \boldsymbol{A} + \frac{1}{2!}\boldsymbol{A}^2 + \cdots + \frac{1}{k!}\boldsymbol{A}^k + \cdots$$

Now, assume that the eigenvalues $\lambda_1, \lambda_2, \cdots, \lambda_n$ of $\boldsymbol{A}$ are distinct, and try to prove $\det(\mathrm{e}^{\boldsymbol{A}}) = \prod_{i=1}^{n} \mathrm{e}^{\lambda_i}$.

1.18 Given the square constant matrix $\boldsymbol{A}$ as

$$\boldsymbol{A} = \begin{bmatrix} 0 & 1 \\ -2 & -3 \end{bmatrix}$$

Compute $\mathrm{e}^{\boldsymbol{A}}$.

1.19 Given the feedback connection system in Figure P1.6, where

$$\boldsymbol{G}_1(s) = \begin{bmatrix} \frac{1}{s+1} & \frac{1}{s+2} \\ 0 & \frac{s+1}{s+2} \end{bmatrix}, \boldsymbol{G}_2(s) = \begin{bmatrix} \frac{1}{s+3} & \frac{1}{s+4} \\ \frac{1}{s+1} & 0 \end{bmatrix}$$

Try to determine the transfer matrix $\boldsymbol{G}(s)$ of the feedback connection system.

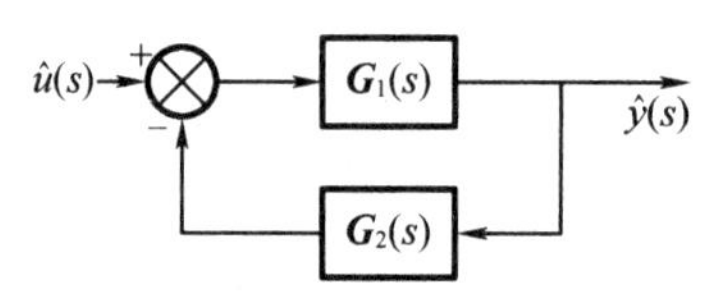

Figure P1.6

1.20 Given the feedback connection system in Figure P1.6, where

$$\boldsymbol{G}_1(s) = \frac{2s+1}{s(s+1)(s+3)}, \boldsymbol{G}_2(s) = \frac{s+2}{s+4}$$

Try to determine the state equation and output equation of the feedback connection system.

1.21 Find the state space description and transfer function of the following parallel system.

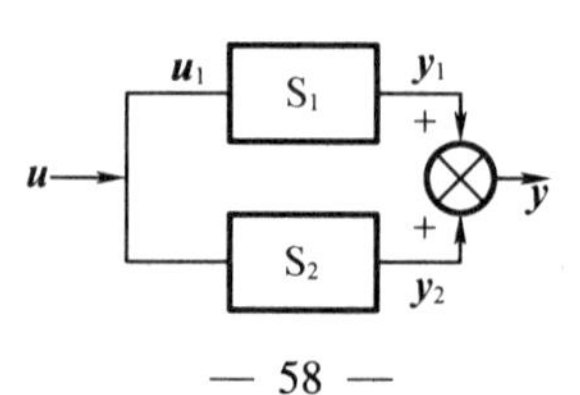

where

$$\mathrm{S}_1: \boldsymbol{A}_1 = \begin{bmatrix} 0 & 1 \\ 1 & 2 \end{bmatrix}, \boldsymbol{B}_1 = \begin{bmatrix} 1 \\ 0 \end{bmatrix}, \boldsymbol{C}_1 = [1 \quad 1]$$

$$\mathrm{S}_2: \boldsymbol{A}_2 = \begin{bmatrix} 0 & 1 \\ -2 & -3 \end{bmatrix}, \boldsymbol{B}_2 = \begin{bmatrix} 1 \\ 1 \end{bmatrix}, \boldsymbol{C}_2 = [0 \quad 1]$$

1.22 Find the state space description and transfer function of the feedback connection system shown in the following figure.

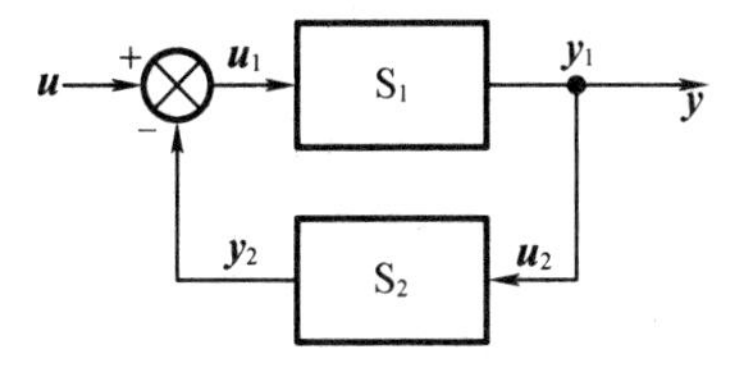

where

$$\mathrm{S}_1: \boldsymbol{A}_1 = \begin{bmatrix} 0 & 1 \\ -2 & -3 \end{bmatrix}, \boldsymbol{B}_1 = \begin{bmatrix} 0 \\ 1 \end{bmatrix}, \boldsymbol{C}_1 = [1 \quad 0]$$

$$\mathrm{S}_2: A_2 = 3, B_2 = 2, C_2 = 1$$

1.23 Find the state space description and transfer function of the following cascade connection system.

where

$$\mathrm{S}_1: \boldsymbol{A}_1 = -3, B_1 = 1, C_1 = 1$$

$$\mathrm{S}_2: \boldsymbol{A}_2 = \begin{bmatrix} 0 & 1 \\ -3 & 1 \end{bmatrix}, \boldsymbol{B}_2 = \begin{bmatrix} 0 \\ 1 \end{bmatrix}, \boldsymbol{C}_2 = [3 \quad 1]$$

Chapter 2 Motion Analysis of Linear Systems

After the mathematical description of linear systems is established, the system's mathematical description can be used to analyze the behavior of systems. The purpose of system analysis is to reveal the motion law and basic characteristics of systems. Its analysis methods mainly include quantitative analysis and qualitative analysis. In quantitative analysis of systems, the motion law of systems is accurately studied, that is, the system response under a given input and an initial state is quantitatively determined, which is mathematically reflected in the solution of the analytical form of the state equation. In the qualitative analysis of systems, we will mainly discuss the basic properties of systems, such as controllability, observability, and stability. In this chapter we will take linear system as the object, discuss the problems of quantitative analysis of the system, determine the motion law of the system, clarify its motion properties, and introduce its analysis methods. In the later chapters, the qualitative analysis of linear systems will be discussed.

2.1 Introduction

2.1.1 Math substance of the motion analysis

For a linear system, its state motion is determined by the state equation. For a LTV system, the state equation is

$$\dot{\boldsymbol{x}} = \boldsymbol{A}(t)\boldsymbol{x} + \boldsymbol{B}(t)\boldsymbol{u},\ \boldsymbol{x}(t_0) = \boldsymbol{x}_0,\ t \in [t_0, t_\alpha] \tag{2.1.1}$$

For a LTI system, the state equation is

$$\dot{\boldsymbol{x}} = \boldsymbol{A}\boldsymbol{x} + \boldsymbol{B}\boldsymbol{u},\ \boldsymbol{x}(0) = \boldsymbol{x}_0,\ t \geqslant 0 \tag{2.1.2}$$

The purpose of system motion analysis is to quantitatively and precisely determine the motion law of systems based on the mathematical model of systems. As can be seen from eq. (2.1.1) and eq. (2.1.2), for a given linear system, its state motion is determined by its initial state $\boldsymbol{x}_0$ and external input $\boldsymbol{u}$. Mathematically, to solve the state motion of the system is to solve the vector differential eq. (2.1.1) or eq. (2.1.2) under the action of given $\boldsymbol{x}_0$ and $\boldsymbol{u}$. After $\boldsymbol{x}(t)$ is obtained, the output of the system can be obtained by substituting $\boldsymbol{x}(t)$ into the output equation which is a algebraic equation. This chapter mainly studies how to solve $\boldsymbol{x}(t)$ and other related problems.

Although the state motion of systems is a response excited by the initial state $\boldsymbol{x}_0$ and external input $\boldsymbol{u}$, the motion law of the system state is mainly determined by the structure and parameters of the system, that is, by $(\boldsymbol{A}(t), \boldsymbol{B}(t))$ or $(\boldsymbol{A}, \boldsymbol{B})$. The state responses of systems with different structures and parameters are different under given $\boldsymbol{x}_0$ and $\boldsymbol{u}$.

2.1.2 Zero-input response and zero-state response

Since the mapping between the input $\boldsymbol{u}(t)$ and output $\boldsymbol{y}(t)$ of a linear system satisfies the superposition principle, and the discussion in Section 1.2 shows that the output variables can be chosen as the state variables, so the mapping between the input $\boldsymbol{u}(t)$ and the state $\boldsymbol{x}(t)$ also satisfies the superposition principle. In addition, according to the discussion in Section 1.1, the non-zero initial condition can be equivalent to adding an impulse input to the input terminal of the system at the initial moment, so we can view $\boldsymbol{x}_0$ as an input. Therefore, the superposition principle is also satisfied between $(\boldsymbol{x}_0, \boldsymbol{u})$ and $\boldsymbol{x}(t)$. Thus, the state motion $\boldsymbol{x}(t)$ of linear systems is equal to the algebraic sum of the responses excited by $\boldsymbol{u}(t)$ and $\boldsymbol{x}_0$, separately.

1. Zero-input response

If the input $\boldsymbol{u}(t)$ is identically zero for $t \geqslant t_0$, the response exclusively excited by the initial state is called zero input response. By this time, the state equation of the system becomes

$$\dot{\boldsymbol{x}} = \boldsymbol{A}(t)\boldsymbol{x}, \ \boldsymbol{x}(t_0) = \boldsymbol{x}_0, \ t \in [t_0, t_\alpha] \tag{2.1.3}$$

The solution of above state equation is the zero-input response which is denoted by $\varphi(t; t_0, \boldsymbol{x}_0, \boldsymbol{0})$.

2. Zero-state response

If the initial state $\boldsymbol{x}_0$ is zero, then the response will be excited exclusively by the input $\boldsymbol{u}(t)$. This response is called the zero-state response. By this time, the state equation of the system becomes

$$\dot{\boldsymbol{x}} = \boldsymbol{A}(t)\boldsymbol{x} + \boldsymbol{B}(t)\boldsymbol{u}, \ \boldsymbol{x}(t_0) = \boldsymbol{0}, \ t \in [t_0, t_\alpha] \tag{2.1.4}$$

The solution of above state equation is the zero-state response which is denoted by $\varphi(t; t_0, \boldsymbol{0}, \boldsymbol{u})$.

Based on above discussion, the total motion response $\varphi(t; t_0, \boldsymbol{x}_0, \boldsymbol{u})$ of the system is the response excited by $\boldsymbol{x}_0$ and $\boldsymbol{u}(t)$ together, which is the sum of zero input response and zero state response, i. e.

$$\varphi(t; t_0, x_0, \boldsymbol{u}) = \varphi(t; t_0, \boldsymbol{x}_0, \boldsymbol{0}) + \boldsymbol{\varphi}(t; t_0, \boldsymbol{0}, \boldsymbol{u}) \tag{2.1.5}$$

2.2 Motion analysis of LTI systems

First, the motion analysis of LTI systems is discussed, then we discuss the motion analysis of LTV systems. In this Section, we mainly analyze the state motion of LTI systems. According to Section 2.1, the state motion response of systems includes zero-input response and zero-state response.

2.2.1 Zero-input response

Let the input of the system $\boldsymbol{u}=\boldsymbol{0}$, and then the state equation of the system becomes

$$\dot{\boldsymbol{x}}=\boldsymbol{A}\boldsymbol{x},\ \boldsymbol{x}(0)=\boldsymbol{x}_0,\ t\geqslant 0 \tag{2.2.1}$$

where $\boldsymbol{x}$ is an n-dimensional state vector, $\boldsymbol{A}$ is an $n\times n$ constant matrix.

In order to compute the zero-input response, the concept of matrix exponential function is introduced first.

1. Matrix exponential function

The matrix exponential function is a matrix function on square matrices analogous to the ordinary exponential function.

For an $n\times n$ matrix $\boldsymbol{A}$, define the $n\times n$ matrix function as

$$\mathrm{e}^{\boldsymbol{A}t}\triangleq\boldsymbol{I}+\boldsymbol{A}t+\frac{1}{2!}\boldsymbol{A}^2t^2+\frac{1}{3!}\boldsymbol{A}^3t^3+\cdots=\sum_{k=0}^{\infty}\frac{1}{k!}\boldsymbol{A}^kt^k \tag{2.2.2}$$

The $\mathrm{e}^{\boldsymbol{A}t}$ or $\exp(\boldsymbol{A}t)$ is called matrix exponential function.

2. Zero-input response

Conclusion 2.2.1 The zero-input response of the LTI system eq. (2.2.1) is expressed as

$$\boldsymbol{x}(t)=\boldsymbol{\varphi}(t;0,\boldsymbol{x}_0,\boldsymbol{0})=\mathrm{e}^{\boldsymbol{A}t}\boldsymbol{x}_0,t\geqslant 0 \tag{2.2.3}$$

Proof Let the solution of eq. (2.2.1) be a power series of undetermined coefficient vectors, i. e.

$$\boldsymbol{x}(t)=\boldsymbol{b}_0+\boldsymbol{b}_1t+\boldsymbol{b}_2t^2+\cdots=\sum_{k=0}^{\infty}\boldsymbol{b}_kt^k,t\geqslant 0 \tag{2.2.4}$$

Substituting eq. (2.2.4) into eq. (2.2.1), yields

$$\boldsymbol{b}_1+2\boldsymbol{b}_2t+3\boldsymbol{b}_3t^2+\cdots=\boldsymbol{A}\boldsymbol{b}_0+\boldsymbol{A}\boldsymbol{b}_1t+\boldsymbol{A}\boldsymbol{b}_2t^2+\cdots \tag{2.2.5}$$

Compare coefficient vectors of $t^k(k=0,1,2,\cdots)$ on both sides of eq. (2.2.5) and get

$$\begin{cases}\boldsymbol{b}_1=\boldsymbol{A}\boldsymbol{b}_0\\ 2\boldsymbol{b}_2=\boldsymbol{A}\boldsymbol{b}_1\\ 3\boldsymbol{b}_3=\boldsymbol{A}\boldsymbol{b}_2\\ \quad\vdots\\ k\boldsymbol{b}_k=\boldsymbol{A}\boldsymbol{b}_{k-1}\\ \quad\vdots\end{cases} \tag{2.2.6}$$

Furthermore, we have

$$\begin{cases} \boldsymbol{b}_1 = \boldsymbol{A}\boldsymbol{b}_0 \\ \boldsymbol{b}_2 = \frac{1}{2}\boldsymbol{A}\boldsymbol{b}_1 = \frac{1}{2}\boldsymbol{A}^2\boldsymbol{b}_0 = \frac{1}{2!}\boldsymbol{A}^2\boldsymbol{b}_0 \\ \boldsymbol{b}_3 = \frac{1}{3}\boldsymbol{A}\boldsymbol{b}_2 = \frac{1}{3}\boldsymbol{A} \cdot \frac{1}{2!}\boldsymbol{A}^2\boldsymbol{b}_0 = \frac{1}{3!}\boldsymbol{A}^3\boldsymbol{b}_0 \\ \vdots \\ \boldsymbol{b}_k = \frac{1}{k!}\boldsymbol{A}^k\boldsymbol{b}_0 \\ \vdots \end{cases} \tag{2.2.7}$$

Substituting eq. (2.2.7) into eq. (2.2.4), yields

$$\boldsymbol{x}(t) = \left(\boldsymbol{I} + \boldsymbol{A}t + \frac{1}{2!}\boldsymbol{A}^2t^2 + \frac{1}{3!}\boldsymbol{A}^3t^3 + \cdots\right)\boldsymbol{b}_0, t \geqslant 0 \tag{2.2.8}$$

Let $t = 0$ in the above equation, we get

$$\boldsymbol{x}_0 = \boldsymbol{b}_0 \tag{2.2.9}$$

From eq. (2.2.2), eq. (2.2.8) and eq. (2.2.9), we get

$$\boldsymbol{x}(t) = e^{\boldsymbol{A}t}\boldsymbol{x}_0 \tag{2.2.10}$$

When $t_0 \neq 0$, by using the similar method we can get the zero-input response of the LTI system as:

$$\boldsymbol{x}(t) = \varphi(t; t_0, \boldsymbol{x}_0, \boldsymbol{0}) = e^{\boldsymbol{A}(t-t_0)}\boldsymbol{x}_0, t \geqslant t_0 \tag{2.2.11}$$

3. Properties of matrix exponential function $e^{\boldsymbol{A}t}$

Based on the definition of matrix exponential function eq. (2.2.2), we can get its some properties as follows:

(1) $\lim_{t \to 0} e^{\boldsymbol{A}t} = \boldsymbol{I}$

(2) $(e^{\boldsymbol{A}t})^{\mathrm{T}} = e^{\boldsymbol{A}^{\mathrm{T}}t}$

(3) Let t and τ are two variables, we have

$$e^{\boldsymbol{A}(t+\tau)} = e^{\boldsymbol{A}t} \cdot e^{\boldsymbol{A}\tau} = e^{\boldsymbol{A}\tau} \cdot e^{\boldsymbol{A}t} \tag{2.2.12}$$

(4) $e^{\boldsymbol{A}t}$ is always nonsingular and its inverse is

$$(e^{\boldsymbol{A}t})^{-1} = e^{-\boldsymbol{A}t} \tag{2.2.13}$$

Obviously, let $\tau = -t$ in eq. (2.2.12), we get eq. (2.2.13).

(5) If the $n \times n$ matrices $\boldsymbol{A}$ and $\boldsymbol{F}$ are commutative, i. e. $\boldsymbol{AF} = \boldsymbol{FA}$, we have

$$e^{(\boldsymbol{A}+\boldsymbol{F})t} = e^{\boldsymbol{A}t} \cdot e^{\boldsymbol{F}t} = e^{\boldsymbol{F}t} \cdot e^{\boldsymbol{A}t} \tag{2.2.14}$$

(6) The derivative of $e^{\boldsymbol{A}t}$ with respect to t is

$$\frac{\mathrm{d}}{\mathrm{d}t}e^{\boldsymbol{A}t} = \boldsymbol{A}e^{\boldsymbol{A}t} = e^{\boldsymbol{A}t}\boldsymbol{A} \tag{2.2.15}$$

(7) For a square matrix $\boldsymbol{A}$, we have

$$(e^{\boldsymbol{A}t})^m = e^{\boldsymbol{A}(mt)}, m = 0, 1, 2, \cdots \tag{2.2.16}$$

4. Matrix exponential functions of some typical matrices

(1) When $\boldsymbol{A}$ is a diagonal matrix, i. e. $\boldsymbol{A} = \mathrm{diag}(\lambda_1, \lambda_2, \cdots, \lambda_n)$,

$$e^{At} = I + At + \frac{1}{2!}A^2t^2 + \frac{1}{3!}A^3t^3 + \cdots$$

$$= I + \begin{bmatrix} \lambda_1 & & & \\ & \lambda_2 & & \\ & & \ddots & \\ & & & \lambda_n \end{bmatrix} t + \frac{1}{2\,!}\begin{bmatrix} \lambda_1^2 & & & \\ & \lambda_2^2 & & \\ & & \ddots & \\ & & & \lambda_n^2 \end{bmatrix} t^2 + \cdots$$

$$= \begin{bmatrix} e^{\lambda_1 t} & & & \\ & e^{\lambda_2 t} & & \\ & & \ddots & \\ & & & e^{\lambda_n t} \end{bmatrix} \tag{2.2.17}$$

(2) When A is a diagonal block matrix, i. e. $A = \mathrm{diag}(A_1, A_2, \cdots, A_l)$,

$$e^{At} = \begin{bmatrix} e^{A_1 t} & & & \\ & e^{A_2 t} & & \\ & & \ddots & \\ & & & e^{A_l t} \end{bmatrix} \tag{2.2.18}$$

(3) When A has the following form:

$$A = \begin{bmatrix} 0 & 1 & 0 \\ 0 & 0 & 1 \\ 0 & 0 & 0 \end{bmatrix}$$

It is called a nilpotent matrix, i. e. after several multiplications, it becomes zero.

$$A^2 = \begin{bmatrix} 0 & 0 & 1 \\ 0 & 0 & 0 \\ 0 & 0 & 0 \end{bmatrix}$$

$$A^k = \mathbf{0}_{3\times 3}, \quad k = 3, 4, \cdots$$

So, applying the definition of e^{At}, yields

$$e^{At} = \sum_{k=0}^{\infty} \frac{1}{k!}A^k t^k = \sum_{k=0}^{2} \frac{1}{k!}A^k t^k = \begin{bmatrix} 1 & t & \frac{t^2}{2} \\ 0 & 1 & t \\ 0 & 0 & 1 \end{bmatrix} \tag{2.2.19}$$

Eq. (2.2.19) can be extended to the square matrix of the following form,

$$A = \begin{bmatrix} 0 & 1 & 0 & & & \\ & 0 & 1 & 0 & & \\ & & 0 & \ddots & \ddots & \\ & & & \ddots & 1 & 0 \\ & & & & 0 & 1 \\ & & & & & 0 \end{bmatrix} \tag{2.2.20}$$

$\boldsymbol{A}$ is the matrix where only the elements on the upper right subdiagonal are 1 and the other elements are zero, we have

$$e^{At}=\begin{bmatrix}1 & t & \frac{t^2}{2!} & \frac{t^3}{3!} & \cdots & \frac{t^{n-1}}{(n-1)!} \\ 0 & 1 & t & \frac{t^2}{2!} & \cdots & \frac{t^{n-2}}{(n-2)!} \\ 0 & 0 & 1 & t & \ddots & \vdots \\ & & 0 & & \ddots & \frac{t^2}{2!} \\ & & & & \ddots & t \\ & & & & \ddots & 1 \\ & & & & & 0\end{bmatrix} \tag{2.2.21}$$

(4) When $\boldsymbol{A}$ has the following form:

$$\boldsymbol{A}=\begin{bmatrix}\lambda & 1 & 0 & & & \\ & \lambda & 1 & & & \\ & & \lambda & 1 & & \\ & & & \ddots & \ddots & \\ & & & & \lambda & 1 \\ & & & & & \lambda\end{bmatrix} \tag{2.2.22}$$

$\boldsymbol{A}$ can be decomposed as the sum of the two matrices like

$$\boldsymbol{A}=\begin{bmatrix}\lambda & & & & & \\ & \lambda & & & & \\ & & \lambda & & & \\ & & & \ddots & & \\ & & & & \lambda & \\ & & & & & \lambda\end{bmatrix}+\begin{bmatrix}0 & 1 & & & & \\ & 0 & 1 & & & \\ & & 0 & 1 & & \\ & & & \ddots & \ddots & \\ & & & & 0 & 1 \\ & & & & & 0\end{bmatrix}=\boldsymbol{A}_1+\boldsymbol{A}_2$$

$\boldsymbol{A}_1$ and $\boldsymbol{A}_2$ are commutative , i. e. $\boldsymbol{A}_1\cdot\boldsymbol{A}_2=\boldsymbol{A}_2\cdot\boldsymbol{A}_1$, so we have $e^{At}=e^{A_1t}\cdot e^{A_2t}$. From eq. (2.2.17) and eq. (2.2.21), we get

$$e^{At}=\begin{bmatrix}e^{\lambda t} & te^{\lambda t} & \frac{t^2}{2!}e^{\lambda t} & \frac{t^3}{3!}e^{\lambda t} & \cdots & \frac{t^{n-1}}{(n-1)!}e^{\lambda t} \\ 0 & e^{\lambda t} & te^{\lambda t} & \frac{t^2}{2!}e^{\lambda t} & \cdots & \frac{t^{n-2}}{(n-2)!}e^{\lambda t} \\ 0 & 0 & e^{\lambda t} & te^{\lambda t} & \ddots & \vdots \\ & & & & \ddots & \frac{t^2}{2!}e^{\lambda t} \\ & & & & \ddots & te^{\lambda t} \\ & & & & & e^{\lambda t}\end{bmatrix} \tag{2.2.23}$$

(5) When $\boldsymbol{A}$ has the following form:

$$A=\begin{bmatrix}0 & \omega\\ -\omega & 0\end{bmatrix} \tag{2.2.24}$$

Based on the definition of e^{At}, we have

$$\begin{aligned} e^{At} &= \begin{bmatrix}1 & 0\\ 0 & 1\end{bmatrix} + t\begin{bmatrix}0 & \omega\\ -\omega & 0\end{bmatrix} + \frac{t^2}{2!}\begin{bmatrix}-\omega^2 & 0\\ 0 & -\omega^2\end{bmatrix} + \frac{t^3}{3!}\begin{bmatrix}0 & -\omega^3\\ -\omega^3 & 0\end{bmatrix} + \cdots \\ &= \begin{bmatrix}\cos\omega t & \sin\omega t\\ -\sin\omega t & \cos\omega t\end{bmatrix} \end{aligned} \tag{2.2.25}$$

(6) When A has the following form:

$$A=\begin{bmatrix}\sigma & \omega\\ -\omega & \sigma\end{bmatrix} \tag{2.2.26}$$

A can be decomposed as the sum of the two matrices like

$$A=\begin{bmatrix}\sigma & 0\\ 0 & \sigma\end{bmatrix}+\begin{bmatrix}0 & \omega\\ -\omega & 0\end{bmatrix}$$

Similar to (4), from eq. (2.2.17) and eq. (2.2.25), we have

$$e^{At}=\begin{bmatrix}e^{\sigma t}\cos\omega t & e^{\sigma t}\sin\omega t\\ -e^{\sigma t}\sin\omega t & e^{\sigma t}\cos\omega t\end{bmatrix} \tag{2.2.27}$$

7) When A has the following form:

$$A=\left[\begin{array}{cc:cc}\sigma & \omega & 1 & 0\\ -\omega & \sigma & 0 & 1\\ \hdashline 0 & 0 & \sigma & \omega\\ 0 & 0 & -\omega & \sigma\end{array}\right]=\begin{bmatrix}A_1 & I\\ 0 & A_1\end{bmatrix}$$

Decompose A into

$$A=\begin{bmatrix}A_1 & 0\\ 0 & A_1\end{bmatrix}+\begin{bmatrix}0 & I\\ 0 & 0\end{bmatrix}=\alpha+\beta$$

Because $\alpha\cdot\beta=\beta\cdot\alpha$ and $e^{At}=e^{(\alpha+\beta)t}=e^{\alpha t}\cdot e^{\beta t}$, using eq. (2.2.17), (2.2.18) and (2.2.27), we get

$$e^{At}=\begin{bmatrix}e^{A_1 t} & 0\\ 0 & e^{A_1 t}\end{bmatrix}\cdot\begin{bmatrix}I & tI\\ 0 & I\end{bmatrix}=\left[\begin{array}{cc:cc}e^{\sigma t}\cos\omega t & e^{\sigma t}\sin\omega t & te^{\sigma t}\cos\omega t & te^{\sigma t}\sin\omega t\\ -e^{\sigma t}\sin\omega t & e^{\sigma t}\cos\omega t & -te^{\sigma t}\sin\omega t & te^{\sigma t}\cos\omega t\\ \hdashline 0 & 0 & e^{\sigma t}\cos\omega t & e^{\sigma t}\sin\omega t\\ 0 & 0 & -e^{\sigma t}\sin\omega t & e^{\sigma t}\cos\omega t\end{array}\right] \tag{2.2.28}$$

5. The calculation methods of matrix exponential function

Here are some calculation methods of e^{At} when A doesn't have a special form.

Method 1 Direct method. Compute e^{At} based on the definition of matrix exponential function:

$$e^{At}=I+At+\frac{1}{2!}A^2t^2+\frac{1}{3!}A^3t^3+\cdots \tag{2.2.29}$$

In general, it is difficult to obtain closed analytical expressions by this method.

Method 2 When the eigenvalues $\lambda_1, \lambda_2 \cdots, \lambda_n$ of $\boldsymbol{A}$ are distinct, choose a suitable matrix $\boldsymbol{P}$ such that $\boldsymbol{A}$ be transformed into a diagonal form, and

$$\boldsymbol{A} = \boldsymbol{P}\begin{bmatrix} \lambda_1 & & & \\ & \lambda_2 & & \\ & & \ddots & \\ & & & \lambda_n \end{bmatrix}\boldsymbol{P}^{-1} \tag{2.2.30}$$

So

$$\mathrm{e}^{\boldsymbol{A}t} = \boldsymbol{P}\begin{bmatrix} \mathrm{e}^{\lambda_1 t} & & & \\ & \mathrm{e}^{\lambda_2 t} & & \\ & & \ddots & \\ & & & \mathrm{e}^{\lambda_n t} \end{bmatrix}\boldsymbol{P}^{-1} \tag{2.2.31}$$

Proof From eq. (2.2.30), we have

$$\boldsymbol{P}^{-1}\boldsymbol{A}\boldsymbol{P} = \begin{bmatrix} \lambda_1 & & & \\ & \lambda_2 & & \\ & & \ddots & \\ & & & \lambda_n \end{bmatrix}$$

Then

$$(\boldsymbol{P}^{-1}\boldsymbol{A}\boldsymbol{P})(\boldsymbol{P}^{-1}\boldsymbol{A}\boldsymbol{P}) = \boldsymbol{P}^{-1}\boldsymbol{A}^2\boldsymbol{P} = \begin{bmatrix} \lambda_1^{\ 2} & & & \\ & \lambda_2^{\ 2} & & \\ & & \ddots & \\ & & & \lambda_n^{\ 2} \end{bmatrix}$$

So

$$\boldsymbol{A}^2 = \boldsymbol{P}\begin{bmatrix} \lambda_1^{\ 2} & & & \\ & \lambda_2^{\ 2} & & \\ & & \ddots & \\ & & & \lambda_n^{\ 2} \end{bmatrix}\boldsymbol{P}^{-1}$$

In a similar way, multiply $\boldsymbol{P}^{-1}\boldsymbol{A}\boldsymbol{P}$ k times,

$$(\boldsymbol{P}^{-1}\boldsymbol{A}\boldsymbol{P})(\boldsymbol{P}^{-1}\boldsymbol{A}\boldsymbol{P})\cdots(\boldsymbol{P}^{-1}\boldsymbol{A}\boldsymbol{P}) = \boldsymbol{P}^{-1}\boldsymbol{A}^k P = \begin{bmatrix} \lambda_1^{\ k} & & & \\ & \lambda_2^{\ k} & & \\ & & \ddots & \\ & & & \lambda_n^{\ k} \end{bmatrix}$$

So

$$\boldsymbol{A}^k=\boldsymbol{P}\begin{bmatrix}\lambda_1{}^k & & & \\ & \lambda_2{}^k & & \\ & & \ddots & \\ & & & \lambda_n{}^k\end{bmatrix}\boldsymbol{P}^{-1},k=2,3,\cdots$$

From the definition of e^{At}, we get

$$\begin{aligned}
e^{At}&=\boldsymbol{I}+\boldsymbol{A}t+\frac{1}{2!}\boldsymbol{A}^2t^2+\frac{1}{3!}\boldsymbol{A}^3t^3+\cdots\\
&=\boldsymbol{P}\boldsymbol{P}^{-1}+\boldsymbol{P}\begin{bmatrix}\lambda_1 & & & \\ & \lambda_2 & & \\ & & \ddots & \\ & & & \lambda_n\end{bmatrix}\boldsymbol{P}^{-1}t+\frac{1}{2!}\boldsymbol{P}\begin{bmatrix}\lambda_1{}^2 & & & \\ & \lambda_2{}^2 & & \\ & & \ddots & \\ & & & \lambda_n{}^2\end{bmatrix}\boldsymbol{P}^{-1}t^2+\\
&\quad\frac{1}{3!}\boldsymbol{P}\begin{bmatrix}\lambda_1{}^3 & & & \\ & \lambda_2{}^3 & & \\ & & \ddots & \\ & & & \lambda_n{}^3\end{bmatrix}\boldsymbol{P}^{-1}t^3+\cdots\\
&=\boldsymbol{P}\left[\boldsymbol{I}+\begin{bmatrix}\lambda_1 & & & \\ & \lambda_2 & & \\ & & \ddots & \\ & & & \lambda_n\end{bmatrix}t+\frac{1}{2!}\begin{bmatrix}\lambda_1{}^2 & & & \\ & \lambda_2{}^2 & & \\ & & \ddots & \\ & & & \lambda_n{}^2\end{bmatrix}t^2+\frac{1}{3!}\begin{bmatrix}\lambda_1{}^3 & & & \\ & \lambda_2{}^3 & & \\ & & \ddots & \\ & & & \lambda_n{}^3\end{bmatrix}t^3+\cdots\right]\boldsymbol{P}^{-1}\\
&=\boldsymbol{P}\begin{bmatrix}e^{\lambda_1t} & & & \\ & e^{\lambda_2t} & & \\ & & \ddots & \\ & & & e^{\lambda_nt}\end{bmatrix}\boldsymbol{P}^{-1}
\end{aligned}$$

Similarly, when $\boldsymbol{A}$ has repeated eigenvalues and does not meet the conditions of transforming the diagonal form, determine matrix $\boldsymbol{Q}$ such that $\boldsymbol{A}$ is transformed into the Jordan form. In order to simplify the discussion, we take an example. Let $\boldsymbol{A}$ has eigenvalues λ_1 (multiplicity 3) and λ_2 (multiplicity 2). The matrix $\boldsymbol{Q}$ transforms $\boldsymbol{A}$ into the Jordan form, and

$$\boldsymbol{A}=\boldsymbol{Q}\begin{bmatrix}\lambda_1 & 1 & 0 & 0 & 0\\ 0 & \lambda_1 & 1 & 0 & 0\\ 0 & 0 & \lambda_1 & 0 & 0\\ 0 & 0 & 0 & \lambda_2 & 1\\ 0 & 0 & 0 & 0 & \lambda_2\end{bmatrix}\boldsymbol{Q}^{-1} \tag{2.2.32}$$

Then the matrix exponential function is

$$e^{At}=Q\begin{bmatrix} e^{\lambda_1 t} & te^{\lambda_1 t} & \dfrac{t^2 e^{\lambda_1 t}}{2\,!} & 0 & 0 \\ 0 & e^{\lambda_1 t} & te^{\lambda_1 t} & 0 & 0 \\ 0 & 0 & e^{\lambda_1 t} & 0 & 0 \\ 0 & 0 & 0 & e^{\lambda_2 t} & te^{\lambda_2 t} \\ 0 & 0 & 0 & 0 & e^{\lambda_2 t} \end{bmatrix} QA^{-1} \tag{2.2.33}$$

Proof let

$$\begin{bmatrix} \lambda_1 & 1 & 0 & 0 & 0 \\ 0 & \lambda_1 & 1 & 0 & 0 \\ 0 & 0 & \lambda_1 & 0 & 0 \\ 0 & 0 & 0 & \lambda_2 & 1 \\ 0 & 0 & 0 & 0 & \lambda_2 \end{bmatrix} = \begin{bmatrix} J_1 & \\ & J_2 \end{bmatrix}$$

and

$$A^k = Q\begin{bmatrix} J_1{}^k & \\ & J_2{}^k \end{bmatrix} Q^{-1}, k=2,3,\cdots$$

Based on the definition of matrix exponential function, we have

$$\begin{aligned} e^{At} &= I + At + \frac{1}{2!}A^2t^2 + \frac{1}{3!}A^3t^3 + \cdots \\ &= QQ^{-1} + Q\begin{bmatrix} J_1 & \\ & J_2 \end{bmatrix}Q^{-1}t + \frac{1}{2!}Q\begin{bmatrix} J_1{}^2 & \\ & J_2{}^2 \end{bmatrix}Q^{-1}t^2 + \frac{1}{3!}Q\begin{bmatrix} J_1{}^3 & \\ & J_2{}^3 \end{bmatrix}Q^{-1}t^3 + \cdots \\ &= Q\left(I + \begin{bmatrix} J_1 & \\ & J_2 \end{bmatrix}t + \frac{1}{2!}\begin{bmatrix} J_1{}^2 & \\ & J_2{}^2 \end{bmatrix}t^2 + \frac{1}{3!}\begin{bmatrix} J_1{}^3 & \\ & J_2{}^3 \end{bmatrix}t^3 + \cdots\right)Q^{-1} \\ &= Q\begin{bmatrix} e^{J_1 t} & \\ & e^{J_2 t} \end{bmatrix}Q^{-1} \end{aligned}$$

From eq. (2.2.18) and eq. (2.2.23), we get

$$e^{At}=Q\begin{bmatrix} e^{\lambda_1 t} & te^{\lambda_1 t} & \dfrac{t^2 e^{\lambda_1 t}}{2\,!} & 0 & 0 \\ 0 & e^{\lambda_1 t} & te^{\lambda_1 t} & 0 & 0 \\ 0 & 0 & e^{\lambda_1 t} & 0 & 0 \\ 0 & 0 & 0 & e^{\lambda_2 t} & te^{\lambda_2 t} \\ 0 & 0 & 0 & 0 & e^{\lambda_2 t} \end{bmatrix} Q^{-1}$$

Method 3 Laplace method. Given an $n \times n$ constant matrix A,

$$e^{At} = L^{-1}(sI - A)^{-1} \tag{2.2.34}$$

Proof For vector equation

$$\dot{x} = Ax \tag{2.2.35}$$

take the Laplace transform on both sides of the above equation, we get

$$sx(s) - x(0) = Ax(s)$$

$$(sI - A)x(s) = x(0) \tag{2.2.36}$$

When the solution of eq. (2.2.35) exists, i. e. inverse of $(sI - A)$ exists, from eq. (2.2.36), we have

$$x(s) = (sI - A)^{-1}x(0) \tag{2.2.37}$$

Take inverse Laplace transform on the above equation, we get

$$x(t) = \mathscr{L}^{-1}[(sI - A)^{-1}]x(0)$$

Compare it with eq. (2.2.3), so $e^{At} = \mathscr{L}^{-1}(sI - A)^{-1}$.

Method 4 Use the following theorem to calculate the matrix exponential function.

Theorem 2.2.1 Given an $n \times n$ matrix A, its characteristic polynomial is

$$\det(\lambda I - A) = \prod_{i=1}^{m} (\lambda - \lambda_i)^{n_i} \tag{2.2.38}$$

For any polynomial $f(\lambda)$, there is an $n-1$ order polynomial

$$g(\lambda) = a_0 + a_1\lambda + \cdots + a_{n-1}\lambda^{n-1} \tag{2.2.39}$$

Such that

$$f^{(l)}(\lambda_i) = g^{(l)}(\lambda_i), i = 1, 2, \cdots, m; l = 0, 1, 2, \cdots, n_i - 1 \tag{2.2.40}$$

where $f^{(l)}(\lambda_i) \triangleq \left.\dfrac{d^l f(\lambda)}{d\lambda^l}\right|_{\lambda=\lambda_i}$, $g^{(l)}(\lambda_i) \triangleq \left.\dfrac{d^l g(\lambda)}{d\lambda^l}\right|_{\lambda=\lambda_i}$

Then

$$f(A) = g(A) = a_0 I + a_1 A + \cdots + a_{n-1}A^{n-1} \tag{2.2.41}$$

By using theorem 2.2.1, let $f(\lambda) = e^{\lambda t}$, from eq. (2.2.39) and eq. (2.2.40), we get the values of $a_0, a_1, a_2, \cdots, a_{n-1}$. Then substitute them into eq. (2.2.41), we can get e^{At}.

Example 2.2.1 Given $A = \begin{bmatrix} 0 & 0 & -2 \\ 0 & 1 & 0 \\ 1 & 0 & 3 \end{bmatrix}$, compute e^{At}.

Answer Let $f(\lambda) = e^{\lambda t}$, $e^{At} = f(A)$

The characteristic polynomial of A is $\det(\lambda I - A) = (\lambda - 1)^2(\lambda - 2)$. So eigenvalues of A are $\lambda_1 = 1$, $\lambda_2 = 2$.

Let $g(\lambda) = a_0 + a_1\lambda + a_2\lambda^2$, from eq. (2.2.40), we have

$$\begin{cases} f(\lambda_1) = g(\lambda_1) \\ f'(\lambda_1) = g'(\lambda_1) \\ f(\lambda_2) = g(\lambda_2) \end{cases} \Rightarrow \begin{cases} e^t = a_0 + a_1 + a_2 \\ te^t = a_1 + 2a_2 \\ e^{2t} = a_0 + 2a_1 + 4a_2 \end{cases}$$

Solve the above equation set, we get

$$\begin{cases} a_0 = -2te^t + e^{2t} \\ a_1 = 3te^t + 2e^t - 2e^{2t} \\ a_2 = e^{2t} - e^t - te^t \end{cases}$$

And

$$\begin{aligned}
e^{At} &= f(A) \\
&= g(A) \\
&= a_0\boldsymbol{I} + a_1\boldsymbol{A} + a_2\boldsymbol{A}^2 \\
&= (-2te^t + e^{2t})\boldsymbol{I} + (3te^t + 2e^t - 2e^{2t})\boldsymbol{A} + (e^{2t} - e^t - te^t)\boldsymbol{A}^2 \\
&= \begin{bmatrix} 2e^t - e^{2t} & 0 & 2e^t - 2e^{2t} \\ 0 & e^t & 0 \\ -e^t + e^{2t} & 0 & -e^t + 2e^{2t} \end{bmatrix}
\end{aligned}$$

2.2.2 Zero-state response

The state equation of a LTI system with zero initial state is:

$$\dot{\boldsymbol{x}} = \boldsymbol{Ax} + \boldsymbol{Bu},\ \boldsymbol{x}(0) = ,\ t \geqslant 0 \tag{2.2.42}$$

where $\boldsymbol{x}$ is an n-dimensional state vector, $\boldsymbol{u}$ is p-dimensional input vector, $\boldsymbol{A}$ and $\boldsymbol{B}$ are $n \times n$ and $n \times p$ constant matrix, respectively. Then, for the zero-state response, there are the following conclusions.

Conclusion 2.2.2 The zero-state response of the LTI system described by eq. (2.2.42) can be expressed as:

$$\boldsymbol{A}(t) = \varphi(t;0,\boldsymbol{0},\boldsymbol{u}) = \int_0^t e^{\boldsymbol{A}(t-\tau)}\boldsymbol{Bu}(\tau)\mathrm{d}\tau, t \geqslant 0 \tag{2.2.43}$$

Proof Consider the following equation:

$$\frac{\mathrm{d}}{\mathrm{d}t}e^{-\boldsymbol{A}t}\boldsymbol{x} = \left(\frac{\mathrm{d}}{\mathrm{d}t}e^{-\boldsymbol{A}t}\right)\boldsymbol{x} + e^{-\boldsymbol{A}t}\dot{\boldsymbol{x}} = e^{-\boldsymbol{A}t}(-\boldsymbol{A})\boldsymbol{x} + e^{-\boldsymbol{A}t}\boldsymbol{Ax} + e^{-\boldsymbol{A}t}\boldsymbol{Bu} = e^{-\boldsymbol{A}t}\boldsymbol{Bu}$$

Integrating the both sides from 0 to t, yields

$$e^{-\boldsymbol{A}t}\boldsymbol{x}(t) - \boldsymbol{x}(0) = \int_0^t e^{-\boldsymbol{A}\tau}\boldsymbol{Bu}(\tau)\mathrm{d}\tau \tag{2.2.44}$$

Multiply both sides of eq. (2.2.44) by $e^{\boldsymbol{A}t}$ from left side, and $\boldsymbol{x}(0) = \boldsymbol{0}$, we get

$$\boldsymbol{x}(t) = \varphi(t;0,\boldsymbol{0},\boldsymbol{u}) = \int_0^t e^{\boldsymbol{A}t}e^{-\boldsymbol{A}\tau}\boldsymbol{Bu}(\tau)\mathrm{d}\tau = \int_0^t e^{\boldsymbol{A}(t-\tau)}\boldsymbol{Bu}(\tau)\mathrm{d}\tau$$

When $t_0 \neq 0$, the zero-state response of the LTI system can be expressed as:

$$\boldsymbol{x}(t) = \varphi(t;t_0,\boldsymbol{0},\boldsymbol{u}) = \int_{t_0}^t e^{\boldsymbol{A}(t-\tau)}\boldsymbol{Bu}(\tau)\mathrm{d}\tau, t \geqslant t_0 \tag{2.2.45}$$

2.2.3 State's motion of a LTI system

The State's motion of a LTI system with initial state and external input, i.e. the solution of the state equation

$$\dot{\boldsymbol{x}} = \boldsymbol{Ax} + \boldsymbol{Bu},\ \boldsymbol{x}(0) = \boldsymbol{x}_0,\ t \geqslant 0 \tag{2.2.46}$$

can be obtained by superposition of eq. (2.2.3) and eq. (2.2.43). We have the following conclusions.

Conclusion 2.2.3 The state motion of the LTI system under the initial state and external input can be expressed as:

$$x(t) = \varphi(t;0,x_0,u) = e^{At}x_0 + \int_0^t e^{A(t-\tau)}Bu(\tau)d\tau,\ t \geqslant 0 \tag{2.2.47}$$

For $t_0 \neq 0$, the state motion of the LTI system can be expressed as:

$$x(t) = \varphi(t;t_0,x_0,u) = e^{A(t-t_0)}x_0 + \int_{t_0}^t e^{A(t-\tau)}Bu(\tau)d\tau,\ t \geqslant t_0 \tag{2.2.48}$$

Conclusion 2.2.4 The state motion of the LTI system under the initial state and external input can be expressed as:

$$x(t) = \varphi(t;t_0,x_0,u) = L^{-1}((sI-A)^{-1})x_0 + L^{-1}((sI-A)^{-1}Bu(s)) \tag{2.2.49}$$

Proof Take the Laplace transform on both sides of eq. (2.2.46), we get

$$sx(s) - x(0) = Ax(s) + Bu(s)$$

$$(sI-A)x(s) = x(0) + Bu(s) \tag{2.2.50}$$

When the solution of eq. (2.2.46) exists, i. e. inverse of $(sI-A)$ exists, multiply both sides of the of eq. (2.2.50) by $(sI-A)^{-1}$ from left side, we have

$$x(s) = (sI-A)^{-1}x(0) + (sI-A)^{-1}Bu(s) \tag{2.2.51}$$

Take the inverse Laplace transform on both sides of the above equation, we get

$$x(t) = L^{-1}((sI-A)^{-1})x_0 + L^{-1}((sI-A)^{-1}Bu(s))$$

Example 2.2.2 Given a LTI system

$$\begin{bmatrix}\dot{x}_1\\ \dot{x}_2\end{bmatrix} = \begin{bmatrix}0 & 1\\ -2 & -3\end{bmatrix}\begin{bmatrix}x_1\\ x_2\end{bmatrix} + \begin{bmatrix}0\\ 1\end{bmatrix}u, t \geqslant 0$$

$$y = \begin{bmatrix}0 & 1\end{bmatrix}\begin{bmatrix}x_1\\ x_2\end{bmatrix}$$

where, $x_1(0) = 1$, $x_2(0) = 0$, $u(t)$ is unit step function $1(t)$, compute the system's output $y(t)$.

Answer

Method 1 Using eq. (2.2.47)

$$x(t) = \varphi(t;0,x_0,u) = e^{At}x_0 + \int_0^t e^{A(t-\tau)}Bu(\tau)d\tau, t \geqslant 0$$

compute $x(t)$, then substitute $x(t)$ into the output equation, we can get $y(t)$. First compute e^{At}. Here, we use 3 methods to compute e^{At}.

Method (1) for calculating e^{At}: Diagonal canonical form method

The characteristic polynomial of A is $\det(sI-A) = s^2 + 3s + 2$, so the eigenvalues of A are $s_1 = -1$, $s_2 = -2$.

First determine a 2×2 matrix P such that $P^{-1}AP = \begin{bmatrix}s_1 & 0\\ 0 & s_2\end{bmatrix} = \begin{bmatrix}-1 & 0\\ 0 & -2\end{bmatrix}$. Let $P = [p_1 \quad p_2]$, where p_1 and p_2 are the eigenvectors of A with s_1 and s_2, respectively.

$$(s_1\boldsymbol{I}-\boldsymbol{A})\boldsymbol{p}_1=\boldsymbol{0},\begin{bmatrix}-1 & -1\\ 2 & 2\end{bmatrix}\boldsymbol{p}_1=\boldsymbol{0},\boldsymbol{p}_1=\begin{bmatrix}1\\ -1\end{bmatrix}$$

$$(s_2\boldsymbol{I}-\boldsymbol{A})\boldsymbol{p}_2=\boldsymbol{0},\begin{bmatrix}-2 & -1\\ 2 & 1\end{bmatrix}\boldsymbol{p}_2=\boldsymbol{0},\boldsymbol{p}_2=\begin{bmatrix}1\\ -2\end{bmatrix}$$

So $\boldsymbol{P}=\begin{bmatrix}1 & 1\\ -1 & -2\end{bmatrix},\boldsymbol{P}^{-1}=\begin{bmatrix}2 & 1\\ -1 & -1\end{bmatrix}$. And

$$\begin{aligned}\mathrm{e}^{\boldsymbol{A}t}&=\boldsymbol{P}\begin{bmatrix}\mathrm{e}^{s_1t} & 0\\ 0 & \mathrm{e}^{s_2t}\end{bmatrix}\boldsymbol{P}^{-1}\\ &=\begin{bmatrix}1 & 1\\ -1 & -2\end{bmatrix}\begin{bmatrix}\mathrm{e}^{s_1t} & 0\\ 0 & \mathrm{e}^{s_2t}\end{bmatrix}\begin{bmatrix}2 & 1\\ -1 & -1\end{bmatrix}\\ &=\begin{bmatrix}2\mathrm{e}^{-t}-\mathrm{e}^{-2t} & \mathrm{e}^{-t}-\mathrm{e}^{-2t}\\ -2\mathrm{e}^{-t}+2\mathrm{e}^{-2t} & -\mathrm{e}^{-t}+2\mathrm{e}^{-2t}\end{bmatrix}\end{aligned}$$

Method (2) for calculating $\mathrm{e}^{\boldsymbol{A}t}$: Laplace method

$$\begin{aligned}(s\boldsymbol{I}-\boldsymbol{A})^{-1}&=\begin{bmatrix}s & -1\\ 2 & s+3\end{bmatrix}^{-1}\\ &=\frac{1}{s^2+3s+2}\begin{bmatrix}s+3 & 1\\ -2 & s\end{bmatrix}\\ &=\begin{bmatrix}\frac{2}{s+1}+\frac{-1}{s+2} & \frac{1}{s+1}+\frac{-1}{s+2}\\ \frac{-2}{s+1}+\frac{2}{s+2} & \frac{-1}{s+1}+\frac{2}{s+2}\end{bmatrix}\end{aligned}$$

The inverse Laplace transform of the above equation yields

$$\mathrm{e}^{\boldsymbol{A}t}=L^{-1}[(s\boldsymbol{I}-\boldsymbol{A})^{-1}]=\begin{bmatrix}2\mathrm{e}^{-t}-\mathrm{e}^{-2t} & \mathrm{e}^{-t}-\mathrm{e}^{-2t}\\ -2\mathrm{e}^{-t}+2\mathrm{e}^{-2t} & -\mathrm{e}^{-t}+2\mathrm{e}^{-2t}\end{bmatrix}$$

Method (3) for calculating $\mathrm{e}^{\boldsymbol{A}t}$:

Let $f(\lambda)=\mathrm{e}^{\lambda t}$, so $\mathrm{e}^{\boldsymbol{A}t}=f(\lambda)|_{\lambda=\boldsymbol{A}}=f(\boldsymbol{A})$.

The characteristic polynomial of $\boldsymbol{A}$ is $\det(\lambda\boldsymbol{I}-\boldsymbol{A})=\lambda^2+3\lambda+2$, so the eigenvalues of $\boldsymbol{A}$ are $\lambda_1=-1,\lambda_2=-2$.

Let $g(\lambda)=a_0+a_1\lambda$, from eq. (2.2.40), we have

$$\begin{cases}f(\lambda_1)=g(\lambda_1)\\ f(\lambda_2)=g(\lambda_2)\end{cases}\Rightarrow\begin{cases}\mathrm{e}^{-t}=a_0-a_1\\ \mathrm{e}^{-2t}=a_0-2a_1\end{cases}$$

Solve the above equation set, we have

$$\begin{cases}a_1=\mathrm{e}^{-t}-\mathrm{e}^{-2t}\\ a_0=2\mathrm{e}^{-t}-\mathrm{e}^{-2t}\end{cases}$$

$$\begin{aligned}\mathrm{e}^{\boldsymbol{A}t}&=f(\boldsymbol{A})\\ &=g(\boldsymbol{A})\end{aligned}$$

$$= a_0\boldsymbol{I} + a_1\boldsymbol{A}$$

$$= (2e^{-t} - e^{-2t})\boldsymbol{I} + (e^{-t} - e^{-2t})\boldsymbol{A}$$

$$= \begin{bmatrix} 2e^{-t} - e^{-2t} & 0 \\ 0 & 2e^{-t} - e^{-2t} \end{bmatrix} + \begin{bmatrix} 0 & e^{-t} - e^{-2t} \\ -2e^{-t} + 2e^{-2t} & -3e^{-t} + 3e^{-2t} \end{bmatrix}$$

$$= \begin{bmatrix} 2e^{-t} - e^{-2t} & e^{-t} - e^{-2t} \\ -2e^{-t} + 2e^{-2t} & -e^{-t} + 2e^{-2t} \end{bmatrix}$$

From eq. (2.2.47), we get

$$\boldsymbol{x}(t) = e^{\boldsymbol{A}t}\boldsymbol{x}_0 + \int_0^t e^{\boldsymbol{A}(t-\tau)}\boldsymbol{B}\boldsymbol{u}(\tau)\,d\tau$$

$$= \begin{bmatrix} 2e^{-t} - e^{-2t} & e^{-t} - e^{-2t} \\ -2e^{-t} + 2e^{-2t} & -e^{-t} + 2e^{-2t} \end{bmatrix}\begin{bmatrix} 1 \\ 0 \end{bmatrix} +$$

$$\int_0^t \begin{bmatrix} 2e^{-(t-\tau)} - e^{-2(t-\tau)} & e^{-(t-\tau)} - e^{-2(t-\tau)} \\ -2e^{-(t-\tau)} + 2e^{-2(t-\tau)} & -e^{-(t-\tau)} + 2e^{-2(t-\tau)} \end{bmatrix}\begin{bmatrix} 0 \\ 1 \end{bmatrix} d\tau$$

$$= \begin{bmatrix} 2e^{-t} - e^{-2t} \\ -2e^{-t} + 2e^{-2t} \end{bmatrix} + \int_0^t \begin{bmatrix} e^{-(t-\tau)} - e^{-2(t-\tau)} \\ -e^{-(t-\tau)} + 2e^{-2(t-\tau)} \end{bmatrix} d\tau$$

$$= \begin{bmatrix} \frac{1}{2} + e^{-t} - \frac{1}{2}e^{-2t} \\ -e^{-t} + e^{-2t} \end{bmatrix}$$

$$y = [0 \quad 1]\begin{bmatrix} x_1 \\ x_2 \end{bmatrix} = -e^{-t} + e^{-2t}$$

Method 2 From eq. (2.2.49)

$$\boldsymbol{x}(t) = L^{-1}((s\boldsymbol{I} - \boldsymbol{A})^{-1})\boldsymbol{x}_0 + L^{-1}((s\boldsymbol{I} - \boldsymbol{A})^{-1}\boldsymbol{B}\boldsymbol{u}(s))$$

$$= L^{-1}((s\boldsymbol{I} - \boldsymbol{A})^{-1}(\boldsymbol{x}_0 + \boldsymbol{B}\boldsymbol{u}(s)))$$

$$= L^{-1}\left((s\boldsymbol{I} - \boldsymbol{A})^{-1}\left(\begin{bmatrix} 1 \\ 0 \end{bmatrix} + \begin{bmatrix} 0 \\ \frac{1}{s} \end{bmatrix}\right)\right)$$

$$= L^{-1}\left(\frac{1}{s^2 + 3s + 2}\begin{bmatrix} s + 3 & 1 \\ -2 & s \end{bmatrix}\begin{bmatrix} 1 \\ \frac{1}{s} \end{bmatrix}\right)$$

$$= L^{-1}\left(\frac{1}{s^2 + 3s + 2}\begin{bmatrix} s + 3 + \frac{1}{s} \\ -1 \end{bmatrix}\right)$$

$$= L^{-1}\left(\begin{bmatrix} \frac{1/2}{s} + \frac{1}{s+1} - \frac{1/2}{s+2} \\ -\frac{1}{s+1} + \frac{1}{s+2} \end{bmatrix}\right) = \begin{bmatrix} \frac{1}{2} + e^{-t} - \frac{1}{2}e^{-2t} \\ -e^{-t} + e^{-2t} \end{bmatrix}$$

and

$$y = [0 \quad 1]\begin{bmatrix} x_1 \\ x_2 \end{bmatrix} = -e^{-t} + e^{-2t}$$

2.3 The state transition matrix of LTI systems

In this section, the state transition matrix is introduced. The state motion excited by $\boldsymbol{x}_0$ and $\boldsymbol{u}(t)$ is a kind of state transition which can be expressed by state transition matrix. In addition, the state transition matrix can be used to establish the uniformly analytical expression for the state motion of linear systems no matter LTV systems or LTI systems.

2.3.1 State transition matrix

1. Definition

Given a LTI system

$$\dot{\boldsymbol{x}} = \boldsymbol{A}\boldsymbol{x} + \boldsymbol{B}\boldsymbol{u},\ \boldsymbol{x}(t_0) = \boldsymbol{x}_0,\ t \geqslant t_0 \tag{2.3.1}$$

where $\boldsymbol{x}$ is an n-dimensional state vector. The $n \times n$ matrix $\boldsymbol{\Phi}(t-t_0)$ which satisfies the following equations

$$\begin{cases} \dot{\boldsymbol{\Phi}}(t-t_0) = \boldsymbol{A}\boldsymbol{\Phi}(t-t_0), t \geqslant t_0 \\ \boldsymbol{\Phi}(0) = \boldsymbol{I} \end{cases} \tag{2.3.2}$$

is called the state transition matrix of the LTI system (2.3.1).

2. The state transition matrix of the LTI system is

$$\boldsymbol{\Phi}(t-t_0) = e^{\boldsymbol{A}(t-t_0)},\ t \geqslant t_0 \tag{2.3.3}$$

or

$$\boldsymbol{\Phi}(t) = e^{\boldsymbol{A}t},\ t \geqslant 0 \tag{2.3.4}$$

Proof Based on the properties of $\exp(\boldsymbol{A}t)$, we can easily prove eq. (2.3.3) and eq. (2.3.4) satisfying the definition of state transition matrix (2.3.2).

Theorem 2.3.1 The set of all solutions of the vector differential equation

$$\dot{\boldsymbol{x}} = \boldsymbol{A}(t)\boldsymbol{x} \tag{2.3.5}$$

forms an n-dimensional linear space over the real number field **R**.

Proof Let $\boldsymbol{p}_1$ and $\boldsymbol{p}_2$ be any two solutions of the vector differential eq. (2.3.5). The $\alpha_1 \boldsymbol{p}_1 + \alpha_2 \boldsymbol{p}_2$ is also a solution of the vector differential eq. (2.3.5) for any two real numbers α_1 and α_2. This is because:

$$\frac{d}{dt}(\alpha_1 \boldsymbol{p}_1 + \alpha_2 \boldsymbol{p}_2) = \alpha_1 \frac{d}{dt}\boldsymbol{p}_1 + \alpha_2 \frac{d}{dt}\boldsymbol{p}_2 = \alpha_1 \boldsymbol{A}(t)\boldsymbol{p}_1 + \alpha_2 \boldsymbol{A}(t)\boldsymbol{p}_2 = \boldsymbol{A}(t)(\alpha_1 \boldsymbol{p}_1 + \alpha_2 \boldsymbol{p}_2)$$

Therefore, the solution space of the vector differential eq. (2.3.5) is linear.

Next, to prove the solution space of the vector differential eq. (2.3.5) is n-dimensional. Let

$e_1, e_2, \cdots, e_n$ be any linearly independent vectors in an n-dimensional vector space $(\mathbf{R}, \mathbf{R}^n)$, and p_i be the solution of the vector differential eq. (2.3.5) with $p_i(t_0) = e_i (i = 1, 2, \cdots, n)$ as initial condition. As long as $p_1(t), p_2(t), \cdots, p_n(t)$ is linearly independent and every solution of the vector differential equation (2.3.5) can be represented by a linear combination of $p_1(t), p_2(t), \cdots, p_n(t)$, the solution space must be n-dimensional.

Now prove it by contradiction. We assume that $p_1(t), p_2(t), \cdots, p_n(t)$ are linearly dependent, then must exist $\boldsymbol{a} = \begin{bmatrix} \alpha_1 \\ \alpha_2 \\ \vdots \\ \alpha_n \end{bmatrix} \neq \mathbf{0}$, such that

$$[p_1(t) \quad p_2(t) \quad \cdots \quad p_n(t)]\boldsymbol{a} = \mathbf{0}, \forall t \in (-\infty, +\infty)$$

Now let $t = t_0$, we have $[p_1(t_0) \quad p_2(t_0) \quad \cdots \quad p_n(t_0)]\boldsymbol{a} = [e_1 \quad e_2 \quad \cdots \quad e_n]\boldsymbol{a} = \mathbf{0}$.

It means that $e_1, e_2, \cdots, e_n$ is linearly dependent, which contradicts the assumption that $e_1, e_2, \cdots, e_n$ is linearly independent. So $p_1(t), p_2(t), \cdots, p_n(t)$ are linearly independent when $t \in (-\infty, +\infty)$.

Let $p(t)$ be any solution of the eq. (2.3.5) and $p(t_0) = e$. Because $e_1, e_2, \cdots, e_n$ is n linearly independent vectors in the n-dimensional vector space $(\mathbf{R}, \mathbf{R}^n)$, e can be uniquely represented by a linear combination of $e_1, e_2, \cdots, e_n$, i.e. $e = \sum_{i=1}^{n} \alpha_i e_i$. Obviously, $\sum_{i=1}^{n} \alpha_i p_i(t)$ is a solution of eq. (2.3.5) with $e = \sum_{i=1}^{n} \alpha_i p_i(t_0)$ as initial condition. From the uniqueness of the solution, we get $p(t) = \sum_{i=1}^{n} \alpha_i p_i(t)$.

3. Fundamental Matrix $\boldsymbol{\Psi}(t)_{n \times n} = [\boldsymbol{\psi}_1 \boldsymbol{\psi}_2 \cdots \boldsymbol{\psi}_n]$

According to theorem 2.3.1, there are n linearly independent solutions in the solution space of vector differential equation $\dot{\boldsymbol{x}} = \boldsymbol{A}(t)\boldsymbol{x}$. When $\boldsymbol{\psi}_1, \boldsymbol{\psi}_2, \cdots, \boldsymbol{\psi}_n$ are linearly independent solutions of $\dot{\boldsymbol{x}} = \boldsymbol{A}(t)\boldsymbol{x}$, call the $n \times n$ matrix $\boldsymbol{\Psi}(t)$ the fundamental matrix of $\dot{\boldsymbol{x}} = \boldsymbol{A}(t)\boldsymbol{x}$, where $\boldsymbol{\Psi}(t) = [\boldsymbol{\psi}_1 \quad \boldsymbol{\psi}_2 \quad \cdots \quad \boldsymbol{\psi}_n]$.

The fundamental matrix $\boldsymbol{\Psi}(t)$ has the following properties:

$$\begin{cases} \dot{\boldsymbol{\Psi}}(t) = \boldsymbol{A}\boldsymbol{\Psi}(t) \\ \boldsymbol{\Psi}(t_0) = \boldsymbol{H}, t \geqslant t_0 \end{cases} \tag{2.3.6}$$

where $\boldsymbol{H}$ is a non-singular constant matrix.

Any $n \times n$ matrix satisfying eq. (2.3.6) is the fundamental matrix of $\dot{\boldsymbol{x}} = \boldsymbol{A}(t)\boldsymbol{x}$.

According to eq. (2.3.2) and eq. (2.3.6), the state transition matrix of the LTI system is constructed by the fundamental matrix $\boldsymbol{\Psi}(t)$ as follow:

$$\boldsymbol{\Phi}(t - t_0) = \boldsymbol{\Psi}(t)\boldsymbol{\Psi}^{-1}(t_0), \ t \geqslant t_0 \tag{2.3.7}$$

This is because $\dot{\boldsymbol{\Phi}}(t-t_0)=\dot{\boldsymbol{\Psi}}(t)\boldsymbol{\Psi}^{-1}(t_0)=\boldsymbol{A}\boldsymbol{\Psi}(t)\boldsymbol{\Psi}^{-1}(t_0)=\boldsymbol{A}\boldsymbol{\Phi}(t-t_0)$ and $\boldsymbol{\Phi}(0)=\boldsymbol{\Psi}(t_0)\boldsymbol{\Psi}^{-1}(t_0)=\boldsymbol{I}$. Therefore, eq. (2.3.7) satisfies the definition of state transition matrix of LTI systems.

In addition, $\frac{\mathrm{d}}{\mathrm{d}t}\mathrm{e}^{At}=A\mathrm{e}^{At}$ and e^{At_0} is a non-singular constant matrix, so e^{At} is a fundamental matrix, i. e. $\boldsymbol{\Psi}(t)=\mathrm{e}^{At}, t\geqslant t_0$. Thus, from eq. (2.3.7), we get

$$\boldsymbol{\Phi}(t-t_0)=\mathrm{e}^{At}(\mathrm{e}^{At_0})^{-1}=\mathrm{e}^{At}\mathrm{e}^{-At_0}=\mathrm{e}^{A(t-t_0)}, t\geqslant t_0$$

when $t_0=0$, $\boldsymbol{\Phi}(t)=\mathrm{e}^{At}, t\geqslant 0$.

2.3.2 State motion expression by using state transition matrix

According to eq. (2.3.3) and eq. (2.3.4), the analytical expression of the state motion of a LTI system discussed in the previous section can be rewritten into the following form using the state transition matrix:

$$\boldsymbol{\Phi}(t;t_0,\boldsymbol{x}_0,\boldsymbol{u}) = \boldsymbol{\Phi}(t-t_0)\boldsymbol{x}_0+\int_{t_0}^{t}\boldsymbol{F}(t-\tau)\boldsymbol{B}\boldsymbol{u}(\tau)\mathrm{d}\tau, t\geqslant t_0 \tag{2.3.8}$$

and

$$\boldsymbol{\Phi}(t;0,\boldsymbol{x}_0,\boldsymbol{u}) = \boldsymbol{\Psi}(t)\boldsymbol{x}_0+\int_{0}^{t}\boldsymbol{\Psi}(t-\tau)\boldsymbol{B}\boldsymbol{u}(\tau)\mathrm{d}\tau, t\geqslant 0 \tag{2.3.9}$$

2.3.3 The properties of transition matrix of LTI systems

Form eq. (2.3.7), some properties of the state transition matrix can be obtained as follows:

(1) $\boldsymbol{\Phi}(0)=\boldsymbol{\Psi}(t_0)\boldsymbol{\Psi}^{-1}(t_0)=\boldsymbol{I}$

(2) $\boldsymbol{\Phi}^{-1}(t-t_0)=\boldsymbol{\Psi}(t_0)\boldsymbol{\Psi}^{-1}(t)=\boldsymbol{\Phi}(t_0-t)$

(3) $\boldsymbol{\Phi}(t_2-t_0)=\boldsymbol{\Psi}(t_2)\boldsymbol{\Psi}^{-1}(t_0)=\boldsymbol{\Psi}(t_2)\boldsymbol{\Psi}^{-1}(t_1)\boldsymbol{\Psi}(t_1)\boldsymbol{\Psi}^{-1}(t_0)=\boldsymbol{\Phi}(t_2-t_1)\boldsymbol{\Phi}(t_1-t_0)$

(4) $\boldsymbol{\Phi}(t_2+t_1)=\boldsymbol{\Phi}(t_2-(-t_1))=\boldsymbol{\Phi}(t_2-0)\boldsymbol{\Phi}(0-(-t_1))=\boldsymbol{\Phi}(t_2)\boldsymbol{\Phi}(t_1)$

(5) $\boldsymbol{\Phi}(mt)=\boldsymbol{\Phi}(t+t+\cdots+t)=(\boldsymbol{\Phi}(t))^m$

(6) $\boldsymbol{\Phi}(t-t_0)$ is uniquely determined by $\boldsymbol{A}$ and independent of the choice of the fundamental matrix.

Proof Let $\boldsymbol{\Psi}_1(t)_{n\times n}$和$\boldsymbol{\Psi}_2(t)_{n\times n}$ are the fundamental matrix of $\dot{\boldsymbol{x}}=\boldsymbol{A}(t)\boldsymbol{x}$, so $\boldsymbol{\Psi}_2(t)=\boldsymbol{\Psi}_1(t)\boldsymbol{P}_{n\times n}$. Where, $\boldsymbol{P}_{n\times n}$ is a non-singular constant matrix.

Therefor, $\boldsymbol{\Phi}(t-t_0)=\boldsymbol{\Psi}_2(t)\boldsymbol{\Psi}_2^{-1}(t_0)=\boldsymbol{\Psi}_1(t)\boldsymbol{P}\boldsymbol{P}^{-1}\boldsymbol{\Psi}_1^{-1}(t_0)=\boldsymbol{\Psi}_1(t)\boldsymbol{\Psi}_1^{-1}(t_0)$. It can be seen that $\boldsymbol{\Phi}(t-t_0)$ is unique and does not change with the different fundamental matrices. In fact, from $\dot{\boldsymbol{\Phi}}(t-t_0)=\boldsymbol{A}\boldsymbol{\Phi}(t-t_0)$, $\boldsymbol{\Phi}(t-t_0)$ is determined uniquely by $\boldsymbol{A}$.

2.4 Motion analysis of LTV systems

The state motion expression of LTV systems is the same as that of the LTI system, but the calculation is complicated and difficult to get the closed-form solution. Usually, it can only be calculated by computer. In this section, we will study on the following LTV systems:

$$\begin{cases} \dot{\boldsymbol{x}} = \boldsymbol{A}(t)\boldsymbol{x} + \boldsymbol{B}(t)\boldsymbol{u}, \ \boldsymbol{x}(t_0) = \boldsymbol{x}_0, \ t \in [t_0, t_\alpha] \\ \boldsymbol{y} = \boldsymbol{C}(t)\boldsymbol{x} + \boldsymbol{D}(t)\boldsymbol{u} \end{cases} \tag{2.4.1}$$

where $\boldsymbol{x}$ is an n-dimensional state vector, $\boldsymbol{u}$ is a p-dimensional input vector, $\boldsymbol{y}$ is a q-dimensional output vector, $\boldsymbol{A}(t)$, $\boldsymbol{B}(t)$, $\boldsymbol{C}(t)$ and $\boldsymbol{D}(t)$ are $n\times n, n\times p, q\times n$ and $q\times p$ time-varying real-valued matrices, respectively.

2.4.1 The state transition matrix of LTV Systems

1. Definition

The $n\times n$ $\boldsymbol{\Phi}(t,t_0)$ which satisfies the following matrix equations

$$\dot{\boldsymbol{\Phi}}(t,t_0) = \boldsymbol{A}\boldsymbol{\Phi}(t,t_0) \tag{2.4.2}$$

$$\boldsymbol{\Phi}(t_0,t_0) = \boldsymbol{I} \tag{2.4.3}$$

is called the state transition matrix of the LTV system (2.4.1).

2. The construction formula of state transition matrix

Similar to the LTI system, the construction formula of $\boldsymbol{\Phi}(t,t_0)$ is

$$\boldsymbol{\Phi}(t,t_0) = \boldsymbol{\psi}(t)\boldsymbol{\psi}^{-1}(t_0), \ t \geqslant t_0 \tag{2.4.4}$$

where $\boldsymbol{\psi}(t)$ is any a fundamental solution matrix of $\dot{\boldsymbol{x}} = \boldsymbol{A}(t)\boldsymbol{x}$, whose, the n columns of $\boldsymbol{\psi}(t)$ are linearly independent solutions of $\dot{\boldsymbol{x}} = \boldsymbol{A}(t)\boldsymbol{x}$. Because $\boldsymbol{\Phi}(t,t_0)$ satisfies eq. (2.4.2) and eq. (2.4.3), we can verify eq. (2.4.4) using $\dot{\boldsymbol{\psi}}(t) = \boldsymbol{A}(t)\boldsymbol{\psi}(t)$.

3. The calculation of state transition matrix $\boldsymbol{\Phi}(t,t_0)$

(1) If $\boldsymbol{A}(t_1)\boldsymbol{A}(t_2) = \boldsymbol{A}(t_2)\boldsymbol{A}(t_1)$, then

$$\boldsymbol{\Phi}(t,t_0) = \mathrm{e}^{\int_{t_0}^{t}\boldsymbol{A}(\tau)\mathrm{d}\tau} = \boldsymbol{I} + \int_{t_0}^{t}\boldsymbol{A}(\tau)\mathrm{d}\tau + \frac{1}{2!}\left(\int_{t_0}^{t}\boldsymbol{A}(\tau)\mathrm{d}\tau\right)^2 + \frac{1}{3!}\left(\int_{t_0}^{t}\boldsymbol{A}(\tau)\mathrm{d}\tau\right)^3 + \cdots \tag{2.4.5}$$

(2) If $\boldsymbol{A}(t_1)\boldsymbol{A}(t_2) \neq \boldsymbol{A}(t_2)\boldsymbol{A}(t_1)$, then

$$\boldsymbol{\Phi}(t,t_0) = \boldsymbol{I} + \int_{t_0}^{t}\boldsymbol{A}(\tau)\mathrm{d}\tau + \int_{t_0}^{t}\boldsymbol{A}(\tau_1)\int_{t_0}^{\tau_1}\boldsymbol{A}(\tau_2)\mathrm{d}\tau_2\mathrm{d}\tau_1 + \cdots \tag{2.4.6}$$

4. The properties of state transition matrix $\boldsymbol{\Phi}(t,t_0)$

(1) $\boldsymbol{\Phi}(t,t) = \boldsymbol{I}$

(2) $\boldsymbol{\Phi}^{-1}(t,t_0) = \boldsymbol{\psi}(t_0)\boldsymbol{\psi}^{-1}(t) = \boldsymbol{\Phi}(t_0,t)$ (2.4.7)

(3) $\boldsymbol{\Phi}(t_2,t_0)=\boldsymbol{\psi}(t_2)\boldsymbol{\psi}^{-1}(t_0)=\boldsymbol{\psi}(t_2)\boldsymbol{\psi}^{-1}(t_1)\boldsymbol{\psi}(t_1)\boldsymbol{\psi}^{-1}(t_0)=\boldsymbol{\Phi}(t_2,t_1)\boldsymbol{\Phi}(t_1,t_0)$ (2.4.8)

(4) $\frac{\mathrm{d}}{\mathrm{d}t}\boldsymbol{\Phi}(t_0,t)=-\boldsymbol{\Phi}(t_0,t)\boldsymbol{A}(t)$ (2.4.9)

Proof From $\boldsymbol{\Phi}(t_0,t)\boldsymbol{\Phi}^{-1}(t_0,t)=I$ and eq. (2.4.7), we have

$$\boldsymbol{\Phi}(t_0,t)\boldsymbol{\Phi}(t,t_0)=\boldsymbol{I}$$

Taking the first derivative of both sides of the above equation with respect to time t, and yields

$$\frac{\mathrm{d}}{\mathrm{d}t}\boldsymbol{F}(t_0,t)\cdot\boldsymbol{\Phi}(t,t_0)+\boldsymbol{\Phi}(t_0,t)\cdot\frac{\mathrm{d}}{\mathrm{d}t}\boldsymbol{\Phi}(t,t_0)=\boldsymbol{O}$$

$$\begin{aligned}\frac{d}{dt}\boldsymbol{\Phi}(t_0,t)&=-\boldsymbol{\Phi}(t_0,t)\cdot\frac{\mathrm{d}}{\mathrm{d}t}\boldsymbol{\Phi}(t,t_0)\cdot\boldsymbol{\Phi}^{-1}(t,t_0)\\&=-\boldsymbol{\Phi}(t_0,t)\cdot\boldsymbol{\Phi}(t)\boldsymbol{\Phi}(t,t_0)\cdot\boldsymbol{\Phi}^{-1}(t,t_0)\\&=-\boldsymbol{\Phi}(t_0,t)\cdot\boldsymbol{\Phi}(t)\end{aligned}$$

2.4.2 The state motion of LTV sytems

The expression of state motion of LTV systems excited by an initial state and a input, namely, the expression of the solution of state eq. (2.4.1) is:

$$\boldsymbol{x}(t)=\boldsymbol{\Phi}(t,t_0)\boldsymbol{x}(t_0)+\int_{t_0}^{t}\boldsymbol{\Phi}(t,\tau)\boldsymbol{B}(\tau)\boldsymbol{u}(\tau)\mathrm{d}\tau \tag{2.4.10}$$

Proof Take the first derivative of both sides of eq. (2.4.10) with respect to time t, we have

$$\begin{aligned}\frac{\mathrm{d}}{\mathrm{d}t}\boldsymbol{x}(t)&=\frac{\partial}{\partial t}\boldsymbol{F}(t,t_0)\boldsymbol{x}(t_0)+\frac{\partial}{\partial t}\int_{t_0}^{t}\boldsymbol{F}(t,\tau)\boldsymbol{B}(\tau)\boldsymbol{u}(\tau)\mathrm{d}\tau\\&=\boldsymbol{A}(t)\boldsymbol{\Phi}(t,t_0)\boldsymbol{x}(t_0)+\boldsymbol{\Phi}(t,t)\boldsymbol{B}(t)\boldsymbol{u}(t)+\int_{t_0}^{t}\boldsymbol{A}(t)\boldsymbol{\Phi}(t,\tau)\boldsymbol{B}(\tau)\boldsymbol{u}(\tau)\mathrm{d}\tau\\&=\boldsymbol{A}(t)\left(\boldsymbol{\Phi}(t,t_0)\boldsymbol{x}(t_0)+\int_{t_0}^{t}\boldsymbol{\Phi}(t,\tau)\boldsymbol{B}(\tau)\boldsymbol{u}(\tau)\mathrm{d}\tau\right)+\boldsymbol{B}(t)\boldsymbol{u}(t)\\&=\boldsymbol{A}(t)\boldsymbol{x}(t)+\boldsymbol{B}(t)\boldsymbol{u}(t)\end{aligned}$$

note: $\frac{\partial}{\partial t}\int_{t_0}^{t}f(t,\tau)\mathrm{d}\tau=f(t,\tau)\big|_{\tau=t}+\int_{t_0}^{t}\frac{\partial}{\partial t}f(t,\tau)\mathrm{d}\tau$

Meanwhile, when $t=t_0$,

$$\boldsymbol{x}(t_0)=\boldsymbol{\Phi}(t_0,t_0)\boldsymbol{x}(t_0)+\int_{t_0}^{t_0}\boldsymbol{\Phi}(t_0,\tau)\boldsymbol{B}(\tau)\boldsymbol{u}(\tau)\mathrm{d}\tau=\boldsymbol{I}\times\boldsymbol{x}(t_0)+\boldsymbol{0}=\boldsymbol{x}(t_0)$$

So, eq. (2.4.10) is the solution of eq. (2.4.1) satisfying initial conditions.

Example 2.4.1 Given $\dot{\boldsymbol{x}}=\boldsymbol{A}(t)\boldsymbol{x}$, $\boldsymbol{A}(t)=\begin{bmatrix}0&\frac{1}{(t+1)^2}\\0&0\end{bmatrix}$, find $\boldsymbol{\Phi}(t,t_0)$.

Answer Because $\boldsymbol{A}(t_1)\boldsymbol{A}(t_2)=\begin{bmatrix}0&0\\0&0\end{bmatrix}=\boldsymbol{A}(t_2)\boldsymbol{A}(t_1)$, so

$$\boldsymbol{\Phi}(t,t_0) = \boldsymbol{I} + \int_{t_0}^{t}\begin{bmatrix} 0 & \dfrac{1}{(\tau+1)^2} \\ 0 & 0 \end{bmatrix}\mathrm{d}\tau + \frac{1}{2!}\left(\int_{t_0}^{t}\begin{bmatrix} 0 & \dfrac{1}{(\tau+1)^2} \\ 0 & 0 \end{bmatrix}\mathrm{d}\tau\right)^2 + \cdots$$

$$= \begin{bmatrix} 1 & 0 \\ 0 & 1 \end{bmatrix} + \begin{bmatrix} 0 & \dfrac{1}{\tau+1}\Big|_{t}^{t_0} \\ 0 & 0 \end{bmatrix} + \boldsymbol{O} + \boldsymbol{O} + \cdots$$

$$= \begin{bmatrix} 1 & \dfrac{t-t_0}{(t+1)(t_0+1)} \\ 0 & 1 \end{bmatrix}$$

Example 2.4.2 Given $\dot{\boldsymbol{x}}(t) = \begin{bmatrix} 0 & 0 \\ t & 0 \end{bmatrix}\boldsymbol{x}(t) + \begin{bmatrix} 1 \\ 1 \end{bmatrix}\boldsymbol{u}$, $u = 1(t-1)$, $x_1(1) = 1$, $x_2(1) = 2$, $t_0 = 1$, find $\boldsymbol{x}(t)$.

Answer Solving vector differential equations $\begin{cases} \dot{x}_1 = 0 \\ \dot{x}_2 = tx_1 \end{cases}$ yields

$$x_1(t) = x_1(t_0)$$

$$x_2(t) = \frac{1}{2}t^2 x_1(t) - \frac{1}{2}t_0{}^2 x_1(t) + x_2(t_0)$$

take $\begin{cases} x_1(t_0) = 0 \\ x_2(t_0) = 1 \end{cases}$, we have the solution of the above differential equations is $\boldsymbol{\psi}_1(t) = \begin{bmatrix} 0 \\ 1 \end{bmatrix}$

take $\begin{cases} x_1(t_0) = 2 \\ x_2(t_0) = 0 \end{cases}$, we have the solution of the above differential equations is $\boldsymbol{\psi}_2(t) = \begin{bmatrix} 2 \\ t^2 - t_0^2 \end{bmatrix}$

So, The fundamental solution matrix of the above vector differential equation is

$$\boldsymbol{\psi}(t) = [\boldsymbol{\psi}_1(t) \quad \boldsymbol{\psi}_1(t)] = \begin{bmatrix} 0 & 2 \\ 1 & t^2 - t_0^2 \end{bmatrix}$$

The state transition matrix of the system is

$$\boldsymbol{\Phi}(t,t_0) = \boldsymbol{\psi}(t)\boldsymbol{\psi}^{-1}(t_0) = \begin{bmatrix} 0 & 2 \\ 1 & t^2 - t_0^2 \end{bmatrix}\begin{bmatrix} 0 & 2 \\ 1 & 0 \end{bmatrix}^{-1} = \begin{bmatrix} 1 & 0 \\ 0.5t^2 - 0.5t_0^2 & 1 \end{bmatrix}$$

Therefor,

$$\boldsymbol{x}(t) = \boldsymbol{\Phi}(t,t_0)\boldsymbol{x}(t_0) + \int_{t_0}^{t}\boldsymbol{\Phi}(t,\tau)\boldsymbol{B}(\tau)\boldsymbol{u}(\tau)\mathrm{d}\tau = \begin{bmatrix} t \\ \dfrac{1}{3}t^3 + t + \dfrac{2}{3} \end{bmatrix}$$

2.4.3 Impulse response matrix of LTV systems

Let $\boldsymbol{G}(t,\tau)$ is the impulse response matrix of the LTV system eq. (2.4.1), the relationship between $\boldsymbol{G}(t,\tau)$ and the state space description is

$$\boldsymbol{G}(t,\tau) = \boldsymbol{C}(t)\boldsymbol{\Phi}(t,\tau)\boldsymbol{B}(\tau) + \boldsymbol{D}(t)\delta(t-\tau) \tag{2.4.11}$$

Proof The input-output description of the system eq. (2.4.1) is

$$\boldsymbol{y}(t) = \int_{t_0}^{t} \boldsymbol{G}(t,\tau)\boldsymbol{u}(\tau)\mathrm{d}\tau \tag{2.4.12}$$

Substitute eq. (2.4.10) into the output equation of the system eq. (2.4.1) under zero condition $\boldsymbol{x}(t_0) = \boldsymbol{0}$, we have

$$\begin{aligned}\boldsymbol{y}(t) &= \boldsymbol{C}(t)\int_{t_0}^{t} \boldsymbol{\Phi}(t,\tau)\boldsymbol{B}(\tau)\boldsymbol{u}(\tau)\mathrm{d}\tau + \boldsymbol{D}(t)\boldsymbol{u}(t) \\ &= \int_{t_0}^{t} (\boldsymbol{C}(t)\boldsymbol{\Phi}(t,\tau)\boldsymbol{B}(\tau) + \boldsymbol{D}(\tau)\delta(t-\tau))\boldsymbol{u}(\tau)\mathrm{d}\tau \end{aligned} \tag{2.4.13}$$

Compare eq. (2.4.12) and eq. (2.4.13), we have

$$\boldsymbol{G}(t,\tau) = \boldsymbol{C}(t)\boldsymbol{\Phi}(t,\tau)\boldsymbol{B}(\tau) + \boldsymbol{D}(t)\delta(t-\tau)$$

Problems

2.1 Given the following constant matrices $\boldsymbol{A}$, compute their exponential matrix $\mathrm{e}^{\boldsymbol{A}t}$.

(1) $\boldsymbol{A} = \begin{bmatrix} -2 & 0 \\ 0 & -3 \end{bmatrix}$

(2) $\boldsymbol{A} = \begin{bmatrix} -2 & 1 \\ 0 & -2 \end{bmatrix}$

(3) $\boldsymbol{A} = \begin{bmatrix} 0 & 0 \\ 1 & 0 \end{bmatrix}$

(4) $\boldsymbol{A} = \begin{bmatrix} 0 & -1 \\ 4 & 0 \end{bmatrix}$

2.2 Compute the exponential matrix $\mathrm{e}^{\boldsymbol{A}t}$ of following matrices $\boldsymbol{A}$:

(1) $\boldsymbol{A} = \begin{bmatrix} 0 & 1 \\ -2 & -3 \end{bmatrix}$

(2) $\boldsymbol{A} = \begin{bmatrix} 0 & 1 & 0 \\ 0 & 0 & 1 \\ -6 & -11 & -6 \end{bmatrix}$

2.3 Compute the solutions of state equations of the following systems:

(1) $\begin{bmatrix} \dot{x}_1 \\ \dot{x}_2 \end{bmatrix} = \begin{bmatrix} 0 & 1 \\ -3 & -2 \end{bmatrix}\begin{bmatrix} x_1 \\ x_2 \end{bmatrix}, \begin{bmatrix} x_1(0) \\ x_2(0) \end{bmatrix} = \begin{bmatrix} 1 \\ 1 \end{bmatrix}$

(2) $\begin{bmatrix} \dot{x}_1 \\ \dot{x}_2 \end{bmatrix} = \begin{bmatrix} 0 & 1 \\ -2 & -3 \end{bmatrix}\begin{bmatrix} x_1 \\ x_2 \end{bmatrix} + \begin{bmatrix} 2 \\ 0 \end{bmatrix}u, \begin{bmatrix} x_1(0) \\ x_2(0) \end{bmatrix} = \begin{bmatrix} 0 \\ 1 \end{bmatrix}, u(t) = \mathrm{e}^{-t}, t \geqslant 0$

2.4 For a LTI system, given by

$\boldsymbol{\Phi}(t) = \begin{bmatrix} \mathrm{e}^{-t} & 0 \\ 0 & \mathrm{e}^{-2t} \end{bmatrix}, \boldsymbol{b} = \begin{bmatrix} 1 \\ 1 \end{bmatrix}, \boldsymbol{x}(0) = \begin{bmatrix} 2 \\ 3 \end{bmatrix}$

compute the state response $\boldsymbol{x}(t)$ when

(1) $u(t)=\delta(t)$ (unit impulse function)

(2) $u(t)=1(t)$ (unit step function)

(3) $u(t)=t$

(4) $u(t)=\sin t$

2.5 Given a LTI system

$$\begin{bmatrix}\dot{x}_1\\ \dot{x}_2\end{bmatrix}=\begin{bmatrix}5 & 9\\ 1 & -1\end{bmatrix}\begin{bmatrix}x_1\\ x_2\end{bmatrix}+\begin{bmatrix}0\\ 1\end{bmatrix}u,\begin{bmatrix}x_1(0)\\ x_2(0)\end{bmatrix}=\begin{bmatrix}1\\ 0\end{bmatrix},u(t)=1(t)$$

compute the state response $\boldsymbol{x}(t)$.

2.6 Given the state transition matrix $\boldsymbol{\Phi}(t)$ of a LTI system $\dot{\boldsymbol{x}}=\boldsymbol{A}\boldsymbol{x},t\geqslant 0$ as

$$\boldsymbol{\Phi}(t)=\begin{bmatrix}\frac{1}{2}(\mathrm{e}^{-t}+\mathrm{e}^{3t}) & \frac{1}{4}(-\mathrm{e}^{-t}+\mathrm{e}^{3t})\\ -\mathrm{e}^{-t}+\mathrm{e}^{3t} & \frac{1}{2}(\mathrm{e}^{-t}+\mathrm{e}^{3t})\end{bmatrix}$$

compute $(\mathrm{e}^{\boldsymbol{A}t})^{-1}$ and $\boldsymbol{A}$.

2.7 Use Laplace transform to prove that the expression of state motion of the LTI system $\dot{\boldsymbol{x}}=\boldsymbol{A}\boldsymbol{x}+\boldsymbol{B}\boldsymbol{u},\boldsymbol{x}(0)=\boldsymbol{x}_0$ is:

$$\boldsymbol{x}(t)=\mathrm{e}^{\boldsymbol{A}t}x_0+\int_0^t \mathrm{e}^{\boldsymbol{A}(t-\tau)}\boldsymbol{B}\boldsymbol{u}(t)\mathrm{d}t$$

2.8 Given the matrix differential equation

$$\dot{\boldsymbol{X}}=\boldsymbol{A}\boldsymbol{X}+\boldsymbol{X}\boldsymbol{A}^{\mathrm{T}},\boldsymbol{X}(0)=\boldsymbol{P}_0$$

where, $\boldsymbol{X}$ is an $n\times n$ matrix, $\boldsymbol{A}$ is a constant matrix, try to prove the solution of the above equation is

$$\boldsymbol{X}(t)=\mathrm{e}^{\boldsymbol{A}t}\boldsymbol{P}_0\mathrm{e}^{\boldsymbol{A}^{\mathrm{T}}t}$$

2.9 Given $\dot{\boldsymbol{x}}=\boldsymbol{A}(t)\boldsymbol{x}$ and its companion matrix $\dot{\boldsymbol{z}}=-\boldsymbol{A}^{\mathrm{T}}(t)\boldsymbol{z}$, let $\boldsymbol{\Phi}(t,t_0)$ and $\boldsymbol{\Phi}_z(t,t_0)$ are their state transition matrices, try to prove $\boldsymbol{\Phi}(t,t_0)\boldsymbol{\Phi}_z^{\mathrm{T}}(t,t_0)=\boldsymbol{I}$.

2.10 For the following LTV system

$$\dot{\boldsymbol{x}}=\begin{bmatrix}\boldsymbol{A}_{11}(t) & \boldsymbol{A}_{12}(t)\\ \boldsymbol{A}_{21}(t) & \boldsymbol{A}_{22}(t)\end{bmatrix}x+\begin{bmatrix}\boldsymbol{B}_1(t)\\ \boldsymbol{B}_2(t)\end{bmatrix}\boldsymbol{u},t\geqslant t_0$$

its state transition matrix is

$$\boldsymbol{\Phi}(t)=\begin{bmatrix}\boldsymbol{\Phi}_{11}(t,t_0) & \boldsymbol{\Phi}_{12}(t,t_0)\\ \boldsymbol{\Phi}_{21}(t,t_0) & \boldsymbol{\Phi}_{22}(t,t_0)\end{bmatrix}$$

try to prove that when $\boldsymbol{A}_{21}(t)=\boldsymbol{O}$, $\boldsymbol{\Phi}_{21}(t,t_0)\equiv\boldsymbol{O}$.

2.11 For a 2-dimensinal LTI system

$$\dot{\boldsymbol{x}}=\boldsymbol{A}\boldsymbol{x},t\geqslant 0$$

when

$$\boldsymbol{x}(0)=\begin{bmatrix}1\\-4\end{bmatrix},\boldsymbol{x}(t)=\begin{bmatrix}e^{-3t}\\-4e^{-3t}\end{bmatrix}$$

and

$$\boldsymbol{x}(0)=\begin{bmatrix}2\\-1\end{bmatrix},\boldsymbol{x}(t)=\begin{bmatrix}2e^{-2t}\\-e^{-2t}\end{bmatrix}$$

try to determine the matrix $\boldsymbol{A}$.

2.12 Let $\boldsymbol{A}$ is a square matrix and its eigenvalues are distinct, try to prove

$$\det(e^{\boldsymbol{A}t})=e^{(tr\,\boldsymbol{A})t}$$

where tr $\boldsymbol{A}$ denotes the trace of $\boldsymbol{A}$.

Chapter 3 Controllability and Observability of Linear Systems

System analysis generally consists of quantitative analysis and qualitative analysis. In chapter 2, we discussed the quantitative analysis of systems, which specifies precisely how the state and output of systems change over time, given certain inputs and initial states. The qualitative analysis does not solve the motion equation of systems but determines the qualitative properties of systems directly from the mathematical model of systems, including the controllability and observability, the stability of systems, etc. Controllability and observability are basic structural properties of systems, which were first proposed by R. E. Kalman in the early 1960s. The development of control theory shows that the two concepts have great significance to the study of control and estimation problems. This chapter mainly discusses the definition and criterion of controllability and observability of systems, controllable canonical form and observable canonical form of linear systems, and structural decomposition of LTI systems, etc.

3.1 Definition of controllability and observability

3.1.1 Intuitive discussion of controllability and observability

We first discuss the meaning of controllability and observability of systems intuitively with some physical systems. This intuitive discussion has great help to understand the concepts of controllability and observability. For a system, input and output constitute its external variables, while state variables are its internal variables. Roughly speaking, controllability studies the possibility of steering the state from the input; observability studies the possibility of estimating the state from the output. Controllability refers to whether the state of the system can be influenced by the input. Observability deal with whether or not the initial state can be observed from the output. A system is said to be controllable, or state controllable, if the motion of every state variable can be influenced or controlled by the input from any starting point to the origin, separately. Otherwise, the system is said to be not fully controllable or uncontrollable. Correspondingly, if the motion of all state variables of the system can be fully reflected by the output, the state of the system is observable, or the system is observable. Otherwise, the system is not completely observable or unobservable. We give some examples to illustrate the concepts.

Example 3.1.1 Consider the circuit shown in Figure 3.1.1. We choose the voltages across the capacitors with capacitance C_1 and C_2 as state variables x_1 and x_2 of the system. The input $u(t)$ is a current source and the output y is the voltage.

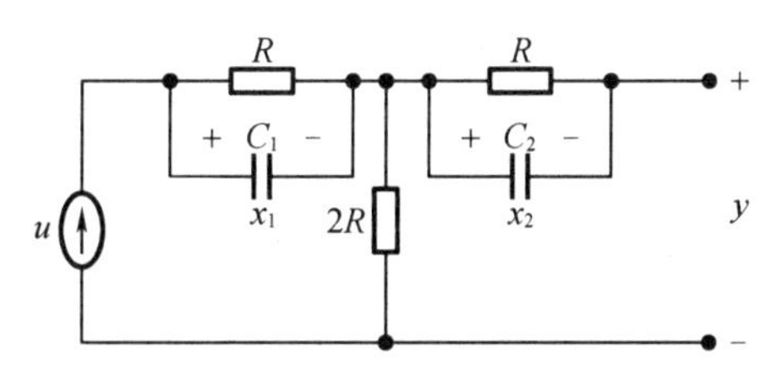

Figure 3.1.1 Incomplete controllable and incomplete observable circuit

We see that, because of the open circuit across y, x_2 is not affected by the input $u(t)$, no matter what input is applied. Therefore, the state variable x_2 cannot be controlled by $u(t)$. But the state variable x_1 can be controlled by the input $u(t)$. Since observability reflects the relationship between the output y and the state, we let the input $u(t)$ be zero when we check the observability. For the current source, this is equivalent to opening the branch where the current source is located. For this case, no matter how the state variable x_1 changes, the output y does not change with it, so the state variable x_1 cannot be observed by y. The state variable x_2 affects the output y, so the state variable x_2 is observable.

Example 3.1.2 Consider the circuit shown in Figure 3.1.2. We choose the voltage across the capacitor as state variable x. The input $u(t)$ is a voltage source and the output y is the voltage.

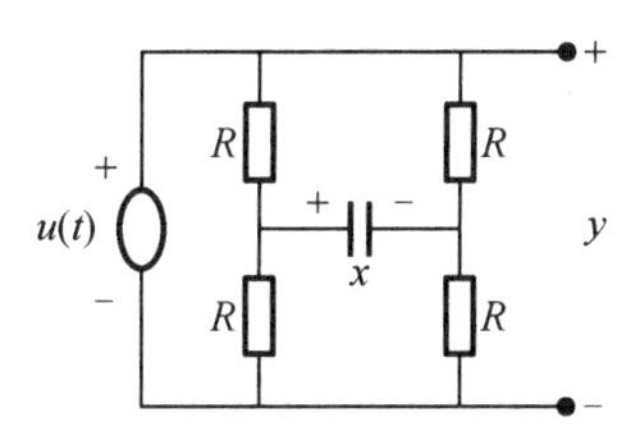

Figure 3.1.2 Uncontrollable and unobservable circuit

We see that, $x(t)$ excited by $u(t)$ is always zero no matter what input $u(t)$ is applied. This is due to the symmetry of the circuit, and the input has no effect on the voltage across the capacitor. Thus the system or more precisely, the state variable x is not controllable. Let $u(t) = 0$, no matter what value of x_0 is, $y(t) = 0$ for all $t \geqslant 0$. So, state motion excited by x_0 is not reflected by y, i. e. state variable x is not observable.

Example 3.1.3 Consider the circuit shown in Figure 3.1.3. It has two state variables x_1 and x_2 as shown. The input $u(t)$ is a voltage source. Input u can transfer x_1 and x_2 to any value at the same time, but it cannot transfer them to different values separately, i. e. no matter how you change u, x_1 always equals x_2, if $x_1(0) = x_2(0)$. Thus the system is not completely controllable.

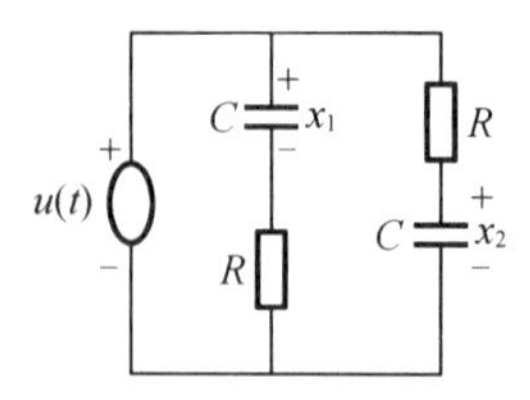

Figure 3.1.3 Not completely controllable circuit

The above descriptions of controllability and observability are only intuitionistic but not rigorous descriptions, and can only be used to explain and judge the controllability and observability of very intuitive and very simple systems. In order to reveal the essential properties of controllability and observability, and to analyze and judge more general and complex systems, it is necessary to establish strict definitions of these two concepts, and derive corresponding criteria and basic properties.

3.1.2 Definition of controllability

Consider a LTV system

$$\dot{\boldsymbol{x}} = \boldsymbol{A}(t)\boldsymbol{x} + \boldsymbol{B}(t)\boldsymbol{u},\ \boldsymbol{x}(t_0) = \boldsymbol{x}_0,\ t \in J \tag{3.1.1}$$

where $\boldsymbol{x}$ is an $n \times 1$ state vector, $\boldsymbol{u}$ is a $p \times 1$ input vector, $\boldsymbol{A}(t)$ is an $n \times n$ time continuous matrix, $\boldsymbol{B}(t)$ is an $n \times p$ time continuous matrix.

The state equation or the pair $(\boldsymbol{A}, \boldsymbol{B})$ is said to be controllable if for any non-zero initial state $\boldsymbol{x}(0) = \boldsymbol{x}_0$ and any final state $\boldsymbol{x}_{\mathrm{f}} = \boldsymbol{0}$, there exists an input u that transfers $\boldsymbol{x}_0$ to $\boldsymbol{x}_{\mathrm{f}}$ in a finite time $[t_0, t_{\mathrm{f}}]$. Otherwise the system is said to be uncontrollable.

The definition of controllability is explained as follows:

(1) Only when each variable x_i of the state vector $\boldsymbol{x}(t)$ can independently reach the final state $x_i(t_{\mathrm{f}}) = 0$ from a non-zero initial state , the system be said to be completely controllable.

(2) The trajectory of $\boldsymbol{x}(t)$ is not specified as long as it arrives at the origin of the state space in a finite time.

(3) There is no constraint imposed on input, but each of components $u_i(t)$ of $\boldsymbol{u}(t)$ is square integrable, i. e. $\int_0^t u_i^2(\tau)\mathrm{d}\tau < \infty, \forall t < +\infty$.

(4) The system is controllable only if, in a finite time $[t_0, t_{\mathrm{f}}]$, the state vector $\boldsymbol{x}(t)$ approaches the origin of the state space. Otherwise, $\boldsymbol{x}(t) \to \boldsymbol{0}$ when $t \to \infty$, it cannot be said that the system is controllable.

(5) For a time-varying system, it does not mean that the systemis controllable at $t = t_1$ if it is controllable at t_0. Whereas, for a LTI system, it is controllable has nothing to do with the selection of t_0.

(6) In practice, a lot of systems are controllable.

(7) In Definition 3.1.1, it is stipulated that the system state is transferred from the non-zero initial state to the origin of the state space. If the system state is transferred to any specified non-zero state from zero initial state, the system is said to be state reachable. For continuous LTI systems, controllability and reachability are equivalent. For discrete systems and time-varying systems, they are not strictly equivalent, and the system may be not completely controllable, but completely reachable.

3.1.3 Definition of observability

Observability describes whether the state of a system can be fully reflected by the output, so the state equation and the output equation of the system (3.1.2) should be considered simultaneously.

$$\begin{cases}\dot{\boldsymbol{x}} = \boldsymbol{A}(t)\boldsymbol{x} + \boldsymbol{B}(t)\boldsymbol{u}, \boldsymbol{x}(t_0) = \boldsymbol{x}_0 \\ \boldsymbol{y} = \boldsymbol{C}(t)\boldsymbol{x} + \boldsymbol{D}(t)\boldsymbol{u}, t \in J\end{cases} \tag{3.1.2}$$

From the discussion in Chapter 2, we know that the state motion of the system isexcited by non-zero initial states and external inputs, and is determined by the parameter matrix $\boldsymbol{A}(t)$ and $\boldsymbol{B}(t)$. From the output equation, it can be seen that the output of the system is related to the state and input, and this relationship is determined by the parameter matrix $\boldsymbol{C}(t)$ and $\boldsymbol{D}(t)$. And since the state observability is an inherent property of the system and has nothing to do with external input, so let $\boldsymbol{u}(t) = \boldsymbol{0}$ when we check the observability. At this point, the output of the system is related only to the state $\boldsymbol{x}$, and furthermore, only to the initial state. This relationship is determined by the parameter matrix $\boldsymbol{A}(t)$ and $\boldsymbol{C}(t)$. For this case, the state space description becomes

$$\begin{cases}\dot{\boldsymbol{x}} = \boldsymbol{A}(t)\boldsymbol{x}, \boldsymbol{x}(t_0) = \boldsymbol{x}_0 \\ \boldsymbol{y} = \boldsymbol{C}(t)\boldsymbol{x}, t_0, t \in J\end{cases} \tag{3.1.3}$$

We give the definition of observability based on eq. (3.1.3).

Definition 3.1.1 $\boldsymbol{x}(t_0)$ is said to be observable at time t_0 if for any unknown initial state $\boldsymbol{x}(t_0)$, there exists a finite $t_f > 0$ such that the knowledge of the output y over $[0, t_f]$ suffices to determine uniquely the initial state $\boldsymbol{x}(t_0)$. Otherwise, the system is said to be unobservable.

Definition 3.1.2 $\boldsymbol{x}(t_0)$ is said to be unobservable if for any non-zero initial state $\boldsymbol{x}(t_0)$, there exists a finite $t_1 > t_0$ such that $\boldsymbol{y}(t) = \boldsymbol{0}$ when $t \in [t_0, t_1]$.

Definition 3.1.3 $\boldsymbol{x}(t_0)$ is said to be not completely observable at time t_0 if there exists a state variable or some state variables are unobservable at t_0.

3.2 Controllability criteria of linear time-continuous systems

It can be seen from the previous discussion that the controllability of linear systems is only related to the state equation of systems, that is, only related to the parameter matrices $\boldsymbol{A}(t)$ and $\boldsymbol{B}(t)$. This section studies the controllability criterion of linear systems. Firstly, we discuss the controllability criterion of LTI systems, then the controllability criterion of LTV systems.

3.2.1 Controllability criteria of LTI systems

Consider the n-dimensional p-input state equation

$$\dot{\boldsymbol{x}} = \boldsymbol{A}\boldsymbol{x} + \boldsymbol{B}\boldsymbol{u}, \boldsymbol{x}(0) = \boldsymbol{x}_0, t \geqslant 0 \tag{3.2.1}$$

where $\boldsymbol{A}$ and $\boldsymbol{B}$ are $n \times n$ and $n \times p$ constant matrices, respectively. The some criteria for controllability of the LTI system are given below.

Conclusion 3.2.1 (Grammian matrix criterion) The system $\{\boldsymbol{A}, \boldsymbol{B}\}$ is completely state-controllable if and only if that Grammian matrix

$$\boldsymbol{W}_c[0,t_1] \triangleq \int_0^{t_1} \mathrm{e}^{-\boldsymbol{A}t}\boldsymbol{B}\boldsymbol{B}^{\mathrm{T}}\mathrm{e}^{-\boldsymbol{A}^{\mathrm{T}}t}\mathrm{d}t \tag{3.2.2}$$

is nonsingular for any $t_1 > 0$, where $\boldsymbol{W}_c[0,t_1]$ is an $n \times n$ matrix.

Proof Sufficiency. Prove the system is completely state-controllable if $\boldsymbol{W}_c[0, t_1]$ is nonsingular.

$\forall \boldsymbol{x}_0 \neq 0$, construct an input $u(t)$ as :

$$\boldsymbol{u}(t) = -\boldsymbol{B}^{\mathrm{T}}\mathrm{e}^{-\boldsymbol{A}^{\mathrm{T}}t}\boldsymbol{W}_c^{-1}[0,t_1]\boldsymbol{x}_0, t \in [t_0, t_1] \tag{3.2.3}$$

At time t_1, $\boldsymbol{x}(t_1)$, excited by $u(t)$, is

$$\begin{aligned}
\boldsymbol{x}(t_1) &= \mathrm{e}^{\boldsymbol{A}t_1}\boldsymbol{x}_0 + \int_0^{t_1}\mathrm{e}^{\boldsymbol{A}(t_1-t)}\boldsymbol{B}\boldsymbol{u}(t)\mathrm{d}t \\
&= \mathrm{e}^{\boldsymbol{A}t_1}\boldsymbol{x}_0 - \int_0^{t_1}\mathrm{e}^{\boldsymbol{A}t_1}\mathrm{e}^{-\boldsymbol{A}t}\boldsymbol{B}\boldsymbol{B}^{\mathrm{T}}\mathrm{e}^{-\boldsymbol{A}^{\mathrm{T}}t}\boldsymbol{W}_c^{-1}[0,t_1]\boldsymbol{x}_0\mathrm{d}t \\
&= \mathrm{e}^{\boldsymbol{A}t_1}\boldsymbol{x}_0 - \mathrm{e}^{\boldsymbol{A}t_1} \cdot \int_0^{t_1}\mathrm{e}^{-\boldsymbol{A}t}\boldsymbol{B}\boldsymbol{B}^{\mathrm{T}}\mathrm{e}^{-\boldsymbol{A}^{\mathrm{T}}t}\mathrm{d}t \cdot \boldsymbol{W}_c^{-1}[0,t_1]\boldsymbol{x}_0 \\
&= \mathrm{e}^{\boldsymbol{A}t_1}\boldsymbol{x}_0 - \mathrm{e}^{\boldsymbol{A}t_1}\boldsymbol{W}_c[0,t_1]\boldsymbol{W}_c^{-1}[0,t_1]\boldsymbol{x}_0 \\
&= \boldsymbol{0}
\end{aligned}$$

Thus, according to the definition of controllability the system is controllable.

Necessity. Prove $\boldsymbol{W}_c[0,t_1]$ is nonsingular, if the system is completely state-controllable.

Adopt proof by contradiction to prove it. We suppose $\boldsymbol{W}_c[0,t_1]$ is singular, i. e. $\exists\ \bar{\boldsymbol{x}}_0 \in \mathbf{R}^n$ and $\bar{\boldsymbol{x}}_0 \neq \boldsymbol{0}$, such that

$$\bar{x}_0^{\mathrm{T}} W_c[0,t_1]\bar{x}_0 = 0 \tag{3.2.4}$$

Then from eq. (3.2.2) and eq. (3.2.4), we get

$$\begin{aligned} 0 &= \bar{x}_0^{\mathrm{T}} W_c[0,t_1]\bar{x}_0 \\ &= \int_0^{t_1} \bar{x}_0^{\mathrm{T}} \mathrm{e}^{-At} BB^{\mathrm{T}} \mathrm{e}^{-A^{\mathrm{T}}t} \bar{x}_0 \mathrm{d}t \\ &= \int_0^{t_1} (B^{\mathrm{T}} \mathrm{e}^{-A^{\mathrm{T}}t} \bar{x}_0)^{\mathrm{T}} (B^{\mathrm{T}} \mathrm{e}^{-A^{\mathrm{T}}t} \bar{x}_0) \mathrm{d}t \\ &= \int_0^{t_1} \| B^{\mathrm{T}} \mathrm{e}^{-A^{\mathrm{T}}t} \bar{x}_0 \|^2 \mathrm{d}t \end{aligned}$$

So

$$B^{\mathrm{T}} \mathrm{e}^{-A^{\mathrm{T}}t} \bar{x}_0 = \mathbf{0}, \quad \forall t \in [0,t_1] \tag{3.2.5}$$

And since the system is completely controllable, so for the non-zero $\bar{x}_0$, we get

$$\mathbf{0} = x(t_1) = \mathrm{e}^{At_1} \bar{x}_0 + \int_0^{t_1} \mathrm{e}^{A(t_1-t)} Bu(t) \mathrm{d}t \tag{3.2.6}$$

So

$$\begin{aligned} \mathrm{e}^{At_1} \bar{x}_0 &= -\int_0^{t_1} \mathrm{e}^{A(t_1-t)} Bu(t) \mathrm{d}t \\ \bar{x}_0 &= -\int_0^{t_1} \mathrm{e}^{-At} Bu(t) \mathrm{d}t \end{aligned}$$

$$\|\bar{x}_0\|^2 = \bar{x}_0^{\mathrm{T}} \bar{x}_0 = \left(-\int_0^{t_1} \mathrm{e}^{-At} Bu(t) \mathrm{d}t\right)^{\mathrm{T}} \bar{x}_0 = -\int_0^{t_1} u^{\mathrm{T}}(t) B^{\mathrm{T}} \mathrm{e}^{-A^{\mathrm{T}}t} \bar{x}_0 \mathrm{d}t \tag{3.2.7}$$

Using eq. (3.2.5) and eq. (3.2.7) yields

$$\|\bar{x}_0\|^2 = 0 \quad \text{i. e.} \quad \bar{x}_0 = \mathbf{0} \tag{3.2.8}$$

The above equation contradicts the assumption $\bar{x}_0 \neq \mathbf{0}$. Thus the assumption that $W_c[0,t_1]$ is singular is not correct if the system is completely state-controllable.

Because of the complexity of calculation, Grammian matrix criterion is mainly used for theoretical analysis and derivation of other controllability criteria.

Conclusion 3.2.2(Rank criterion) The system $\{A, B\}$ is completely state-controllable if and only if its $n \times np$ controllability matrix $[B \vdots AB \vdots \cdots \vdots A^{n-1}B]$ has rank n (full row rank).

i. e.

$$\operatorname{rank}[B \vdots AB \vdots \cdots \vdots A^{n-1}B] = n \tag{3.2.9}$$

where n is the order of the system, $Q_c \triangleq [B \vdots AB \vdots \cdots \vdots A^{n-1}B]$ is called controllability matrix.

Proof Sufficiency. Prove the system is completely state-controllable if rank $Q_c = n$.

Adopt proof by contradiction to prove it. Suppose the system is uncontrollable. Then the Grammian matrix

$$W_c[0,t_1] \triangleq \int_0^{t_1} \mathrm{e}^{-At} BB^{\mathrm{T}} \mathrm{e}^{-A^{\mathrm{T}}t} \mathrm{d}t, \forall t_1 > 0$$

is singular according to conclusion 3.2.1. That means there exists a non-zero $n \times 1$ vector $\boldsymbol{\alpha}$

such that

$$\begin{aligned} 0 &= \boldsymbol{\alpha}^{\mathrm{T}} \boldsymbol{W}_c[0,t_1]\boldsymbol{\alpha} \\ &= \int_0^{t_1} \boldsymbol{\alpha}^{\mathrm{T}} \mathrm{e}^{-\boldsymbol{A}t}\boldsymbol{B}\boldsymbol{B}^{\mathrm{T}} \mathrm{e}^{-\boldsymbol{A}^{\mathrm{T}}t}\boldsymbol{\alpha}\mathrm{d}t \\ &= \int_0^{t_1} [\boldsymbol{\alpha}^{\mathrm{T}} \mathrm{e}^{-\boldsymbol{A}t}\boldsymbol{B}][\boldsymbol{\alpha}^{\mathrm{T}} \mathrm{e}^{-\boldsymbol{A}t}\boldsymbol{B}]^{\mathrm{T}}\mathrm{d}t \end{aligned}$$

Thus

$$\boldsymbol{\alpha}^{\mathrm{T}} \mathrm{e}^{-\boldsymbol{A}t}\boldsymbol{B} = \boldsymbol{0},\ \forall t \in [0,t_1] \tag{3.2.10}$$

Take the derivative of the aboveequation until $n-1$ times, and then set $t=0$ in the result, we get

$$\boldsymbol{\alpha}^{\mathrm{T}}\boldsymbol{B} = \boldsymbol{0},\ \boldsymbol{\alpha}^{\mathrm{T}}\boldsymbol{A}\boldsymbol{B} = \boldsymbol{0},\ \boldsymbol{\alpha}^{\mathrm{T}}\boldsymbol{A}^2\boldsymbol{B} = \boldsymbol{0}, \cdots,\ \boldsymbol{\alpha}^{\mathrm{T}}\boldsymbol{A}^{n-1}\boldsymbol{B} = \boldsymbol{0} \tag{3.2.11}$$

They can arranged as

$$\boldsymbol{\alpha}^{\mathrm{T}}[\boldsymbol{B} \vdots \boldsymbol{A}\boldsymbol{B} \vdots \cdots \vdots \boldsymbol{A}^{n-1}\boldsymbol{B}] = \boldsymbol{\alpha}^{\mathrm{T}}\boldsymbol{Q}_c = \boldsymbol{0} \tag{3.2.12}$$

Because $\boldsymbol{\alpha} \neq \boldsymbol{0}$, n rows of $\boldsymbol{Q}_c$ are linearly dependent. This contradicts the hypothesis that rank $\boldsymbol{Q}_c = n$. So we finished the proof of sufficiency.

Necessity. Prove rank $\boldsymbol{Q}_c = n$, if the system is completely state-controllable.

Adopt proof by contradiction to prove it. Suppose rank $\boldsymbol{Q}_c < n$. Thus n rows of $\boldsymbol{Q}_c$ are linearly dependent, and there exists a non-zero $n \times 1$ vector $\boldsymbol{\alpha}$ such that

$$\boldsymbol{\alpha}^{\mathrm{T}}\boldsymbol{Q}_c = \boldsymbol{\alpha}^{\mathrm{T}}[\boldsymbol{B} \vdots \boldsymbol{A}\boldsymbol{B} \vdots \cdots \vdots \boldsymbol{A}^{n-1}\boldsymbol{B}] = \boldsymbol{0}$$

So

$$\boldsymbol{\alpha}^{\mathrm{T}}\boldsymbol{A}^i\boldsymbol{B} = \boldsymbol{0}, \quad i = 0,1,2,\cdots,n-1 \tag{3.2.13}$$

According to the Cayley-Hamilton theorem, we know that if characteristic polynomial of $\boldsymbol{A}$ is

$$\det(s\boldsymbol{I} - \boldsymbol{A}) = s^n + a_{n-1}s^{n-1} + \cdots + a_1 s + a_0$$

Then there exists

$$\boldsymbol{A}^n = -a_{n-1}\boldsymbol{A}^{n-1} - \cdots - a_1\boldsymbol{A} - a_0\boldsymbol{I} \tag{3.2.14}$$

It indicates $\boldsymbol{A}^n$ is a combination of $\boldsymbol{I}, \boldsymbol{A}, \boldsymbol{A}^2, \cdots, \boldsymbol{A}^{n-1}$. From eq. (3.2.13), we have

$$\boldsymbol{\alpha}^{\mathrm{T}}\boldsymbol{A}^n\boldsymbol{B} = \boldsymbol{0} \tag{3.2.15}$$

From eq. (3.2.14), $\boldsymbol{A}^{n+1}$ also is combination of $\boldsymbol{A}, \boldsymbol{A}^2, \cdots, \boldsymbol{A}^n$. From eq. (3.2.15) and eq. (3.2.13), we have

$$\boldsymbol{\alpha}^{\mathrm{T}}\boldsymbol{A}^{n+1}\boldsymbol{B} = \boldsymbol{0} \tag{3.2.16}$$

And so on, we have

$$\boldsymbol{\alpha}^{\mathrm{T}}\boldsymbol{A}^i\boldsymbol{B} = \boldsymbol{0}, i = 0,1,2,\cdots \tag{3.2.17}$$

So, for any $t_1 > 0$ there is

$$\pm\boldsymbol{\alpha}^{\mathrm{T}}\frac{\boldsymbol{A}^i t^i}{i!}\boldsymbol{B} = \boldsymbol{0},\ \forall t \in [0,t_1], \mathrm{i} = 0,1,2,\cdots \tag{3.2.18}$$

Add all the terms of eq. (3.2.18), we get

$$\boldsymbol{0} = \boldsymbol{\alpha}^{\mathrm{T}}[\boldsymbol{I} - \boldsymbol{A}t + \frac{1}{2!}\boldsymbol{A}^2t^2 - \frac{1}{3!}\boldsymbol{A}^3t^3 + \cdots]\boldsymbol{B} = \boldsymbol{\alpha}^{\mathrm{T}}\mathrm{e}^{-\boldsymbol{A}t}\boldsymbol{B},\ \forall t \in [0,t_1] \tag{3.2.19}$$

and

$$\alpha^{\mathrm{T}} W_c[0,t_1]\alpha = \alpha^{\mathrm{T}}\int_0^{t_1} \mathrm{e}^{-At}BB^{\mathrm{T}}\mathrm{e}^{-A^{\mathrm{T}}t}\mathrm{d}t \cdot \alpha = \int_0^{t_1}(\alpha^{\mathrm{T}}\mathrm{e}^{-At}B)(\alpha^{\mathrm{T}}\mathrm{e}^{-At}B)^{\mathrm{T}}\mathrm{d}t = 0 \tag{3.2.20}$$

It shows that $W_c[0,t_1]$ is singular. According to the conclusion 3.2.1, the system is uncontrollable. This contradicts the assumption that the system is completely state-controllable. Thus if rank $Q_c = n$, the system is completely state-controllable.

Example 3.2.1 Consider a 2-order system

$$\begin{bmatrix}\dot{x}_1\\ \dot{x}_2\end{bmatrix} = \begin{bmatrix}-2 & 1\\ 0 & -3\end{bmatrix}\begin{bmatrix}x_1\\ x_2\end{bmatrix} + \begin{bmatrix}0 & 1\\ 1 & -1\end{bmatrix}\begin{bmatrix}u_1\\ u_2\end{bmatrix}$$

Check whether the system is controllable.

Answer $Q_c = [B \vdots AB] = \begin{bmatrix}0 & 1 & 1 & -3\\ 1 & -1 & -3 & 3\end{bmatrix}$

rank $Q_c = 2$. According to conclusion 3.2.2, the system is controllable. (we can use matlab function rank (ctrb (A, b)) *to compute rank* Q_c.

Conclusion 3.2.3(PBH Rank criterion) The system $\{A, B\}$ is controllable if and only if $n\times(n+p)$ matrix $[\lambda_i I - A, B]$ has full row rank at every eigenvalue $\lambda_i (i=1,2,\cdots,n)$ of A. i.e.

$$\mathrm{rank}[\lambda_i I - A, B] = n, i = 1,2,\cdots,n \tag{3.2.21}$$

or

$$\mathrm{rank}[sI - A, B] = n, \forall s \in \text{complex number} \tag{3.2.22}$$

Proof For eq. (3.2.22), if s is not the eigenvalue of A, then $\det(sI - A) \neq 0$, thus rank $(sI - A) = n$ and rank$[sI - A, B] = n$. So we just need to prove eq. (3.2.21).

Necessity. Prove that eq. (3.2.21) holds, if the system is completely state-controllable. Adopt proof by contradiction to prove it. Suppose there exists a eigenvalue λ_i such that rank$(\lambda_i I - A, B) < n$. This means the rows of $[\lambda_i I - A, B]$ are linearly dependent and there exists a non-zero $n\times 1$ vector α such that

$$\alpha^{\mathrm{T}}[\lambda_i I - A, B] = 0 \tag{3.2.23}$$

From the above equation, we get

$$\alpha^{\mathrm{T}}A = \lambda_i \alpha^{\mathrm{T}}, \quad \alpha^{\mathrm{T}}B = 0 \tag{3.2.24}$$

Furthermore,

$$\alpha^{\mathrm{T}}B = 0, \quad \alpha^{\mathrm{T}}AB = \lambda_i \alpha^{\mathrm{T}}B = 0, \quad \cdots, \quad \alpha^{\mathrm{T}}A^{n-1}B = 0 \tag{3.2.25}$$

From eq. (3.2.25), we have

$$\alpha^{\mathrm{T}}[B \vdots AB \vdots \cdots \vdots A^{n-1}B] = \alpha^{\mathrm{T}}Q_c = 0$$

Because $\alpha \neq 0$, so rank $Q_c < n$. This shows the system is uncontrollable which contradicts the assumption that the system is completely state-controllable. Thus if eq. (3.2.21) holds, he system is completely state-controllable.

Sufficiency. Prove the system is completely state-controllable if eq. (3.2.21) holds. Adopt proof by contradiction to prove it. Suppose the system is uncontrollable. Then the rows of $\boldsymbol{Q}_c = [\boldsymbol{B} \vdots \boldsymbol{AB} \vdots \cdots \vdots \boldsymbol{A}^{n-1}\boldsymbol{B}]$ are linearly dependent. Thus there exists a non-zero $n \times 1$ vector $\boldsymbol{\alpha}$ such that

$$\boldsymbol{\alpha}^{\mathrm{T}} \boldsymbol{Q}_c = \boldsymbol{\alpha}^{\mathrm{T}}[\boldsymbol{B} \vdots \boldsymbol{AB} \vdots \cdots \vdots \boldsymbol{A}^{n-1}\boldsymbol{B}] = \boldsymbol{0}$$

i. e.

$$\boldsymbol{\alpha}^{\mathrm{T}}\boldsymbol{B} = \boldsymbol{0}, \quad \boldsymbol{\alpha}^{\mathrm{T}}\boldsymbol{AB} = \boldsymbol{0}, \quad \cdots, \quad \boldsymbol{\alpha}^{\mathrm{T}}\boldsymbol{A}^{n-1}\boldsymbol{B} = \boldsymbol{0} \tag{3.2.26}$$

Let $\lambda_i (i=1,2,\cdots,n)$ is i^{th} eigenvalue of $\boldsymbol{A}$. From rank$[\lambda_1\boldsymbol{I} - \boldsymbol{A}, \boldsymbol{B}] = n$ and eq. (3.2.26), we have

$$\boldsymbol{\alpha}^{\mathrm{T}}[\lambda_1\boldsymbol{I} - \boldsymbol{A}, \boldsymbol{B}] = [\boldsymbol{\alpha}^{\mathrm{T}}(\lambda_1\boldsymbol{I} - \boldsymbol{A}), \boldsymbol{0}] \neq \boldsymbol{0}$$

Thus,

$$\boldsymbol{\alpha}^{\mathrm{T}}(\lambda_1\boldsymbol{I} - \boldsymbol{A}) \neq \boldsymbol{0} \tag{3.2.27}$$

From rank$[\lambda_2\boldsymbol{I} - \boldsymbol{A}, \boldsymbol{B}] = n$ which shows the rows of $[\lambda_2\boldsymbol{I} - \boldsymbol{A}, \boldsymbol{B}]$ are linearly independent, we get $\boldsymbol{\alpha}^{\mathrm{T}}(\lambda_1\boldsymbol{I} - \boldsymbol{A})[\lambda_2\boldsymbol{I} - \boldsymbol{A}, \boldsymbol{B}] \neq 0$. From eq. (3.2.26), we have

$$\boldsymbol{\alpha}^{\mathrm{T}}(\lambda_1\boldsymbol{I} - \boldsymbol{A})(\lambda_2\boldsymbol{I} - \boldsymbol{A}) \neq \boldsymbol{0} \tag{3.2.28}$$

Similarly, we can get

$$\boldsymbol{\alpha}^{\mathrm{T}}(\lambda_1\boldsymbol{I} - \boldsymbol{A})(\lambda_2\boldsymbol{I} - \boldsymbol{A})(\lambda_3\boldsymbol{I} - \boldsymbol{A}) \neq \boldsymbol{0}$$

$$\vdots$$

$$\boldsymbol{\alpha}^{\mathrm{T}}(\lambda_1\boldsymbol{I} - \boldsymbol{A})(\lambda_2\boldsymbol{I} - \boldsymbol{A})\cdots(\lambda_n\boldsymbol{I} - \boldsymbol{A}) \neq \boldsymbol{0} \tag{3.2.29}$$

Thus,

$$(\lambda_1\boldsymbol{I} - \boldsymbol{A})(\lambda_2\boldsymbol{I} - \boldsymbol{A})\cdots(\lambda_n\boldsymbol{I} - \boldsymbol{A}) \neq \boldsymbol{0} \tag{3.2.30}$$

According to the Cayley-Hamilton theorem, we know

$$\prod_{i=1}^{n}(\lambda_i\boldsymbol{I} - \boldsymbol{A}) = \boldsymbol{0} \tag{3.2.31}$$

Eq. (3.2.30) contradicts eq. (3.2.31). Thus if eq. (3.2.21) holds, the system is completely state-controllable.

Conclusion 3.2.4 (Jordan canonical form criterion) (1) When all eigenvalues of $\boldsymbol{A}$ are distinct, the state equation has diagonal canonical form:

$$\dot{\bar{\boldsymbol{x}}} = \begin{bmatrix} \lambda_1 & & & \\ & \lambda_2 & & \\ & & \ddots & \\ & & & \lambda_n \end{bmatrix} \bar{\boldsymbol{x}} + \begin{bmatrix} \bar{\boldsymbol{b}}_1 \\ \bar{\boldsymbol{b}}_2 \\ \vdots \\ \bar{\boldsymbol{b}}_n \end{bmatrix} \boldsymbol{u} \tag{3.2.32}$$

its state is completely controllable if and only if the all row vectors of input matrix, $\bar{\boldsymbol{b}}_1, \bar{\boldsymbol{b}}_2, \cdots, \bar{\boldsymbol{b}}_n$, are not zero.

(2) When the $n \times n$ matrix $\boldsymbol{A}$ has repeated eigenvalues:

λ_1 (multiplicity σ_1), λ_2 (multiplicity σ_2), $\cdots$, λ_l (multiplicity σ_l) and $\sigma_1 + \sigma_2 + \cdots + \sigma_l = n$, the Jordan canonical form is

$$\dot{\hat{\boldsymbol{x}}} = \hat{\boldsymbol{A}}\hat{\boldsymbol{x}} + \hat{\boldsymbol{B}}\boldsymbol{u} \tag{3.2.33}$$

where

$$\hat{\boldsymbol{A}}_{n\times n} = \begin{bmatrix} \boldsymbol{J}_1 & & & & \\ & \ddots & & & \\ & & \boldsymbol{J}_i & & \\ & & & \ddots & \\ & & & & \boldsymbol{J}_l \end{bmatrix}, \quad \hat{\boldsymbol{B}}_{n\times p} = \begin{bmatrix} \hat{\boldsymbol{B}}_1 \\ \vdots \\ \hat{\boldsymbol{B}}_i \\ \vdots \\ \hat{\boldsymbol{B}}_l \end{bmatrix} \tag{3.2.34}$$

$$\boldsymbol{J}_{i(\sigma_i\times\sigma_i)} = \begin{bmatrix} \boldsymbol{J}_{i1} & & & \\ & \boldsymbol{J}_{i2} & & \\ & & \ddots & \\ & & & \boldsymbol{J}_{i\alpha_i} \end{bmatrix}, \quad \hat{\boldsymbol{B}}_{i(\sigma_i\times p)} = \begin{bmatrix} \hat{\boldsymbol{B}}_{i1} \\ \hat{\boldsymbol{B}}_{i2} \\ \vdots \\ \hat{\boldsymbol{B}}_{i\alpha_i} \end{bmatrix} \tag{3.2.35}$$

$$\boldsymbol{J}_{ik(r_{ik}\times r_{ik})} = \begin{bmatrix} \lambda_i & 1 & & & \\ & \lambda_i & 1 & & \\ & & \ddots & \ddots & \\ & & & \lambda_i & 1 \\ & & & & \lambda_i \end{bmatrix}, \quad \hat{\boldsymbol{B}}_{ik(r_{ik}\times p)} = \begin{bmatrix} \hat{\boldsymbol{b}}_{1ik} \\ \hat{\boldsymbol{b}}_{2ik} \\ \vdots \\ \hat{\boldsymbol{b}}_{(r-1)ik} \\ \hat{\boldsymbol{b}}_{rik} \end{bmatrix} \tag{3.2.36}$$

where $\boldsymbol{J}_{ik}$ is the kth block of $\boldsymbol{J}_i$ corresponding to λ_i ($k = 1, 2, \cdots, \alpha_i$) which is also called the kth small Jordan block corresponding to λ_i, and $r_{i1} + r_{i2} + \cdots + r_{i\alpha_i} = \sigma_i$.

The LTI system eq. (3.2.33) is completely controllable if and only if the matrix

$$\begin{bmatrix} \hat{\boldsymbol{b}}_{ri1} \\ \hat{\boldsymbol{b}}_{ri2} \\ \vdots \\ \hat{\boldsymbol{b}}_{ri\alpha_i} \end{bmatrix} \tag{3.2.37}$$

composed by last rows of $\hat{\boldsymbol{B}}_{ik}$ ($k = 1, 2, \cdots, \alpha_i$), has full row rank, where $i = 1, 2, \cdots, l$.

Proof (1) Form eq. (3.2.32), we have

rank $\boldsymbol{Q}_c$ = rank$[\overline{\boldsymbol{B}} \vdots \overline{\boldsymbol{A}}\,\overline{\boldsymbol{B}} \vdots \cdots \vdots \overline{\boldsymbol{A}}^{n-1}\overline{\boldsymbol{B}}]$

$$
=\operatorname{rank}\begin{bmatrix} \bar{\boldsymbol{b}}_1 & \lambda_1 \bar{\boldsymbol{b}}_1 & \lambda_1^2 \bar{\boldsymbol{b}}_1 & \cdots & \lambda_1^{n-1} \bar{\boldsymbol{b}}_1 \\ \bar{\boldsymbol{b}}_2 & \lambda_2 \bar{\boldsymbol{b}}_2 & \lambda_2^2 \bar{\boldsymbol{b}}_2 & \cdots & \lambda_2^{n-1} \bar{\boldsymbol{b}}_2 \\ \vdots & \vdots & \vdots & & \vdots \\ \bar{\boldsymbol{b}}_n & \lambda_n \bar{\boldsymbol{b}}_n & \lambda_n^2 \bar{\boldsymbol{b}}_n & \cdots & \lambda_n^{n-1} \bar{\boldsymbol{b}}_n \end{bmatrix}
$$

$$
=\operatorname{rank}\begin{bmatrix} \bar{\boldsymbol{b}}_1 & & & \\ & \bar{\boldsymbol{b}}_2 & & \\ & & \ddots & \\ & & & \bar{\boldsymbol{b}}_n \end{bmatrix}\begin{bmatrix} 1 & \lambda_1 & \lambda_1^2 & \cdots & \lambda_1^{n-1} \\ 1 & \lambda_2 & \lambda_2^2 & \cdots & \lambda_2^{n-1} \\ \vdots & \vdots & \vdots & & \vdots \\ 1 & \lambda_n & \lambda_n^2 & \cdots & \lambda_n^{n-1} \end{bmatrix}
$$

$$
=\operatorname{rank}\begin{bmatrix} \bar{\boldsymbol{b}}_1 & & & \\ & \bar{\boldsymbol{b}}_2 & & \\ & & \ddots & \\ & & & \bar{\boldsymbol{b}}_n \end{bmatrix} \tag{3.2.38}
$$

Because $\bar{\boldsymbol{b}}_i(i=1,2,\cdots,n)$ is row vector, so rank $\boldsymbol{Q}_c=n$ if only if $\bar{\boldsymbol{b}}_i\neq 0(i=1,2,\cdots,n)$.

(2) Here we don't give the proof.

Example 3.2.2 Consider the LTI system

$$
\dot{\bar{\boldsymbol{x}}}=\begin{bmatrix} -7 & 0 & 0 \\ 0 & -2 & 0 \\ 0 & 0 & 1 \end{bmatrix}\bar{\boldsymbol{x}}+\begin{bmatrix} 0 & 2 \\ 4 & 0 \\ 0 & 1 \end{bmatrix}\boldsymbol{u}
$$

Check its controllability.

Answer Because $\boldsymbol{A}$ has distinct eigenvalues and is diagonal form, and all rows of $\bar{\boldsymbol{B}}$ are not zero, the system is controllable.

Example 3.2.3 Consider the LTI system

$$
\dot{\hat{\boldsymbol{x}}}=\begin{bmatrix} -2 & 1 & & & & & \\ 0 & -2 & & & & & \\ & & -2 & & & & \\ & & & -2 & & & \\ & & & & 3 & 1 & \\ & & & & 0 & 3 & \\ & & & & & & 3 \end{bmatrix}\hat{\boldsymbol{x}}+\begin{bmatrix} 0 & 0 & 0 \\ 1 & 0 & 0 \\ 0 & 4 & 0 \\ 0 & 0 & 7 \\ 0 & 0 & 0 \\ 1 & 1 & 0 \\ 0 & 4 & 1 \end{bmatrix}\boldsymbol{u}
$$

Check its controllability.

Answer The matrix $\boldsymbol{A}$ has different eigenvalues $\lambda_1=-2$ and $\lambda_2=3$. For $\lambda_1=-2$, there exists $\begin{bmatrix} \hat{b}_{r11} \\ \hat{b}_{r12} \\ \hat{b}_{r13} \end{bmatrix} = \begin{bmatrix} 1 & 0 & 0 \\ 0 & 4 & 0 \\ 0 & 0 & 7 \end{bmatrix}$, where its all rows are linearly independent.

So, the state variable x_1, x_2, x_3 and x_4 are controllable.

For $\lambda_2 = 3$, there exists $\begin{bmatrix} \hat{b}_{r21} \\ \hat{b}_{r22} \end{bmatrix} = \begin{bmatrix} 1 & 1 & 0 \\ 0 & 4 & 1 \end{bmatrix}$, where all rows are linearly independent.

So, the state variable x_5, x_6 and x_7 are controllable. Thus the system is controllable.

3.2.2 Controllability index

For the LTI system eq. (3.2.1), $\boldsymbol{A}$ and $\boldsymbol{B}$ are $n \times n$ and $n \times p$ constant matrices. We define an $n \times (k+1)p$ matrix

$$\boldsymbol{U}_k \triangleq [\boldsymbol{B} \vdots \boldsymbol{AB} \vdots \cdots \vdots \boldsymbol{A}^k B], k = 1, 2, \cdots \tag{3.2.39}$$

If $(\boldsymbol{A}, \boldsymbol{B})$ is controllable, then

$$\text{rank}\, \boldsymbol{U}_{n-1} = n$$

i. e. there are n linearly independent columns among the np columns of $\boldsymbol{U}_{n-1}$. Let $\boldsymbol{b}_i$ denote the ith column of $\boldsymbol{B}$. Then $\boldsymbol{B}$ can written as

$$\boldsymbol{B} = [\boldsymbol{b}_1 \quad \boldsymbol{b}_2 \quad \cdots \quad \boldsymbol{b}_p]$$

So $\boldsymbol{U}_k$ can written as

$$\boldsymbol{U}_k = [\boldsymbol{b}_1 \quad \boldsymbol{b}_2 \quad \cdots \quad \boldsymbol{b}_p \vdots \boldsymbol{Ab}_1 \quad \boldsymbol{Ab}_2 \quad \cdots \quad \boldsymbol{Ab}_p \vdots \cdots \vdots \boldsymbol{A}^k\boldsymbol{b}_1 \quad \boldsymbol{A}^k\boldsymbol{b}_2 \quad \cdots \quad \boldsymbol{A}^k\boldsymbol{b}_p] \tag{3.2.40}$$

Lemma 3.2.1 For $\boldsymbol{U}_k$, if $\boldsymbol{A}^j \boldsymbol{b}_i (j = 0, 1, 2, \cdots, k;\ i = 1, 2, \cdots, p)$ depends on its left-hand-side (LHS) columns, then $\boldsymbol{A}^{j_1}\boldsymbol{b}_i (j_1 > j)$ will also depend on LHS columns.

Proof If $\boldsymbol{Ab}_i$ depends on its LHS columns, then $\boldsymbol{Ab}_i$ can be expressed a linearly combination of LHS columns of $\boldsymbol{Ab}_i$, i. e.

$$\boldsymbol{Ab}_i = \alpha_1 \boldsymbol{b}_1 + \alpha_2 \boldsymbol{b}_2 + \cdots + \alpha_p \boldsymbol{b}_p + \alpha_{p+1}\boldsymbol{Ab}_1 + \cdots + \alpha_{p+i-1}\boldsymbol{Ab}_{i-1}$$

where the constants $\alpha_1, \alpha_2, \cdots, \alpha_{p+i-1}$ are not all zeros. Multiplying both sides of the above equation by $\boldsymbol{A}$ yields

$$\boldsymbol{A}^2 \boldsymbol{b}_i = \boldsymbol{A}(\boldsymbol{Ab}_i) = \alpha_1\boldsymbol{Ab}_1 + \alpha_2\boldsymbol{Ab}_2 + \cdots + \alpha_p\boldsymbol{Ab}_p + \alpha_{p+1}\boldsymbol{A}^2 \boldsymbol{b}_1 + \cdots + \alpha_{p+i-1}\boldsymbol{A}^2 \boldsymbol{b}_{i-1}$$

It shows $\boldsymbol{A}^2 \boldsymbol{b}_i$ is also a linearly combination of its LHS columns, i. e. $\boldsymbol{A}^2 \boldsymbol{b}_i$ dependents its LHS columns. Lemma 3.2.1 can be proved by repeating the above process.

Let the number of linearly dependent columns of $[\boldsymbol{A}^j \boldsymbol{b}_1 \boldsymbol{A}^j \boldsymbol{b}_2 \cdots \boldsymbol{A}^j \boldsymbol{b}_p] (j = 0, 1, 2, \cdots)$ is r_j, then

$$\begin{aligned} \text{rank}\boldsymbol{B} &= \text{rank}[\boldsymbol{b}_1 \quad \boldsymbol{b}_2 \quad \cdots \quad \boldsymbol{b}_p] = p - r_0 \\ \text{rank}\boldsymbol{AB} &= \text{rank}[\boldsymbol{Ab}_1 \quad \boldsymbol{Ab}_2 \quad \cdots \quad \boldsymbol{Ab}_p] = p - r_1 \\ &\vdots \end{aligned} \tag{3.2.41}$$

For the matrices $\boldsymbol{E}$ and $\boldsymbol{F}$, there exists

$$\text{rank } \boldsymbol{EF} \leqslant \min(\text{rank } \boldsymbol{E}, \text{rank } \boldsymbol{F}) \tag{3.2.42}$$

So, from eq. (3.2.41) and eq. (3.2.42), we have

$$0 \leqslant r_0 \leqslant r_1 \leqslant r_2 \leqslant \cdots \tag{3.2.43}$$

From eq. (3.2.42) and lemma 3.2.1, we have

$$\text{rank } \boldsymbol{U}_0 < \text{rank } \boldsymbol{U}_1 < \cdots < \text{rank } \boldsymbol{U}_{\mu-1} = \text{rank } \boldsymbol{U}_{\mu} = \text{rank } \boldsymbol{U}_{\mu+1} = \cdots \tag{3.2.44}$$

The above equation shows the rank of $\boldsymbol{U}_k$ increases as k increases until $k = \mu - 1$. When $k \geqslant \mu$, p columns of $\boldsymbol{A}^k\boldsymbol{B}$ will depend on its LHS columns. So rank of $\boldsymbol{U}_k$ don't increase again, i. e.

$$\text{rank}[\boldsymbol{B} \quad \boldsymbol{AB} \quad \cdots \quad \boldsymbol{A}^{\mu-1}\boldsymbol{B}] = \text{rank}[\boldsymbol{B} \quad \boldsymbol{AB} \quad \cdots \quad \boldsymbol{A}^{\mu-1}\boldsymbol{B} \quad \boldsymbol{A}^{\mu}\boldsymbol{B}] = \cdots \leqslant n \tag{3.2.45}$$

μ is called the controllability index of $(\boldsymbol{A}, \boldsymbol{B})$.

Example 3.2.4 Compute the controllability index of the following system

$$\dot{\boldsymbol{x}} = \begin{bmatrix} 2 & 1 & 0 \\ 0 & 2 & 0 \\ 0 & 0 & 2 \end{bmatrix} \boldsymbol{x} + \begin{bmatrix} 1 & 0 \\ 1 & 1 \\ 2 & 2 \end{bmatrix} \boldsymbol{u}$$

Answer Because $\text{rank}[\boldsymbol{B} \quad \boldsymbol{AB} \quad \cdots] = \text{rank} \begin{bmatrix} 1 & 0 & 3 & 1 & \cdots \\ 1 & 1 & 2 & 2 & \cdots \\ 2 & 2 & 4 & 4 & \cdots \end{bmatrix} = 2$, the two columns of $\boldsymbol{AB}$ depends on its LHS columns. The columns of $\boldsymbol{A}^2\boldsymbol{B}$ are also depends on its LHS columns according to lemma 3.2.1. Thus, the columns of $\boldsymbol{A}^2\boldsymbol{B}$ do not contribute columns to increase of the rank of the controllability matrix, and the controllability index of the system is 1.

The range of controllability index μ of a controllable system is given by the following theorem.

Theorem 3.2.1 For a controllable LTI system, its controllability index satisfies

$$\frac{n}{p} \leqslant \mu \leqslant \min(\bar{n}, n - \bar{p} + 1) \tag{3.2.46}$$

where $\bar{n}$ is the degree of the minimal polynomial of $\boldsymbol{A}$, $\bar{p} \triangleq \text{rank } \boldsymbol{B}$, n is the order of the system.

(Note: minimal polynomial $\psi(\lambda) = \lambda^{\bar{n}} + a_{\bar{n}-1}\lambda^{\bar{n}} + \cdots + a_1\lambda + a_0$ is the monic polynomial of minimal degree that satisfies $\psi(\boldsymbol{A}) = \boldsymbol{0}$.)

Proof From the definition of minimal polynomial, we get

$$\boldsymbol{A}^{\bar{n}} = -a_{\bar{n}-1}\boldsymbol{A}^{\bar{n}-1} + \cdots + a_1\boldsymbol{A} + a_0\boldsymbol{I}$$

Multiply both sides ofthe above equation by $\boldsymbol{B}$, we get

$$\boldsymbol{A}^{\bar{n}}\boldsymbol{B} = -a_{\bar{n}-1}\boldsymbol{A}^{\bar{n}-1}\boldsymbol{B} + \cdots + a_1\boldsymbol{AB} + a_0\boldsymbol{B}$$

which implies that $\boldsymbol{A}^{\bar{n}}\boldsymbol{B}$ can be written as a linearly combination of $\boldsymbol{B}, \boldsymbol{AB}, \cdots, \boldsymbol{A}^{\bar{n}-1}\boldsymbol{B}$. Thus we have

$$\mu \leqslant \bar{n} \tag{3.2.47}$$

On the other side, as k increases by 1, the rank of $\boldsymbol{U}_k$ increases by at least 1 until it stops increasing, so there are

$$\mu - 1 \leqslant n - \bar{p} \tag{3.2.48}$$

From eq. (3.2.47) and eq. (3.2.48), we have

$$\mu \leqslant \min(\bar{n}, n - \bar{p} + 1) \tag{3.2.49}$$

In order to ensure the rank of $\boldsymbol{U}_{\mu-1}$ is n, its column number must not less then its row number, i. e. $\mu p \geqslant n$. So we have

$$\mu \geqslant \frac{n}{p} \tag{3.2.50}$$

From eq. (3.2.49) and eq. (3.2.50), we have

$$\frac{n}{p} \leqslant \mu \leqslant \min(\bar{n}, n - \bar{p} + 1)$$

Based on above discussion, the conclusion 3.2.2 can be written as the following theorem.

Theorem 3.2.2 The n-dimensional pair $(\boldsymbol{A}, \boldsymbol{B})$ is controllable if and only if

$$\operatorname{rank} \boldsymbol{U}_{\bar{n}-1} = \operatorname{rank}[\boldsymbol{B} \quad \boldsymbol{AB} \quad \cdots \quad \boldsymbol{A}^{\bar{n}-1}\boldsymbol{B}] = n \tag{3.2.51}$$

or

$$\operatorname{rank} \boldsymbol{U}_{n-\bar{p}} = \operatorname{rank}[\boldsymbol{B} \quad \boldsymbol{AB} \quad \cdots \quad \boldsymbol{A}^{n-\bar{p}}\boldsymbol{B}] = n \tag{3.2.52}$$

There is also the following theorem about the controllability index of LTI systems.

Theorem 3.2.3 The controllability index of the LTI system eq. (3.2.1) is invariant under any equivalence transformation.

Proof The matrix $\boldsymbol{P}$ is an $n \times n$ non-singular matrix and

$$\bar{\boldsymbol{A}} = \boldsymbol{PAP}^{-1} \quad \bar{\boldsymbol{B}} = \boldsymbol{PB} \tag{3.2.53}$$

Let $\bar{\boldsymbol{U}}_k = [\bar{\boldsymbol{B}} \quad \bar{\boldsymbol{A}}\bar{\boldsymbol{B}} \quad \cdots \quad \bar{\boldsymbol{A}}^k\bar{\boldsymbol{B}}]$. Substituting eq. (3.2.53) into eq. (3.2.53) yields

$$\bar{\boldsymbol{U}}_k = [\boldsymbol{PB} \quad \boldsymbol{PAB} \quad \cdots \quad \boldsymbol{PA}^k\boldsymbol{B}] = \boldsymbol{P}[\boldsymbol{B} \quad \boldsymbol{AB} \quad \cdots \quad \boldsymbol{A}^k\boldsymbol{B}] = \boldsymbol{PU}_k,\ k = 0, 1, 2, \cdots$$

Because rank $\boldsymbol{P} = n$, so

$$\operatorname{rank} \bar{\boldsymbol{U}}_k = \operatorname{rank} \boldsymbol{U}_k \tag{3.2.54}$$

It shows that the controllability index of the system remains unchanged under any equivalence transformation.

3.2.3 Controllability criteria of LTV systems

Consider a LTV system

$$\dot{\boldsymbol{x}} = \boldsymbol{A}(t)\boldsymbol{x} + \boldsymbol{B}(t)\boldsymbol{u}, \quad \boldsymbol{x}(t_0) = \boldsymbol{x}_0, \quad t, t_0 \in J \tag{3.2.55}$$

where $\boldsymbol{x}$ is an n-dimensional column vector, $\boldsymbol{u}$ is a p-dimensional column vector, J is the time interval of the system, $\boldsymbol{A}(t)$ and $\boldsymbol{B}(t)$ are $n \times n$ and $n \times p$ time varying matrix.

Conclusion 3.2.5 (Grammian matrix criterion) The LTV system eq. (3.2.55) is completely state-observable at t_0 if and only if there exists a time t_1 ($t_1 \in J, t_1 > t_0$) such that the

$n \times n$ Grammian matrix

$$W_c[t_0,t_1] \triangleq \int_{t_0}^{t_1} \boldsymbol{\Phi}(t_0,t)\boldsymbol{B}(t)\boldsymbol{B}^{\mathrm{T}}(t)\boldsymbol{\Phi}^{\mathrm{T}}(t_0,t)\mathrm{d}t \tag{3.2.56}$$

is nonsingular. Where, $\boldsymbol{\Phi}(t,t_0)$ is the state transition matrix of the system (3.2.55).

Proof Sufficiency. Choose the input $u(t)$ as

$$\boldsymbol{u}(t) = -\boldsymbol{B}^{\mathrm{T}}(t)\boldsymbol{\Phi}^{\mathrm{T}}(t_0,t)\boldsymbol{W}_c^{-1}[t_0,t_1]\boldsymbol{x}_0 \tag{3.2.57}$$

At t_1,

$$\begin{aligned}
\boldsymbol{x}(t_1) &= \boldsymbol{\Phi}(t_1,t_0)\boldsymbol{x}_0 + \int_{t_0}^{t_1} \boldsymbol{\Phi}(t_1,\tau)\boldsymbol{B}(\tau)\boldsymbol{u}(\tau)\mathrm{d}\tau \\
&= \boldsymbol{\Phi}(t_1,t_0)\boldsymbol{x}_0 - \int_{t_0}^{t_1} \boldsymbol{\Phi}(t_1,\tau)\boldsymbol{B}(\tau)\boldsymbol{B}^{\mathrm{T}}(\tau)\boldsymbol{\Phi}^{\mathrm{T}}(t_0,\tau)\boldsymbol{W}_c^{-1}[t_0,t_1]\boldsymbol{x}_0\mathrm{d}\tau \\
&= \boldsymbol{\Phi}(t_1,t_0)\boldsymbol{x}_0 - \int_{t_0}^{t_1} \boldsymbol{\Phi}(t_1,t_0)\boldsymbol{\Phi}(t_0,\tau)\boldsymbol{B}(\tau)\boldsymbol{B}^{\mathrm{T}}(\tau)\boldsymbol{\Phi}^{\mathrm{T}}(t_0,\tau)\mathrm{d}\tau \cdot \boldsymbol{W}_c^{-1}[t_0,t_1]\boldsymbol{x}_0 \\
&= \boldsymbol{\Phi}(t_1,t_0)\boldsymbol{x}_0 - \boldsymbol{\Phi}(t_1,t_0)\int_{t_0}^{t_1} \boldsymbol{\Phi}(t_0,\tau)\boldsymbol{B}(\tau)\boldsymbol{B}^{\mathrm{T}}(\tau)\boldsymbol{\Phi}^{\mathrm{T}}(t_0,\tau)\mathrm{d}\tau \cdot \boldsymbol{W}_c^{-1}[t_0,t_1]\boldsymbol{x}_0 \\
&= \boldsymbol{\Phi}(t_1,t_0)\boldsymbol{x}_0 - \boldsymbol{\Phi}(t_1,t_0)\boldsymbol{W}_c[t_0,t_1]\boldsymbol{W}_c^{-1}[t_0,t_1]\boldsymbol{x}_0 \\
&= \boldsymbol{\Phi}(t_1,t_0)\boldsymbol{x}_0 - \boldsymbol{\Phi}(t_1,t_0)\boldsymbol{x}_0 \\
&= \boldsymbol{0}
\end{aligned}$$

According to the definition of controllability, the system is controllable.

Necessity. Prove that $\boldsymbol{W}_c[t_0,t_1]$ is nonsingular, if the system is controllable.

Adopt proof by contradiction to prove it. Suppose $\boldsymbol{W}_c[t_0,t_1]$ is nsingular, so there exists an non-zero constant vector $x_0 \in \mathbf{R}^n$ such that

$$\boldsymbol{x}_0^{\mathrm{T}}\boldsymbol{W}_c[t_0,t_1]\boldsymbol{x}_0 = 0 \tag{3.2.58}$$

Substituting eq. (3.2.56) into the above equation yields

$$\begin{aligned}
0 &= \int_{t_0}^{t_1} \boldsymbol{x}_0^{\mathrm{T}}\boldsymbol{\Phi}(t_0,t)\boldsymbol{B}(t)B^{\mathrm{T}}(t)\boldsymbol{\Phi}^{\mathrm{T}}(t_0,t)\boldsymbol{x}_0\mathrm{d}t \\
&= \int_{t_0}^{t_1} [\boldsymbol{x}_0^{\mathrm{T}}\boldsymbol{\Phi}(t_0,t)\boldsymbol{B}(t)] \cdot [\boldsymbol{x}_0^{\mathrm{T}}\boldsymbol{\Phi}(t_0,t)\boldsymbol{B}(t)]^{\mathrm{T}}\mathrm{d}t \\
&= \int_{t_0}^{t_1} \| \boldsymbol{x}_0^{\mathrm{T}}\boldsymbol{\Phi}(t_0,t)\boldsymbol{B}(t) \|^2\mathrm{d}t
\end{aligned}$$

which implies

$$\boldsymbol{x}_0^{\mathrm{T}}\boldsymbol{\Phi}(t_0,t)\boldsymbol{B}(t) = \boldsymbol{0}, t \in [t_0,t_1] \tag{3.2.59}$$

On the other hand, because the system is controllable, so there exists $\boldsymbol{u}(t)$, such that the state motion under initial condition $\boldsymbol{x}_0$ is

$$\boldsymbol{0} = \boldsymbol{x}(t_1) = \boldsymbol{\Phi}(t_1,t_0)\boldsymbol{x}_0 + \int_{t_0}^{t_1} \boldsymbol{\Phi}(t_1,t)\boldsymbol{B}(t)\boldsymbol{u}(t)\mathrm{d}t$$

Thus

$$\begin{aligned}
\boldsymbol{x}_0 &= -\boldsymbol{\Phi}^{-1}(t_1,t_0)\int_{t_0}^{t_1} \boldsymbol{\Phi}(t_1,t)\boldsymbol{B}(t)\boldsymbol{u}(t)\mathrm{d}t \\
&= -\int_{t_0}^{t_1} \boldsymbol{\Phi}(t_0,t_1)\boldsymbol{\Phi}(t_1,t)\boldsymbol{B}(t)\boldsymbol{u}(t)\mathrm{d}t
\end{aligned}$$

$$= -\int_{t_0}^{t_1} \boldsymbol{\Phi}(t_0,t)\boldsymbol{B}(t)\boldsymbol{u}(t)\mathrm{d}t \quad \boldsymbol{x}_0 \tag{3.2.60}$$

$$\|\boldsymbol{x}_0\|^2 = \boldsymbol{x}_0^{\mathrm{T}}\boldsymbol{x}_0 = -\boldsymbol{x}_0^{\mathrm{T}}\int_{t_0}^{t_1} \boldsymbol{\Phi}(t_0,t)\boldsymbol{B}(t)\boldsymbol{u}(t)\mathrm{d}t = -\int_{t_0}^{t_1} \boldsymbol{x}_0^{\mathrm{T}}\boldsymbol{\Phi}(t_0,t)\boldsymbol{B}(t)\boldsymbol{u}(t)\mathrm{d}t \tag{3.2.61}$$

From eq. (3.2.59) and eq. (3.2.61), we have

$\|\boldsymbol{x}_0\| = 0$, i.e. $\boldsymbol{x}_0 = \boldsymbol{0}$. It contradicts that $\boldsymbol{x}_0$ is not zero. Thus $\boldsymbol{W}_c[t_0,t_1]$ is nonsingular, if the system is controllable.

It should be pointed out that although the Grammian matrix criterion has a simple form, it is difficult to apply in practice and only has theoretical significance because it is very difficult to solve the state transition matrix of time-varying systems. In order to make it possible to determine the controllability directly according to $\boldsymbol{A}(t)$ and $\boldsymbol{B}(t)$, a sufficient criterion theorem is given below.

Conclusion 3.2.6 (Rank criterion) Let $\boldsymbol{A}(t)$ and $\boldsymbol{B}(t)$ be $n-1$ times continuously differentiable. The n-dimensional LTV system $\{\boldsymbol{A}(t), \boldsymbol{B}(t)\}$ is controllable at t_0 if there is a finite time $t_1 > t_0$ such that

$$\mathrm{rank}[\boldsymbol{M}_0(t_1) \vdots \boldsymbol{M}_1(t_1) \vdots \cdots \vdots \boldsymbol{M}_{n-1}(t_1)] = n \tag{3.2.62}$$

where,

$$\begin{cases} \boldsymbol{M}_0(t) = \boldsymbol{B}(t) \\ \boldsymbol{M}_1(t) = -\boldsymbol{A}(t)\boldsymbol{M}_0(t) + \dfrac{\mathrm{d}}{\mathrm{d}t}\boldsymbol{M}_0(t) \\ \boldsymbol{M}_2(t) = -\boldsymbol{A}(t)\boldsymbol{M}_1(t) + \dfrac{\mathrm{d}}{\mathrm{d}t}\boldsymbol{M}_1(t) \\ \vdots \\ \boldsymbol{M}_{n-1}(t) = -\boldsymbol{A}(t)\boldsymbol{M}_{n-2}(t) + \dfrac{\mathrm{d}}{\mathrm{d}t}\boldsymbol{M}_{n-2}(t) \end{cases} \tag{3.2.63}$$

Proof (1) Consider $\boldsymbol{\Phi}(t_0,t_1)\boldsymbol{B}(t_1) = \boldsymbol{\Phi}(t_0,t_1)\boldsymbol{M}_0(t_1)$ and

$$\frac{\partial}{\partial t_1}[\boldsymbol{\Phi}(t_0,t_1)\boldsymbol{B}(t_1)] = \left[\frac{\partial}{\partial t}\boldsymbol{\Phi}(t_0,t)\boldsymbol{B}(t)\right]_{t=t_1}$$

From eq. (2.4.9), we have

$$\left[\boldsymbol{\Phi}(t_0,t_1)\boldsymbol{B}(t_1) \vdots \frac{\partial}{\partial t_1}\boldsymbol{\Phi}(t_0,t_1)\boldsymbol{B}(t_1) \vdots \cdots \vdots \frac{\partial^{n-1}}{\partial t_1^{n-1}}\boldsymbol{\Phi}(t_0,t_1)\boldsymbol{B}(t_1)\right]$$

$$= \boldsymbol{\Phi}(t_0,t_1)[\boldsymbol{M}_0(t_1) \vdots \boldsymbol{M}_1(t_1) \vdots \cdots \vdots \boldsymbol{M}_{n-1}(t_1)] \tag{3.2.64}$$

Because $\boldsymbol{\Phi}(t_0,t_1)$ is nonsingular, so from eq. (3.2.62) and eq. (3.2.64), we have

$$\mathrm{rank}\left[\boldsymbol{\Phi}(t_0,t_1)\boldsymbol{B}(t_1) \vdots \frac{\partial}{\partial t_1}\boldsymbol{\Phi}(t_0,t_1)\boldsymbol{B}(t_1) \vdots \cdots \vdots \frac{\partial^{n-1}}{\partial t_1^{n-1}}\boldsymbol{\Phi}(t_0,t_1)\boldsymbol{B}(t_1)\right] = n \tag{3.2.65}$$

(2) Next prove that for $t_1 > t_0$, $\boldsymbol{\Phi}(t_0,t)\boldsymbol{B}(t)$ is row linearly independent in $[t_0,t_1]$. Adopt proof by contradiction to prove it. Suppose $\boldsymbol{\Phi}(t_0,t)\boldsymbol{B}(t)$ is row linearly dependent in $[t_0,t_1]$,

i. e. $\exists \boldsymbol{\alpha} \in \mathbf{R}^n$ and $\boldsymbol{\alpha} \neq \mathbf{0}$, such that

$$\boldsymbol{\alpha}^{\mathrm{T}} \boldsymbol{\Phi}(t_0, t) \boldsymbol{B}(t) = \mathbf{0} \tag{3.2.66}$$

Then for $t \in [t_0, t_1]$ and $k = 1, 2, \cdots, n-1$, we have

$$\boldsymbol{\alpha}^{\mathrm{T}} \frac{\partial^k}{\partial t^k} \boldsymbol{\Phi}(t_0, t) \boldsymbol{B}(t) = \mathbf{0} \tag{3.2.67}$$

i. e. for $t \in [t_0, t_1]$, we have

$$\boldsymbol{\alpha}^{\mathrm{T}} \left[\boldsymbol{\Phi}(t_0, t) \boldsymbol{B}(t) \vdots \frac{\partial}{\partial t} \boldsymbol{\Phi}(t_0, t) \boldsymbol{B}(t) \vdots \cdots \vdots \frac{\partial^{n-1}}{\partial t^{n-1}} \boldsymbol{\Phi}(t_0, t) \boldsymbol{B}(t) \right] = \mathbf{0} \tag{3.2.68}$$

It shows

$$\left[\boldsymbol{\Phi}(t_0, t) \boldsymbol{B}(t) \vdots \frac{\partial}{\partial t} \boldsymbol{\Phi}(t_0, t) \boldsymbol{B}(t) \vdots \cdots \vdots \frac{\partial^{n-1}}{\partial t^{n-1}} \boldsymbol{\Phi}(t_0, t) \boldsymbol{B}(t) \right]$$

are row linearly dependent for $t \in [t_0, t_1]$. This contradicts eq. (3.2.65). So $\boldsymbol{\Phi}(t_0, t) \boldsymbol{B}(t)$ is row linearly independent in $[t_0, t_1]$.

(3) From $\boldsymbol{\Phi}(t_0, t) \boldsymbol{B}(t)$ is row linearly independent in $[t_0, t_1]$, prove $\boldsymbol{W}_c[t_0, t_1]$ is nonsingular. Adopt proof by contradiction to prove it. Suppose $\boldsymbol{W}_c[t_0, t_1]$ is singular. So there exists a $1 \times n$ non-zero constant vector $\boldsymbol{\alpha}$ such that

$$0 = \boldsymbol{\alpha} \boldsymbol{W}_c[t_0, t_1] \boldsymbol{\alpha}^{\mathrm{T}} = \int_{t_0}^{t_1} (\boldsymbol{\alpha} \boldsymbol{\Phi}(t_0, t) \boldsymbol{B}(t)) (\boldsymbol{\alpha} \boldsymbol{\Phi}(t_0, t) \boldsymbol{B}(t))^{\mathrm{T}} \mathrm{d}t \tag{3.2.69}$$

From the above equation, we have

$$\boldsymbol{\alpha} \boldsymbol{\Phi}(t_0, t) \boldsymbol{B}(t) = \mathbf{0}, t \in [t_0, t_1] \tag{3.2.70}$$

This contradicts that $\boldsymbol{\Phi}(t_0, t) \boldsymbol{B}(t)$ is row independent in $[t_0, t_1]$. Thus $\boldsymbol{W}_c[t_0, t_1]$ is nonsingular.

(4) Because $\boldsymbol{W}_c[t_0, t_1]$ is nonsingular, following conclusion 3.2.5, the system is controllable at t_0.

Example 3.2.5 Given a LTV system

$$\dot{\boldsymbol{x}} = \begin{bmatrix} t & 1 & 0 \\ 0 & 0 & t \\ 0 & 0 & t^2 \end{bmatrix} \boldsymbol{x} + \begin{bmatrix} 0 \\ 1 \\ 1 \end{bmatrix} \boldsymbol{u}, t > 0$$

Check its controllability.

Answer

$$\boldsymbol{M}_0(t) = \boldsymbol{B}(t) = \begin{bmatrix} 0 \\ 1 \\ 1 \end{bmatrix}$$

$$\boldsymbol{M}_1(t) = -\boldsymbol{A}(t) \boldsymbol{M}_0(t) + \frac{\mathrm{d}}{\mathrm{d}t} \boldsymbol{M}_0(t) = \begin{bmatrix} -1 \\ -t \\ -t^2 \end{bmatrix}$$

$$\boldsymbol{M}_2(t) = -\boldsymbol{A}(t) \boldsymbol{M}_1(t) + \frac{\mathrm{d}}{\mathrm{d}t} \boldsymbol{M}_1(t) = \begin{bmatrix} 2t \\ t^2 \\ t^4 \end{bmatrix} + \begin{bmatrix} 0 \\ -1 \\ -2t \end{bmatrix} = \begin{bmatrix} 2t \\ t^2 - 1 \\ t^4 - 2t \end{bmatrix}$$

when $t>0$, we have

$$\operatorname{rank}[\boldsymbol{M}_0(t)\ \boldsymbol{M}_1(t)\ \boldsymbol{M}_2(t)]=\operatorname{rank}\begin{bmatrix}0 & -1 & 2t\\ 1 & -t & t^2-1\\ 1 & -t^2 & t^4-2t\end{bmatrix}=3$$

Thus the system is controllable at t_0.

Conclusion 3.2.7 System eq. (3.2.55) is controllable at t_0 if only if there exists $t_1>t_0$, such that $\boldsymbol{\Phi}(t_0,t)\boldsymbol{B}(t)$ is row linearly independent in $[t_0,t_1]$.

Proof Necessity. Prove $\boldsymbol{\Phi}(t_0,t)\boldsymbol{B}(t)$ is row linearly independent in $[t_0,t_1]$, if system eq. (3.2.55) is controllable at t_0. The step (1) and (2) in the proof of conclusion 3.2.6 finish the proof of necessity.

Sufficiency. Proof the system eq. (3.2.55) is controllable at t_0, if $\boldsymbol{\Phi}(t_0,t)\boldsymbol{B}(t)$ is row linearly independent in $[t_0,t_1]$. Because $\boldsymbol{\Phi}(t_0,t)\boldsymbol{B}(t)$ is row linearly independent in $[t_0,t_1]$, the Grammian matrix $\boldsymbol{W}_c[t_0,t_1]$ is nonsingular. According to conclusion 3.2.6, the system is controllable.

3.3 Observability criteria of linear time-continuous systems

Observability of systems is usually studied under $\boldsymbol{u}=0$. In this section, we discuss the observability criterion of linear time-varying and time-invariant systems. Because the observability and controllability of systems have duality form in concepts and analysis methods, we only give conclusions and no proof for most conclusion.

3.3.1 Observability criteria of LTI systems

When $\boldsymbol{u}=\boldsymbol{0}$, the state equation and output equation are

$$\begin{cases}\dot{\boldsymbol{x}}=\boldsymbol{A}\boldsymbol{x},\boldsymbol{x}(0)=\boldsymbol{x}_0,t\geqslant 0\\ \boldsymbol{y}=\boldsymbol{C}\boldsymbol{x}\end{cases}\tag{3.3.1}$$

where $\boldsymbol{x}$ is an n-dimensional state vector, $\boldsymbol{y}$ is a q-dimensional output vector, $\boldsymbol{A}$ and $\boldsymbol{C}$ are $n\times n$ and $q\times n$ constant matrix.

Conclusion 3.3.1 (Grammian matrix criterion) The LTI system eq. (3.3.1) is observable if and only if there exists $t_1>0$, such that the $n\times n$ Grammian matrix

$$\boldsymbol{W}_o[0,t_1]\triangleq\int_0^{t_1}\mathrm{e}^{\boldsymbol{A}^{\mathrm{T}}t}\boldsymbol{C}^{\mathrm{T}}\boldsymbol{C}\mathrm{e}^{\boldsymbol{A}t}\mathrm{d}t\tag{3.3.2}$$

is nonsingular.

Proof Sufficiency. Prove the system is observable, if $\boldsymbol{W}_o[0,t_1]$ is nonsingular.

Because $\boldsymbol{W}_o[0,t_1]$ is nonsingular, there exists $\boldsymbol{W}_o^{-1}$. The following formula can be constructed by using $\boldsymbol{y}(t)$ in $[0,t_1]$:

$$\boldsymbol{W}_o^{-1}[0,t_1]\int_0^{t_1}\mathrm{e}^{\boldsymbol{A}^{\mathrm{T}}t}\boldsymbol{C}^{\mathrm{T}}\boldsymbol{y}(t)\mathrm{d}t = \boldsymbol{W}_o^{-1}[0,t_1]\int_0^{t_1}\mathrm{e}^{\boldsymbol{A}^{\mathrm{T}}t}\boldsymbol{C}^{\mathrm{T}}\boldsymbol{C}\mathrm{e}^{\boldsymbol{A}t}\mathrm{d}t\cdot\boldsymbol{x}_0$$

$$= \boldsymbol{W}_o^{-1}[0,t_1]\,\boldsymbol{W}_o[0,t_1]\,\boldsymbol{x}_0 = \boldsymbol{x}_0 \tag{3.3.3}$$

It shows that when $\boldsymbol{W}_o[0,t_1]$ is nonsingular, nonzero initial state $\boldsymbol{x}_0$ can be constructed by $\boldsymbol{y}(t)$ in $[0,t_1]$. Thus the system is observable.

Necessity. Prove $\boldsymbol{W}_o[0,t_1]$ is nonsingular, if the system is observable.

Adopt proof by contradiction to prove it. We suppose $\boldsymbol{W}_o[0,t_1]$ is singular, i. e. $\exists\ \bar{\boldsymbol{x}}_0\in\mathbf{R}^n$ and $\bar{\boldsymbol{x}}_0\neq 0$, such that

$$\begin{aligned}
0 &= \bar{\boldsymbol{x}}_0^{\mathrm{T}}\boldsymbol{W}_o[0,t_1]\bar{\boldsymbol{x}}_0\\
&= \int_0^{t_1}\bar{x}_0^{\mathrm{T}}\mathrm{e}^{\boldsymbol{A}^{\mathrm{T}}t}\boldsymbol{C}^{\mathrm{T}}\boldsymbol{C}\mathrm{e}^{\boldsymbol{A}t}\bar{\boldsymbol{x}}_0\mathrm{d}t\\
&= \int_0^{t_1}\boldsymbol{y}^{\mathrm{T}}(t)\boldsymbol{y}(t)\mathrm{d}t\\
&= \int_0^{t_1}\|\boldsymbol{y}(t)\|^2\mathrm{d}t
\end{aligned}\tag{3.3.4}$$

It implies

$$\boldsymbol{y}(t) = \boldsymbol{C}\mathrm{e}^{\boldsymbol{A}t}\bar{\boldsymbol{x}}_0\equiv\boldsymbol{0},\ \forall t\in[0,t_1] \tag{3.3.5}$$

According to the definition of observability, $\bar{x}_0$ is not observable which contradicts that the state is observable. Thus, $\boldsymbol{W}_o[0,t_1]$ is non-singular, if the system is observable.

In order to calculate the rank of the Grammian matrix, it is necessary to calculate the matrix exponent $\mathrm{e}^{\boldsymbol{A}t}$, and when the system dimension is large, the calculation is very time consuming. Therefore, grammian matrix criterion is mainly used for theoretical analysis and proof of other criteria.

Conclusion 3.3.2 (Rank criterion) The LTI system eq. (3.3.1) is observable if and only if its $qn\times n$ observability matrix

$$\boldsymbol{Q}_o = \begin{bmatrix}\boldsymbol{C}\\ \boldsymbol{C}\boldsymbol{A}\\ \vdots\\ \boldsymbol{C}\boldsymbol{A}^{n-1}\end{bmatrix}$$

has rank n (full column rank), i. e.

$$\mathrm{rank}\begin{bmatrix}\boldsymbol{C}\\ \boldsymbol{C}\boldsymbol{A}\\ \vdots\\ \boldsymbol{C}\boldsymbol{A}^{n-1}\end{bmatrix} = n \tag{3.3.6}$$

We can use matlab function rank(obsv ($\boldsymbol{A}$, $\boldsymbol{C}$)) *to compute* rank $\boldsymbol{Q}_o$.

Example 3.3.1 Given a LTI system

$$\dot{x} = \begin{bmatrix} 0 & -2 \\ 1 & -3 \end{bmatrix} x$$
$$y = \begin{bmatrix} 0 & 1 \end{bmatrix} x$$

Check its observability.

Answer

$$\operatorname{rank}\begin{bmatrix} c \\ cA \end{bmatrix} = \operatorname{rank}\begin{bmatrix} 0 & 1 \\ 1 & -3 \end{bmatrix} = 2 = n$$

Thus the system is observable.

Conclusion 3.3.3 (PBH rank criterion) The LTI system eq. (3.3.1) is observable if and only if $(n+q)\times n$ matrix

$$\begin{bmatrix} C \\ \lambda_i I - A \end{bmatrix}$$

has full column rank at every eigenvalue $\lambda_i (i=1,2,\cdots,n)$ of A, i.e.

$$\begin{bmatrix} C \\ \lambda_i I - A \end{bmatrix} = n, i=1,2,\cdots,n \tag{3.3.7}$$

or

$$\operatorname{rank}\begin{bmatrix} C \\ sI - A \end{bmatrix} = n, \forall s \in \text{Complex number}$$

Conclusion 3.3.4 (Jordan canonical form criterion)

(1) When all eigenvalues of A, $\lambda_1, \lambda_2, \cdots, \lambda_n$, are distinct, A has diagonal form and

$$\begin{cases} \dot{\bar{x}} = \begin{bmatrix} \lambda_1 & & & \\ & \lambda_2 & & \\ & & \ddots & \\ & & & \lambda_n \end{bmatrix} \bar{x} \\ y = \bar{C}\bar{x} \end{cases}$$

then its state is completely observable if and only if all column vectors of $\bar{C}$ are not zero.

(2) When the $n \times n$ matrix A has repeated eigenvalues:

λ_1 (multiplicity σ_1), λ_2 (multiplicity σ_2), $\cdots$, λ_l (multiplicity σ_l) and $\sigma_1 + \sigma_2 + \cdots + \sigma_l = n$, the Jordan canonical form is

$$\begin{cases} \dot{\hat{x}} = \hat{A}\hat{x} \\ y = \hat{C}\hat{x} \end{cases} \tag{3.3.8}$$

where

$$\hat{A}_{n\times n} = \begin{bmatrix} J_1 & & & & \\ & \ddots & & & \\ & & J_i & & \\ & & & \ddots & \\ & & & & J_l \end{bmatrix}, \quad \hat{C}_{q\times n} = \begin{bmatrix} \hat{C}_1 & \hat{C}_2 & \cdots & \hat{C}_l \end{bmatrix} \tag{3.3.9}$$

$$J_{i(\sigma_i \times \sigma_i)} = \begin{bmatrix} J_{i1} & & & \\ & J_{i2} & & \\ & & \ddots & \\ & & & J_{i\alpha_i} \end{bmatrix}, \quad \hat{C} = [\hat{C}_{i1} \quad \hat{C}_{i2} \quad \cdots \quad \hat{C}_{i\alpha_i}] \tag{3.3.10}$$

$$J_{ik(r_{ik} \times r_{ik})} = \begin{bmatrix} \lambda_i & 1 & & & \\ & \lambda_i & 1 & & \\ & & \ddots & \ddots & \\ & & & \lambda_i & 1 \\ & & & & \lambda_i \end{bmatrix}, \hat{C}_{ik(q \times r_{ik})} = [\hat{c}_{1ik} \quad \hat{c}_{2ik} \quad \cdots \quad \hat{c}_{rik}] \tag{3.3.11}$$

where J_{ik} is the kth block of J_i corresponding to $\lambda_i (k = 1, 2, \cdots, \alpha_i)$ which is also called the kth small Jordan block corresponding to λ_i, and $r_{i1} + r_{i2} + \cdots + r_{i\alpha_i} = \sigma_i$.

The LTI system eq. (3.3.1) is completely observable if and only if the matrix

$$[\hat{c}_{1i1} \quad \hat{c}_{1i2} \quad \cdots \quad \hat{c}_{1i\alpha_i}]$$

composed by first column of $\hat{B}_{ik} (k = 1, 2, \cdots, \alpha_i)$, has full column rank, where $i = 1, 2, \cdots, l$.

Example 3.3.2 Given a LTI system

$$\dot{\bar{x}} = \begin{bmatrix} -1 & 0 & 0 \\ 0 & -2 & 0 \\ 0 & 0 & -3 \end{bmatrix} \bar{x}$$

$$y = \begin{bmatrix} 1 & 2 & 3 \\ 0 & 2 & 5 \end{bmatrix} \bar{x}$$

Check its observability.

Answer Because all columns of $\bar{C}$ are not zero, the system is observable.

Example 3.3.3 The state space description of a LTI system is

$$\dot{\hat{x}} = \left[\begin{array}{cc|c|c|ccc} \lambda_1 & 1 & & & & & \\ 0 & \lambda_1 & & & & & \\ \hline & & \lambda_1 & & & & \\ \hline & & & \lambda_1 & & & \\ \hline & & & & \lambda_2 & 1 & \\ & & & & & \lambda_2 & 1 \\ & & & & & & \lambda_2 \end{array}\right] \hat{x}$$

$$y = \left[\begin{array}{cc|c|c|ccc} 1 & 1 & 2 & 0 & 0 & 2 & 0 \\ 1 & 0 & 1 & 2 & 0 & 1 & 1 \\ 1 & 0 & 2 & 3 & 0 & 2 & 2 \end{array}\right] \hat{x}$$

where $\lambda_1 \neq \lambda_2$. Check its observability.

Answer A has two different eigenvalues λ_1 and λ_2. For λ_1, For $\lambda_1 = -2$, there exists

$[\hat{\boldsymbol{c}}_{111}\quad \hat{\boldsymbol{c}}_{112}\quad \hat{\boldsymbol{c}}_{113}] = \begin{bmatrix} 1 & 2 & 0 \\ 1 & 1 & 2 \\ 1 & 2 & 3 \end{bmatrix}$, where its all columns are linearly independent.

Thus the state variables x_1, x_2, x_3 and x_4 are observable.

For λ_2, there exists $\hat{\boldsymbol{c}}_{121} = \begin{bmatrix} 0 \\ 0 \\ 0 \end{bmatrix}$. Thus the x_5, x_6 and x_7 are not completely observable.

3.3.2 Observability index

For the LTI system eq. (3.3.1), $\boldsymbol{A}$ and $\boldsymbol{C}$ are $n\times n$ and $q\times n$ constant matrices. We define a $(k+1)q\times n$ matrix

$$\boldsymbol{V}_k \triangleq \begin{bmatrix} \boldsymbol{C} \\ \boldsymbol{CA} \\ \vdots \\ \boldsymbol{CA}^k \end{bmatrix},\quad k=1,2,\cdots \tag{3.3.12}$$

If $(\boldsymbol{A},\boldsymbol{C})$ is observable, then

$$\text{rank } \boldsymbol{V}_{n-1} = n$$

i. e. there are n linearly independent rows among the qn rows of $\boldsymbol{V}_{n-1}$. $\boldsymbol{C}$ can written as

$$\boldsymbol{C} = \begin{bmatrix} \boldsymbol{c}_1 \\ \boldsymbol{c}_2 \\ \vdots \\ \boldsymbol{c}_q \end{bmatrix} \tag{3.3.13}$$

where $\boldsymbol{C}_i\,(i=1,2,\cdots,q)$ is $1\times n$ row vector. Thus $\boldsymbol{V}_k$ can written as

$$\boldsymbol{V}_k = \begin{bmatrix} \boldsymbol{c}_1 \\ \boldsymbol{c}_2 \\ \vdots \\ \boldsymbol{c}_q \\ \hdashline \boldsymbol{c}_1\boldsymbol{A} \\ \boldsymbol{c}_2\boldsymbol{A} \\ \vdots \\ \boldsymbol{c}_q\boldsymbol{A} \\ \hdashline \vdots \\ \hdashline \boldsymbol{c}_1\boldsymbol{A}^k \\ \boldsymbol{c}_2\boldsymbol{A}^k \\ \vdots \\ \boldsymbol{c}_q\boldsymbol{A}^k \end{bmatrix} \tag{3.3.14}$$

Let $r_i\,(i=1,2,\cdots,n)$ is the number of linearly dependent rows of $\boldsymbol{CA}^i_{q\times n}$, similar to analysis in controllability index, we have

$$0<r_0\leqslant r_1\leqslant\cdots\leqslant q \tag{3.3.15}$$

Since there are at most n linearly independent rows in $\boldsymbol{V}_k$, there exists an integer v, so that

$$0<r_0\leqslant r_1\leqslant\cdots\leqslant r_v=r_{v+1}=\cdots \tag{3.3.16}$$

or equivalent to

$$\text{rank } \boldsymbol{V}_0<\text{rank } \boldsymbol{V}_1<\cdots<\text{rank } \boldsymbol{V}_{v-1}=\text{rank } \boldsymbol{V}_v=\text{rank}\boldsymbol{V}_{v+1}=\cdots \tag{3.3.17}$$

v is called the observability index of $(\boldsymbol{A},\boldsymbol{C})$.

Theorem 3.3.1 For a observable LTI system, its observability index satisfies

$$\frac{n}{q}\leqslant v\leqslant\min(\bar{n},n-\bar{q}+1) \tag{3.3.18}$$

where $\bar{n}$ is the degree of the minimal polynomial of $\boldsymbol{A}$, $\bar{q}\triangleq\text{rank }\boldsymbol{C}$, n is the order of the system.

Proof

Let the minimal polynomial of A is $\psi(\lambda)=\lambda^{\bar{n}}+a_{\bar{n}-1}\lambda^{\bar{n}-1}+\cdots+a_1\lambda+a_0$, we have

$$\boldsymbol{A}^{\bar{n}}=-a_{\bar{n}-1}\boldsymbol{A}^{\bar{n}-1}+\cdots+a_1\boldsymbol{A}+a_0\boldsymbol{I}$$

Left multiplying both sides of the above equation by $\boldsymbol{C}$, yields

$$\boldsymbol{CA}^{\bar{n}}=-a_{\bar{n}-1}\boldsymbol{CA}^{\bar{n}-1}+\cdots+a_1\boldsymbol{CA}+a_0\boldsymbol{C}$$

which implies that $\boldsymbol{CA}^{\bar{n}}$ can be written as a linearly combination of $\boldsymbol{C},\boldsymbol{CA},\cdots,\boldsymbol{CA}^{\bar{n}-1}$. Thus we have

$$v\leqslant\bar{n} \tag{3.3.19}$$

On the other side, as k increases by 1, the rank of $\boldsymbol{V}_k$ increases by at least 1 until it stops increasing, so there are

$$v-1\leqslant n-\bar{q} \tag{3.3.20}$$

From eq. (3.3.19) and eq. (3.3.20), we have

$$v\leqslant\min(\bar{n},n-\bar{q}+1) \tag{3.3.21}$$

In order to ensure the rank of $\boldsymbol{V}_{v-1}$ is n, its row number must not less then its column number, i. e. $vq\geqslant n$, So we have

$$v\geqslant\frac{n}{q} \tag{3.3.22}$$

From eq. (3.3.21) and eq. (3.3.22), we have

$$\frac{n}{q}\leqslant v\leqslant\min(\bar{n},n-\bar{q}+1)$$

Based on above discussion, the conclusion 3.3.2 can be written as the following theorem.

Theorem 3.3.2 The n-dimensional pair $(\boldsymbol{A},\boldsymbol{C})$ is observable if and only if

$$\text{rank } \boldsymbol{V}_{\bar{n}-1}=\text{rank}\begin{bmatrix}\boldsymbol{C}\\ \boldsymbol{CA}\\ \vdots\\ \boldsymbol{CA}^{\bar{n}-1}\end{bmatrix}=n \tag{3.3.23}$$

or

$$\operatorname{rank} \boldsymbol{V}_{n-\bar{q}} = \operatorname{rank}\begin{bmatrix} \boldsymbol{C} \\ \boldsymbol{CA} \\ \vdots \\ \boldsymbol{CA}^{n-\bar{q}} \end{bmatrix} = n \tag{3.3.24}$$

Theorem 3.3.3 The observability index of the LTI system eq. (3.3.1) is invariant under any equivalence transformation.

3.3.3 Observability criteria of LTV systems

Consider a LTV system

$$\begin{cases} \dot{\boldsymbol{x}} = \boldsymbol{A}(t)\boldsymbol{x}, \boldsymbol{x}(t_0) = \boldsymbol{x}_0, t, t_0 \in J \\ \boldsymbol{y} = \boldsymbol{C}(t)x \end{cases} \tag{3.3.25}$$

where J is the time interval of the system, $\boldsymbol{A}(t)$ and $\boldsymbol{C}(t)$ are $n \times n$ and $q \times n$ time varying matrix.

Conclusion 3.3.5 (Grammian matrix criterion) The LTV system eq. (3.3.25) is completely state-observable at time t_0 if and only if there exists a time $t_1 \in J, t_1 > t_0$ such that $n \times n$ Grammian matrix

$$\boldsymbol{W}_o[t_0, t_1] \triangleq \int_{t_0}^{t_1} \boldsymbol{\Phi}^T(t,t_0)\boldsymbol{C}^T(t)\boldsymbol{C}(t)\boldsymbol{\Phi}(t,t_0)\mathrm{d}t \tag{3.3.26}$$

is nonsingular. Where $\boldsymbol{\Phi}(t,t_0)$ is the state transition matrix of the LTV system eq. (3.3.25).

Proof Sufficiency. Prove the system is observable at t_0 if $\boldsymbol{W}_o[t_0, t_1]$ is nonsingular.

For any $\boldsymbol{x}(t_0) = \boldsymbol{x}_0$, we have

$$\boldsymbol{y}(t) = \boldsymbol{C}(t)\boldsymbol{\Phi}(t,t_0)\boldsymbol{x}_0$$

Left multiply both sides of the above equation by $\boldsymbol{\Phi}^T(t,t_0)\boldsymbol{C}^T(t)$, then integrate, we have

$$\int_{t_0}^{t_1} \boldsymbol{\Phi}^T(t,t_0)\boldsymbol{C}^T(t)y(t)\mathrm{d}t = \int_{t_0}^{t_1} \boldsymbol{\Phi}^T(t,t_0)\boldsymbol{C}^T(t)\boldsymbol{C}(t)\boldsymbol{\Phi}(t,t_0)\mathrm{d}t \cdot \boldsymbol{x}_0 = \boldsymbol{W}_o[t_0, t_1]\boldsymbol{x}_0$$

So

$$\boldsymbol{x}_0 = \boldsymbol{W}_o^{-1}[t_0, t_1]\int_{t_0}^{t_1} \boldsymbol{\Phi}^T(t,t_0)\boldsymbol{C}^T(t)y(t)\mathrm{d}t$$

It shows $\boldsymbol{x}_0$ can be uniquely determined by $\boldsymbol{y}(t)$ in $[t_0, t_1]$.

Necessity. Prove $\boldsymbol{W}_o[t_0, t_1]$ is nonsingular, if the system is observable at t_0.

Adopt proof by contradiction to prove it. We suppose $\boldsymbol{W}_o[t_0, t_1]$ is singular. So there exists $\boldsymbol{X}$ non-zero $n \times 1$ vector, such that

$$\boldsymbol{\alpha}^T \boldsymbol{W}_o[t_0, t_1]\boldsymbol{\alpha} = 0$$

i. e.

$$\int_{t_0}^{t_1} (\boldsymbol{\alpha}^T \boldsymbol{\Phi}^T(t,t_0)\boldsymbol{C}^T(t))(\boldsymbol{\alpha}^T \boldsymbol{\Phi}^T(t,t_0)\boldsymbol{C}^T(t))^T\mathrm{d}t = 0$$

It implies

$$C(t)\boldsymbol{\Phi}(t,t_0)\boldsymbol{\alpha}=0,\quad t\in[t_0,t_1]$$

If let $\boldsymbol{\alpha}$ equals initial state, i. e. $\boldsymbol{x}(t_0)=\boldsymbol{\alpha}$, we have

$$\boldsymbol{y}(t)=\boldsymbol{C}(t)\boldsymbol{\Phi}(t,t_0)\boldsymbol{x}_0=\mathbf{0}$$

It shows the system is not observable that contradicts thatthe system is observable. Thus $\boldsymbol{W}_o[t_0,t_1]$ is nonsingular, if the system is observable at t_0.

Conclusion 3.3.6 (Rank criterion) Let $\boldsymbol{A}(t)$ and $\boldsymbol{C}(t)$ be $n-1$ times continuously differentiable. Then the n-dimensional LTV system eq. (3.3.25) is observable at t_0 if there exists a finite $t_1\in J$, $t_1>t_0$, such that

$$\operatorname{rank}\begin{bmatrix}\boldsymbol{N}_0(t_1)\\ \boldsymbol{N}_2(t_1)\\ \vdots\\ \boldsymbol{N}_{n-1}(t_1)\end{bmatrix}=n \tag{3.3.27}$$

where

$$\begin{cases}\boldsymbol{N}_0(t)=\boldsymbol{C}(t)\\ \boldsymbol{N}_1(t)=\boldsymbol{N}_0(t)\boldsymbol{A}(t)+\dfrac{\mathrm{d}}{\mathrm{d}t}\boldsymbol{N}_0(t)\\ \boldsymbol{N}_2(t)=\boldsymbol{N}_1(t)\boldsymbol{A}(t)+\dfrac{\mathrm{d}}{\mathrm{d}t}\boldsymbol{N}_1(t)\\ \quad\vdots\\ \boldsymbol{N}_{n-1}(t)=\boldsymbol{N}_{n-2}(t)\boldsymbol{A}(t)+\dfrac{\mathrm{d}}{\mathrm{d}t}\boldsymbol{N}_{n-2}(t)\end{cases} \tag{3.3.28}$$

3.4 Duality theorem

Controllability and observability are dual in concept and form. The duality of the criterion of controllability and observability reflects the duality of the control problem and the estimation problem in essence. In this section, we discuss the main conclusions of this dual relationship.

3.4.1 Dual system

Consider a LTV system

$$\Sigma:\begin{cases}\dot{\boldsymbol{x}}=\boldsymbol{A}(t)\boldsymbol{x}+\boldsymbol{B}(t)\boldsymbol{u}\\ \boldsymbol{y}=\boldsymbol{C}(t)\boldsymbol{x}\end{cases} \tag{3.4.1}$$

where $\boldsymbol{x}$ is an n-dimensional column vector, $\boldsymbol{u}$ is a p-dimensional column vector, $\boldsymbol{y}$ is a q-dimensional column vector, $\boldsymbol{A}(t)$, $\boldsymbol{B}(t)$ and $\boldsymbol{C}(t)$ are $n\times n$, $n\times p$ and $q\times n$ matrices.

The state space description of the dual system Σ_d of the system Σ is

$$\Sigma_d:\begin{cases}\dot{\boldsymbol{\psi}}^{\mathrm{T}} = -\boldsymbol{A}^{\mathrm{T}}(t)\,\boldsymbol{\psi}^{\mathrm{T}} + \boldsymbol{C}^{\mathrm{T}}(t)\boldsymbol{\eta}^{\mathrm{T}}\\ \boldsymbol{\varphi}^{\mathrm{T}} = \boldsymbol{B}^{\mathrm{T}}(t)\,\boldsymbol{\psi}^{\mathrm{T}}\end{cases} \tag{3.4.2}$$

where $\boldsymbol{\psi}$ is an n-dimensional row vector, η is a q-dimensional input row vector, $\boldsymbol{\varphi}$ is a p-dimensional output row vector,

The LTV system Σ and its dual system Σ_d have the following relationships:

1. The state transition matrices of both systems satisfy

$$\boldsymbol{\Phi}_d(t,t_0) = \boldsymbol{\Phi}^{\mathrm{T}}(t_0,t) \tag{3.4.3}$$

where $\boldsymbol{\Phi}(t,t_0)$ is state transition matrix of system eq. (3.4.1), $\boldsymbol{\Phi}_d(t,t_0)$ is state transition matrix of system eq. (3.4.2).

Proof Because

$$\boldsymbol{\Phi}(t,t_0)\boldsymbol{\Phi}^{-1}(t,t_0) = \boldsymbol{I}$$

taking the derivative of both sides of the above equation with respect to t, yields

$$\begin{aligned}\boldsymbol{0} &= \frac{\mathrm{d}}{\mathrm{d}t}(\boldsymbol{\Phi}(t,t_0)\boldsymbol{\Phi}^{-1}(t,t_0))\\ &= \frac{\mathrm{d}}{\mathrm{d}t}(\boldsymbol{\Phi}(t,t_0))\cdot\boldsymbol{\Phi}^{-1}(t,t_0) + \boldsymbol{\Phi}(t,t_0)\frac{\mathrm{d}}{\mathrm{d}t}(\boldsymbol{\Phi}^{-1}(t,t_0))\\ &= \boldsymbol{A}(t)\boldsymbol{\Phi}(t,t_0)\boldsymbol{\Phi}^{-1}(t,t_0) + \boldsymbol{\Phi}(t,t_0)\dot{\boldsymbol{\Phi}}(t_0,t)\\ &= \boldsymbol{A}(t) + \boldsymbol{\Phi}(t,t_0)\dot{\boldsymbol{\Phi}}(t_0,t)\end{aligned}$$

So

$$\dot{\boldsymbol{\Phi}}(t_0,t) = -\boldsymbol{\Phi}^{-1}(t,t_0)\boldsymbol{A}(t) = -\boldsymbol{\Phi}(t_0,t)\boldsymbol{A}(t),\ \boldsymbol{\Phi}(t_0,t_0) = \boldsymbol{I} \tag{3.4.4}$$

or

$$\dot{\boldsymbol{\Phi}}^{\mathrm{T}}(t_0,t) = -\boldsymbol{A}^{\mathrm{T}}(t)\boldsymbol{\Phi}^{\mathrm{T}}(t_0,t),\ \boldsymbol{\Phi}(t_0,t_0) = \boldsymbol{I} \tag{3.4.5}$$

According to the definition of state transition matrix, eq. (3.4.3) holds.

2. The block diagrams of system eq. (3.4.1) and its dual system eq. (3.4.2) are dual as shown in Figure 3.4.1.

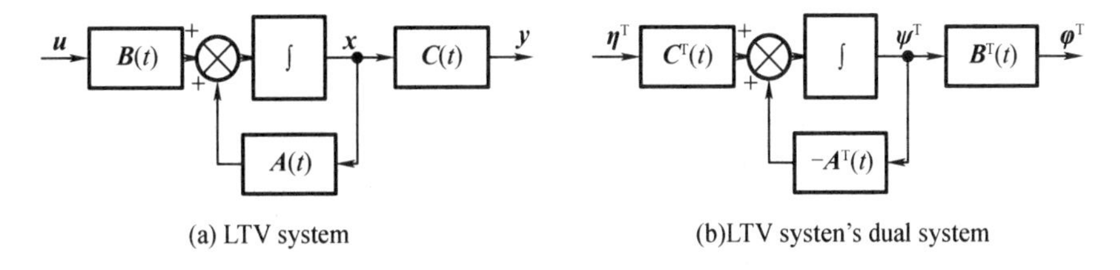

Figure 3.4.1 LTV system(a) and its dual system(b)

3.4.2 Duality theorem of linear systems

Theorem 3.4.1 The LTV system Σ_d is controllable if and only if its dual system Σ is observable. The LTV system Σ_d is observable if and only if its dual system Σ is controllable.

Proof The system Σ is controllable at t_0, there exists $t_1 > t_0$, such that

$$n = \operatorname{rank}\left(\int_{t_0}^{t_1} \boldsymbol{\Phi}^{\mathrm{T}}(t,t_0)\,\boldsymbol{C}^{\mathrm{T}}(t)\boldsymbol{C}(t)\boldsymbol{\Phi}(t,t_0)\,\mathrm{d}t\right)$$

$$= \operatorname{rank}\left(\int_{t_0}^{t_1} \boldsymbol{\Phi}_d(t_0,t)\,\boldsymbol{C}^{\mathrm{T}}(t)\,[\boldsymbol{C}^{\mathrm{T}}(t)]^{\mathrm{T}}\,\boldsymbol{\Phi}_d^{\mathrm{T}}(t_0,t)\,\mathrm{d}t\right)$$

It indicates the system Σ_d is observable at t_0. By using similar method, we can prove that the system Σ is controllable, the system Σ_d is observable.

Example 3.4.1 Check the following system if it is observable.

$$\dot{\boldsymbol{x}} = \begin{bmatrix} -3 & 0 & 0 & 0 \\ 1 & -3 & 0 & 0 \\ 0 & 1 & -3 & 0 \\ 0 & 0 & 0 & -3 \end{bmatrix}\boldsymbol{x},\ \boldsymbol{y} = \begin{bmatrix} 0 & 1 & 1 & 2 \\ 0 & 2 & 1 & 2 \end{bmatrix}\boldsymbol{x}$$

Answer Because $\operatorname{rank}[-\boldsymbol{A}^{\mathrm{T}}, \boldsymbol{C}^{\mathrm{T}}] = \operatorname{rank}[\boldsymbol{A}^{\mathrm{T}}, \boldsymbol{C}^{\mathrm{T}}]$, according to duality theorem, the pair $(\boldsymbol{A}, \boldsymbol{C})$ is observable if only if the pair $(\boldsymbol{A}^{\mathrm{T}}, \boldsymbol{C}^{\mathrm{T}})$ is controllable.

$$\boldsymbol{A}^{\mathrm{T}} = \begin{bmatrix} -3 & 1 & 0 & 0 \\ 0 & -3 & 1 & 0 \\ 0 & 0 & -3 & 0 \\ 0 & 0 & 0 & -3 \end{bmatrix},\ \boldsymbol{C}^{\mathrm{T}} = \begin{bmatrix} 0 & 0 \\ 1 & 2 \\ 1 & 1 \\ 2 & 2 \end{bmatrix}$$

The eigenvalue -3 of $\boldsymbol{A}^{\mathrm{T}}$ has two small Jordan blocks. The last rows of $\boldsymbol{C}^{\mathrm{T}}$ corresponding the last rows of each small Jordan blocks are $[1 \quad 1]$ and $[2 \quad 2]$, which are linearly dependent. Thus according Jordan canonical form criterion, $(\boldsymbol{A}^{\mathrm{T}}, \boldsymbol{C}^{\mathrm{T}})$ is uncontrollable. According to duality theorem, $(\boldsymbol{A}, \boldsymbol{C})$ is unobservable .

3.5 Controllable canonical form and observable canonical form of SISO LTI systems

If the state space description of a system has controllable canonical form or observable canonical form, the system must be controllable or observable without using controllability and observability criteria. For the LTI system which is completely controllable or completely observable, its state space description can be transformed into a controllable canonical form and observable canonical form by introducing linear non-singular transformation. As will be seen in later chapters, the design of systems such as state feedback and state observation should be based on controllable or observable canonical forms. In this section, we discuss the case of single input-single output systems, give the definition of the canonical form and corresponding conversion methods.

3.5.1 Controllable canonical form

For a SISO LTI system, if its state space description has following form

$$\begin{cases} \dot{\bar{x}} = A_c \bar{x} + b_c u \\ y = c_c \bar{x} \end{cases} \tag{3.5.1}$$

where

$$A_c = \left[\begin{array}{c|cccc} 0 & 1 & & & \\ \vdots & & \ddots & & \\ 0 & 0 & & 1 & \\ \hline -a_0 & -a_1 & \cdots & -a_{n-1} & \end{array}\right], b_c = \left[\begin{array}{c} 0 \\ \vdots \\ 0 \\ \hline 1 \end{array}\right] \tag{3.5.2}$$

then the state space description eq. (3.5.1) is called controllable canonical form.

Conclusion 3. 5. 1 If (A_c, b_c) has controllable canonical form, then (A_c, b_c) is controllable.

Proof According to the rank criterion,

$$\text{rank}[b_c \vdots A_c b_c \vdots A_c^2 b_c \vdots \cdots \vdots A_c^{n-1} b_c] = \text{rank}\begin{bmatrix} 0 & 0 & \cdots & \cdots & 1 \\ 0 & 0 & \cdots & ⋰ & * \\ \vdots & \vdots & \vdots & \vdots & \vdots \\ 0 & 0 & 1 & \cdots & * \\ 0 & 1 & -a_{n-1} & \cdots & * \\ 1 & -a_{n-1} & \cdots & \cdots & * \end{bmatrix} = n$$

Thus the system is controllable.

Conclusion 3.5.2 Consider a SISO LTI system

$$\begin{cases} \dot{x} = Ax + bu \\ y = cx \end{cases} \tag{3.5.3}$$

If it is controllable, by using the linearly nonsingular transformation $\bar{x} = P^{-1}x$, we get its controllable canonical form:

$$\begin{cases} \dot{\bar{x}} = A_c \bar{x} + b_c u \\ y = c_c \bar{x} \end{cases} \tag{3.5.4}$$

where,

$$A_c = P^{-1}AP = \left[\begin{array}{c|cccc} 0 & 1 & 0 & \cdots & 0 \\ 0 & 0 & 1 & \cdots & 0 \\ & & & \ddots & \\ 0 & 0 & \cdots & & 1 \\ \hline -a_0 & -a_1 & -a_2 & \cdots & -a_{n-1} \end{array}\right], b_c = P^{-1}b = \begin{bmatrix} 0 \\ 0 \\ \vdots \\ 0 \\ 1 \end{bmatrix}$$

$$c_c = cP = [\beta_0 \quad \beta_1 \quad \cdots \quad \beta_{n-1}] \tag{3.5.5}$$

From eq. (3.5.5), the characteristic polynomial of A is

$$\det(sI - A) = \alpha(s) = s^n + a_{n-1}s^{n-1} + \cdots + a_1 s + a_0 \tag{3.5.6}$$

P can be computed as

$$P = [e_1 \quad e_2 \quad \cdots \quad e_n] = [A^n b \quad \cdots \quad A^2 b \quad Ab]\begin{bmatrix} 1 & & & & \\ a_{n-1} & 1 & & & \\ a_{n-2} & a_{n-1} & 1 & & \\ \vdots & & \ddots & \ddots & \\ a_1 & \cdots & & a_{n-1} & 1 \end{bmatrix} \tag{3.5.7}$$

Proof (1) Prove A_c. Let $P = [e_1 \quad e_2 \quad \cdots \quad e_n]$, from $A_c = P^{-1}AP$, we have

$$PA_c = AP = [A^n b \quad \cdots \quad A^2 b \ Ab]\begin{bmatrix} 1 & & & & \\ a_{n-1} & 1 & & & \\ a_{n-2} & a_{n-1} & 1 & & \\ \vdots & & \ddots & \ddots & \\ a_1 & \cdots & & a_{n-1} & 1 \end{bmatrix} \tag{3.5.8}$$

From Hamilton-Cayley theorem, $\alpha(A) = 0$, and eq. (3.5.8), we have

$$Ae_1 = (A^n b + a_{n-1}A^{n-1}b + \cdots + a_1 Ab + a_0 b) - a_0 b = -a_0 e_n$$

$$Ae_2 = (A^{n-1} b + a_{n-1}A^{n-1}b + \cdots + a_2 Ab + a_1 b) - a_1 b = e_1 - a_1 e_n$$

$$\cdots$$

$$Ae_{n-1} = (A^2 b + a_{n-1}Ab + a_{n-2}b) - a_{n-2}b = e_{n-2} - a_{n-2}e_n$$

$$Ae_n = (Ab + a_{n-1}b) - a_{n-1}b = e_{n-1} - a_{n-1}e_n \tag{3.5.9}$$

Substituting eq. (3.5.9) into eq. (3.5.8), yields

$$PA_c = [-a_0 e_n \quad e_1 - a_1 e_n \quad \cdots \quad e_{n-2} - a_{n-2}e_n \quad e_{n-1} - a_{n-1}e_n]$$

$$= [e_1 \quad e_2 \quad \cdots \quad e_n]\begin{bmatrix} 0 & 1 & & \\ \vdots & & \ddots & \\ 0 & & & 1 \\ -a_0 & -a_2 & \cdots & -a_{n-1} \end{bmatrix}$$

Left multiply both sides of the above equation by P^{-1}, we get the expression of A_c.

(2) Prove b_c. From $b_c = P^{-1}b$ and eq. (3.6.7), we have

$$Pb_c = b = [e_1 \quad e_2 \quad \cdots \quad e_n]\begin{bmatrix} 0 \\ \vdots \\ 0 \\ 1 \end{bmatrix} = P\begin{bmatrix} 0 \\ \vdots \\ 0 \\ 1 \end{bmatrix}$$

Left multiply both sides of the above equation by P^{-1}, we get the expression of b_c.

(3) Prove c_c. From $c_c = cP$ and eq. (3.6.7), we have

$$cP = c[A^{n-1}b \quad \cdots \quad Ab \quad b]\begin{bmatrix} 1 & & & & \\ a_{n-1} & 1 & & & \\ a_{n-2} & a_{n-1} & 1 & & \\ \vdots & & \ddots & \ddots & \\ a_1 & \cdots & & a_{n-1} & 1 \end{bmatrix}$$

$$= [\beta_0 \quad \beta_1 \quad \cdots \quad \beta_{n-1}]$$

Example 3.5.1 Determine the controllable canonical form of the following SISO LTI system:

$$\begin{cases} \dot{x} = \begin{bmatrix} 1 & 2 & 0 \\ 3 & -1 & 1 \\ 0 & 2 & 0 \end{bmatrix} x + \begin{bmatrix} 2 \\ 1 \\ 1 \end{bmatrix} u \\ y = [0 \quad 1 \quad 1] x \end{cases}$$

Answer (1) $\text{rank}[b \quad Ab \quad A^2 b] = \text{rank}\begin{bmatrix} 2 & 4 & 16 \\ 1 & 6 & 8 \\ 1 & 2 & 12 \end{bmatrix} = 3$, so the system is controllable.

(2) $\det(sI - A) = \det\begin{bmatrix} s-1 & -2 & 0 \\ -3 & s+1 & -1 \\ 0 & -2 & s \end{bmatrix} = s^3 - 9s + 2$, $\therefore a_2 = 0, a_1 = -9, a_0 = 2$

So, P is

$$P = [A^{n-1}b \quad \cdots \quad Ab \; b]\begin{bmatrix} 1 & 0 & 0 \\ a_2 & 1 & 0 \\ a_1 & a_2 & 1 \end{bmatrix} = \begin{bmatrix} 16 & 4 & 2 \\ 8 & 6 & 1 \\ 12 & 2 & 1 \end{bmatrix}\begin{bmatrix} 1 & 0 & 0 \\ 0 & 1 & 0 \\ -9 & 0 & 1 \end{bmatrix} = \begin{bmatrix} -2 & 4 & 2 \\ -1 & 6 & 1 \\ 3 & 2 & 1 \end{bmatrix}$$

Introducing $\bar{x} = P^{-1}x$, yields

$$A_c = \begin{bmatrix} 0 & 1 & 0 \\ 0 & 0 & 1 \\ -a_0 & -a_1 & -a_2 \end{bmatrix} = \begin{bmatrix} 0 & 1 & 0 \\ 0 & 0 & 1 \\ -2 & 9 & 0 \end{bmatrix}$$

$$b_c = \begin{bmatrix} 0 \\ 0 \\ 1 \end{bmatrix}$$

$$c_c = cP = [0 \quad 1 \quad 1]\begin{bmatrix} -2 & 4 & 2 \\ -1 & 6 & 1 \\ 3 & 2 & 1 \end{bmatrix} = [2 \quad 8 \quad 2]$$

3.5.2 Observable canonical form

For a SISO LTI system, if its state space description has following form

$$\begin{cases} \dot{\hat{x}} = A_o \hat{x} + b_o u \\ y = c_o \hat{x} \end{cases} \tag{3.5.10}$$

where,

$$A_o = \begin{bmatrix} 0 & \cdots & 0 & -a_0 \\ 1 & & & -a_1 \\ & \ddots & & \vdots \\ & & 1 & -a_{n-1} \end{bmatrix}, c_o = [0 \quad \cdots \quad 0 \quad 1] \tag{3.5.11}$$

then the description eq. (3.5.10) is called observable canonical form.

Conclusion 3.5.3 If (A_o, c_o) has observable canonical form, then (A_o, c_o) must be observable.

Proof The dual system of system eq. (3.5.10) is

$$\begin{cases} \dot{\psi}^{\mathrm{T}} = -A_o^{\mathrm{T}} \psi^{\mathrm{T}} + c_o^{\mathrm{T}} \eta^{\mathrm{T}} \\ \varphi^{\mathrm{T}} = b_o^{\mathrm{T}} \psi^{\mathrm{T}} \end{cases} \tag{3.5.12}$$

Because $(A_0^{\mathrm{T}}, c_0^{\mathrm{T}})$ is controllable canonical form, the system eq. (3.5.12) is controllable. Thus its dual system eq. (3.5.10) is observable.

Conclusion 3.5.4 For the SISO LTI system eq. (3.5.3), if it is observable, it will be transformed into the observable canonical form eq. (3.5.10) by using the linear nonsingular transformation $\hat{x} = Qx$. where

$$A_o = QAQ^{-1} = \begin{bmatrix} 0 & \cdots & 0 & -a_0 \\ 1 & & & -a_1 \\ & \ddots & & \vdots \\ & & 1 & -a_{n-1} \end{bmatrix}, \ b_o = Qb, \ c_o = cQ^{-1} = [0 \quad \cdots \quad 0 \quad 1] \tag{3.5.13}$$

From eq. (3.5.13), the characteristic of A is eq. (3.5.6), we computer Q as the as the following equation

$$Q = \begin{bmatrix} 1 & a_{n-1} & \cdots & a_1 \\ & \ddots & \ddots & \vdots \\ & & \ddots & a_{n-1} \\ & & & 1 \end{bmatrix} \begin{bmatrix} cA^{n-1} \\ \vdots \\ cA \\ c \end{bmatrix} \tag{3.5.14}$$

Proof The proof process is similar to the controllable canonical form, so it is omitted.

Example 3.5.2 Given a LTI system

$$\dot{x} = \begin{bmatrix} 1 & 0 & 2 \\ 2 & 1 & 1 \\ 1 & 0 & -2 \end{bmatrix} x + \begin{bmatrix} 1 \\ 2 \\ 1 \end{bmatrix} u$$

$$y = [0 \quad 1 \quad 1] x$$

Determine its observable canonical form.

Answer (1) Check observability of the system:

$$\operatorname{rank}\begin{bmatrix} c \\ cA \\ cA^2 \end{bmatrix} = \operatorname{rank}\begin{bmatrix} 0 & 1 & 1 \\ 3 & 1 & -1 \\ 4 & 1 & 9 \end{bmatrix} = 3, \text{ thus the system is observable.}$$

(2) The characteristic polynomial of the system is

$$\det(s\boldsymbol{I}-\boldsymbol{A}) = \det\begin{bmatrix} s-1 & 0 & -2 \\ -2 & s-1 & -1 \\ -1 & 0 & s+2 \end{bmatrix} = s^3 - 5s + 4$$

So, $\boldsymbol{Q}$ is

$$\boldsymbol{Q} = \begin{bmatrix} 1 & a_2 & a_1 \\ 0 & 1 & a_2 \\ 0 & 0 & 1 \end{bmatrix}\begin{bmatrix} cA^2 \\ cA \\ c \end{bmatrix} = \begin{bmatrix} 1 & 0 & -5 \\ 0 & 1 & 0 \\ 0 & 0 & 1 \end{bmatrix}\begin{bmatrix} 4 & 1 & 9 \\ 3 & 1 & -1 \\ 0 & 1 & 1 \end{bmatrix} = \begin{bmatrix} 4 & -4 & 4 \\ 3 & 1 & -1 \\ 0 & 1 & 1 \end{bmatrix}$$

$$\boldsymbol{b}_o = \boldsymbol{Q}\boldsymbol{b} = \begin{bmatrix} 0 \\ 4 \\ 3 \end{bmatrix}, \boldsymbol{c}_o = \boldsymbol{c}\boldsymbol{Q}^{-1} = [0 \ \cdots \ 0 \ 1]$$

Thus observable canonical form of the system is

$$\dot{\hat{\boldsymbol{x}}} = \begin{bmatrix} 0 & 0 & -4 \\ 1 & 0 & 5 \\ 0 & 1 & 0 \end{bmatrix}\hat{\boldsymbol{x}} + \begin{bmatrix} 0 \\ 4 \\ 3 \end{bmatrix}u$$

$$y = [0 \quad 0 \quad 1]\hat{\boldsymbol{x}}$$

For controllable canonical form and observable canonical form, we get the following conclusions:

(1) The controllable canonical form $(\boldsymbol{A}_c, \boldsymbol{b}_c)$ and observable canonical form $(\boldsymbol{A}_o, \boldsymbol{c}_o)$ are determined by the coefficients $a_i\,(i=1,2,\cdots,n)$ of the characteristic polynomial of the system matrix which reflects the inherent characteristics of systems.

(2) Algebraically equivalent completely controllable systems have the same controllable canonical form and algebraically equivalent completely observable systems have the same observable canonical form.

Proof Take a observable system for example to prove it. It needs to prove that algebraically equivalent systems have the same $\boldsymbol{\alpha}(s)$ and $\beta_i\,(i=0,1,2,\cdots,n-1)$.

Let $(\boldsymbol{A},\boldsymbol{b},\boldsymbol{c})$ and $(\bar{\boldsymbol{A}},\bar{\boldsymbol{b}},\bar{\boldsymbol{c}})$ are algebraically equivalent, so there exist

$$\bar{\boldsymbol{A}} = \boldsymbol{T}\boldsymbol{A}\boldsymbol{T}^{-1}, \ \bar{\boldsymbol{b}} = \boldsymbol{T}\boldsymbol{b}, \ \bar{\boldsymbol{c}} = \boldsymbol{c}\boldsymbol{T}^{-1}$$

where, $\boldsymbol{T}$ is nonsingular constant matrix. The characteristic polynomial of the system is

$$\bar{\alpha}(s) = \det(s\boldsymbol{I}-\bar{\boldsymbol{A}}) = \det(\boldsymbol{T}(s\boldsymbol{I}-\boldsymbol{A})\boldsymbol{T}^{-1}) = \det(s\boldsymbol{I}-\boldsymbol{A}) = \alpha(s)$$

In the observable canonical form, $\boldsymbol{b}$ is

$$\bar{b}_s = Q\bar{b} = \begin{bmatrix} 1 & \bar{a}_{n-1} & \cdots & \bar{a}_1 \\ & \ddots & \ddots & \vdots \\ & & \ddots & \bar{a}_{n-1} \\ & & & 1 \end{bmatrix} \begin{bmatrix} \bar{c}\bar{A}^{n-1} \\ \vdots \\ \bar{c}\bar{A} \\ \bar{c} \end{bmatrix} \bar{b} = \begin{bmatrix} \bar{\beta}_0 \\ \bar{\beta}_1 \\ \vdots \\ \bar{\beta}_{n-1} \end{bmatrix}$$

where,

$$\begin{aligned} \bar{\beta}_{i-1} &= \overline{cA}^{n-i}\bar{b} + \bar{a}_{n-1}\overline{cA}^{n-i-1}\bar{b} + \cdots + \bar{a}_i\,\overline{cb} \\ &= c\,T^{-1}T\,A^{n-i}T^{-1}Tb + a_{n-1}c\,T^{-1}T\,A^{n-i-1}T^{-1}Tb + \cdots + a_i c\,T^{-1}Tb \\ &= c\,A^{n-i}b + a_{n-1}c\,A^{n-i-1}b + \cdots + a_i cb \\ &= \beta_{i-1}, \quad i = 1, 2, \cdots, n \end{aligned}$$

3.6 Controllable canonical form and observable canonical form of MIMO LTI systems

In this section, we discuss the controllable canonical form and observable canonical form of multiple input-multiple output LTI systems. Compared with single input-single output systems, the canonical forms of multiple input-multiple output systems are more complex both in form and construction methods. The Wonham and Luenberger canonical forms, which are widely used, are discussed below.

3.6.1 Scheme of searching linearly independent columns or rows

For multiple input-multiple output systems, no matter what canonical form is constructed, a common problem is to find linearly independent columns in controllability matrix or linearly independent rows in observability matrix. Usually, this is a search process.

Consider an n-dimensional multiple input-multiple output system

$$\begin{cases} \dot{x} = Ax + Bu \\ y = Cx \end{cases} \tag{3.6.1}$$

where, A is an $n \times n$ constant matrix, B and C are $n \times p$ and $q \times n$ constant matrices. The controllability matrix Q_c is

$$Q_c = [B \vdots AB \vdots \cdots \vdots A^{n-1}B] \tag{3.6.2}$$

observability matrix Q_o is

$$Q_o = \begin{bmatrix} C \\ CA \\ \vdots \\ CA^{n-1} \end{bmatrix} \tag{3.6.3}$$

Obviously, when the system is controllable, rank $\boldsymbol{Q}_c = n$, i. e. there are n linearly independent columns in $n \times pn$ $\boldsymbol{Q}_c$. When the system is observable, *rank* $\boldsymbol{Q}_o = n$, i. e. there are n linearly independent rows in $qn \times n$ $\boldsymbol{Q}_o$. Therefore, in order to determine controllable canonical form or observable canonical form, it is necessary to find the linearly independent columns or rows in $\boldsymbol{Q}_c$ or $\boldsymbol{Q}_o$, and then construct the corresponding transformation matrices. Below, the controllability matrix $\boldsymbol{Q}_c$ is taken as an example to illustrate the steps of searching for linearly independent columns in Q_c. We can use a similar manner to search linearly independent rows in $\boldsymbol{Q}_o$.

For $\{\boldsymbol{A}, \boldsymbol{B}\}$, to make the searching n linearly independent column vectors more intuitive, we establish grid graphs, Figure 3.6.1 and Figure 3.6.2. The grid graphs consists of a number of rows and columns. The columns of $\boldsymbol{B}, \boldsymbol{b}_1 \boldsymbol{b}_2 \cdots$, are on the top of the grid graphs. The several powers of $\boldsymbol{A}$, $\boldsymbol{A}^0$, $\boldsymbol{A}^1$, $\boldsymbol{A}^2 \cdots$, are on the left side of the grid graphs. The grid ji represents the product of $\boldsymbol{A}^j$ and $\boldsymbol{b}_i$, the column vector $\boldsymbol{A}^j \boldsymbol{b}_i$. With different search directions in the grid graph, there are row-first searching scheme and column-first searching scheme.

1. Column-first searching scheme

We take searching n linearly independent column vectors of $\boldsymbol{Q}_c$ as an example. As shown in Figure 3.6.1, column-first searching method first finds all linearly independent column vectors in the first column of the grid graph sequentially from the upper left lattice, $\boldsymbol{A}^0 \boldsymbol{b}_1$, to bottom. Next, go to the second column and find all linearly independent column vectors from the lattice $\boldsymbol{A}^0 \boldsymbol{b}_2$ to bottom. And so on until we find n linearly independent column vectors.

	b_1	b_2	b_3	b_4
A^0	×	×	×	o
A^1	×	×	o	
A^2	×	o		
A^3	o			
A^4				
A^5				

Figure 3.6.1 Column-first searching

	b_1	b_2	b_3	b_4
A^0	×	×	×	o
A^1	×	o	×	
A^2	×		o	
A^3	o			
A^4				
A^5				

Figure 3.6.2 Row-first searching

The procedure of column-first searching is explained below.

Step 1. First look at the left first column in Figure 3.6.1. If the left top vector $\boldsymbol{b}_1$ is not zero, then draw a × in the grid. Then check next vector $\boldsymbol{A}\boldsymbol{b}_1$ in the second row, if $\boldsymbol{A}\boldsymbol{b}_1$ and $\boldsymbol{b}_1$ are dependent, than draw a × inside the grid of $\boldsymbol{A}\boldsymbol{b}_1$ and so on. Until there is a vector $\boldsymbol{A}^{\gamma_1}\boldsymbol{b}_1$ in first column which is dependent on previous vectors $\{\boldsymbol{b}_1, \boldsymbol{A}\boldsymbol{b}_1, \cdots, \boldsymbol{A}^{\gamma_1 - 1}\boldsymbol{b}_1\}$. Draw a o inside this grid and stop searching first column. So we get independent vectors $\boldsymbol{b}_1, \boldsymbol{A}\boldsymbol{b}_1, \cdots, \boldsymbol{A}^{\gamma_1 - 1}\boldsymbol{b}_1$ whose length is γ_1.

Step 2. Look at the second column. If $\boldsymbol{b}_2$ is independent on

$$\{\boldsymbol{b}_1, \boldsymbol{A}\boldsymbol{b}_1, \cdots, \boldsymbol{A}^{\gamma_1 - 1}\boldsymbol{b}_1\}$$

then draw × inside this grid. Next, if vector $\boldsymbol{A}\boldsymbol{b}_2$ in the second row is independent on

$$\{\boldsymbol{b}_1, \boldsymbol{A}\boldsymbol{b}_1, \cdots, \boldsymbol{A}^{\gamma_1 - 1}\boldsymbol{b}_1; \boldsymbol{b}_2\}$$

draw a × inside this grid and so on. Until appears a vector $\boldsymbol{A}^{\gamma_2}\boldsymbol{b}_2$ which is dependent on

$$\{\boldsymbol{b}_1, \boldsymbol{A}\boldsymbol{b}_1, \cdots, \boldsymbol{A}^{\gamma_1 - 1}\boldsymbol{b}_1; \boldsymbol{b}_2, \boldsymbol{A}\boldsymbol{b}_2, \cdots, \boldsymbol{A}^{\gamma_2 - 1}\boldsymbol{b}_2\}$$

Now stop searching second column and draw a o inside the grid of $\boldsymbol{A}^{\gamma_2}\boldsymbol{b}_2$. So we get independent vectors $\boldsymbol{b}_2, \boldsymbol{A}\boldsymbol{b}_2, \cdots, \boldsymbol{A}^{\gamma_2 - 1}\boldsymbol{b}_2$ in the second column. Their length is γ_2.

...

Step l. Rightwards look at the lth column. If $\boldsymbol{b}_l$ is independent on

$$\{\boldsymbol{b}_1, \boldsymbol{A}\boldsymbol{b}_1, \cdots, \boldsymbol{A}^{\gamma_1 - 1}\boldsymbol{b}_1; \boldsymbol{b}_2, \boldsymbol{A}\boldsymbol{b}_2, \cdots, \boldsymbol{A}^{\gamma_2 - 1}\boldsymbol{b}_2; \cdots; \boldsymbol{b}_{l-1}, \boldsymbol{A}\boldsymbol{b}_{l-1}, \cdots, \boldsymbol{A}^{\gamma_{l-1} - 1}\boldsymbol{b}_{l-1}\}$$

then draw a × insidethe grid of $\boldsymbol{b}_l$ and so on. Until appears a vector $\boldsymbol{A}^{\gamma_l}\boldsymbol{b}_l$ which is dependent on

$$\{\boldsymbol{b}_1, \boldsymbol{A}\boldsymbol{b}_1, \cdots, \boldsymbol{A}^{\gamma_1 - 1}\boldsymbol{b}_1; \cdots; \boldsymbol{b}_{l-1}, \boldsymbol{A}\boldsymbol{b}_{l-1}, \cdots, \boldsymbol{A}^{\gamma_{l-1} - 1}\boldsymbol{b}_{l-1}; \boldsymbol{b}_l, \boldsymbol{A}\boldsymbol{b}_l, \cdots, \boldsymbol{A}^{\gamma_l - 1}\boldsymbol{b}_l\}$$

draw a o inside the grid of $\boldsymbol{A}^{\gamma_l}\boldsymbol{b}_l$ and stop searching the lth column. So we get independent vectors $\boldsymbol{b}_l, \boldsymbol{A}\boldsymbol{b}_l, \cdots, \boldsymbol{A}^{\gamma_l - 1}\boldsymbol{b}_l$in the lth column. Their length is γ_l.

Step $l + 1$. If $\gamma_1 + \gamma_2 + \cdots + \gamma_l = n$, stop searching. So far, the above obtained vectors

$$\{\boldsymbol{b}_1, \cdots, \boldsymbol{A}^{\gamma_1 - 1}\boldsymbol{b}_1; \boldsymbol{b}_2, \cdots, \boldsymbol{A}^{\gamma_2 - 1}\boldsymbol{b}_2; \cdots; \boldsymbol{b}_l, \cdots, \boldsymbol{A}^{\gamma_l - 1}\boldsymbol{b}_l\}$$

are linearly independent.

For the example in Figure 3.6.1, $n = 6$, $l = 3$, and we can obtain 6 independent vectors in $\boldsymbol{Q}_c$, which are

$$\boldsymbol{b}_1, \boldsymbol{A}\boldsymbol{b}_1, \boldsymbol{A}^2\boldsymbol{b}_1; \boldsymbol{b}_2, \boldsymbol{A}\boldsymbol{b}_2; \boldsymbol{b}_3$$

2. Row-first searching scheme

We take searching n linearly independent column vectors of $\boldsymbol{Q}_c$ as an example. As shown in Figure 3.6.2, the row-first searching method first finds all linearly independent column vectors in the first row of the grid graph sequentially from the upper left lattice, $\boldsymbol{A}^0\boldsymbol{b}_1$, to right. Next, go to the second row and find all linearly independent column vectors in the second row from the lattice $\boldsymbol{A}\boldsymbol{b}_1$ to right. And so on until we find n linearly independent column vectors.

The procedure are listed below.

Step1. Let rank$\boldsymbol{B} = r \leqslant p$, i. e. $\boldsymbol{B}$ has r independent columns. First look at the first row in Figure 3.6.2. If the left top vector $\boldsymbol{b}_1$ is not zero, then draw a × inside the grid of $\boldsymbol{A}^0\boldsymbol{b}_1$. Then find r independent columns $\boldsymbol{b}_1, \boldsymbol{b}_2, \cdots, \boldsymbol{b}_r$ in the first row from left to right and draw × inside corresponding grids. If not, this can be done by swapping the position of the columns of $\boldsymbol{B}$.

Step 2. Look at the second row. Search independent columns from the grid of $\boldsymbol{A}\boldsymbol{b}_1$ to $\boldsymbol{A}\boldsymbol{b}_r$ (from left to right). If find an independent vector, draw a × inside the grid, otherwise draw a o. Note that If a grid is filled by o, then all column vectors under the grid in the same column must be linearly dependent on the previously obtained linearly independent column vectors. So we don't check these columns.

…

Step l. Go down to the lth row. Search independent columns from the grid of $\boldsymbol{A}^l\boldsymbol{b}_1$ to $\boldsymbol{A}^l\boldsymbol{b}_r$ (from left to right). If find an independent vector, draw × inside this grid, otherwise draw o.

Step $l+1$. If find n independent vectors, stop searching. The column vectors corresponding to the grid with × in Figure 3.6.2 are the n linearly independent column vectors in $\boldsymbol{Q}_c$ obtained by the row-first searching scheme.

For the example in Figure 3.6.2, $n=6$ and $l=3$. The independent column vectors in $\boldsymbol{Q}_c$ are

$$\boldsymbol{b}_1,\boldsymbol{b}_2,\boldsymbol{b}_3;\boldsymbol{Ab}_1,\boldsymbol{Ab}_3;\boldsymbol{A}^2\boldsymbol{b}_1$$

3.6.2 Wonham controllable form

Consider a MIMO LTI system eq. (3.6.1). Assume it is controllable. First we need find n independent column vectors in $\boldsymbol{Q}_c=[\boldsymbol{B} \vdots \boldsymbol{AB} \vdots \cdots \vdots \boldsymbol{A}^{n-1}\boldsymbol{B}]$. Let $\boldsymbol{B}=[\boldsymbol{b}_1,\boldsymbol{b}_2,\cdots,\boldsymbol{b}_p]$. Using column-first searching method, we get n independent vectors

$$\boldsymbol{b}_1,\boldsymbol{Ab}_1,\cdots,\boldsymbol{A}^{v_1-1}\boldsymbol{b}_1;\boldsymbol{b}_2,\boldsymbol{Ab}_2,\cdots,\boldsymbol{A}^{v_2-1}\boldsymbol{b}_2;\cdots;\boldsymbol{b}_l,\boldsymbol{Ab}_l,\cdots,\boldsymbol{A}^{v_l-1}\boldsymbol{b}_l \tag{3.6.4}$$

where

$$v_1+v_2+\cdots+v_l=n$$

$\boldsymbol{A}^{v_1}\boldsymbol{b}_1$ is a linearly combination of $\boldsymbol{b}_1$, $\boldsymbol{Ab}_1,\cdots$, $\boldsymbol{A}^{v_1-1}\boldsymbol{b}_1$, $\boldsymbol{A}^{v_2}\boldsymbol{b}_2$ is a linearly combination of $\boldsymbol{b}_1$, $\boldsymbol{Ab}_1,\cdots,\boldsymbol{A}^{v_1-1}\boldsymbol{b}_1;\boldsymbol{b}_2,\boldsymbol{Ab}_2,\cdots,\boldsymbol{A}^{v_2-1}\boldsymbol{b}_2$, $\cdots$, $\boldsymbol{A}^{v_l}b_l$ is a linearly combination of $\boldsymbol{b}_1,\boldsymbol{Ab}_1,\cdots,\boldsymbol{A}^{v_1-1}\boldsymbol{b}_1;\cdots;\boldsymbol{b}_l,\boldsymbol{Ab}_l,\cdots,\boldsymbol{A}^{v_l-1}\boldsymbol{b}_l$. So, we have

$$\boldsymbol{A}^{v_1}\boldsymbol{b}_1=-\sum_{j=0}^{v_1-1}\alpha_{1j}\boldsymbol{A}^j\boldsymbol{b}_1 \tag{3.6.5}$$

Define base set

$$\begin{cases}\boldsymbol{e}_{11}\triangleq\boldsymbol{A}^{v_1-1}\boldsymbol{b}_1+\alpha_{1,v_1-1}\boldsymbol{A}^{v_1-2}\boldsymbol{b}_1+\cdots+\alpha_{11}\boldsymbol{b}_1\\ \boldsymbol{e}_{12}\triangleq\boldsymbol{A}^{v_1-2}\boldsymbol{b}_1+\alpha_{1,v_1-1}\boldsymbol{A}^{v_1-3}\boldsymbol{b}_1+\cdots+\alpha_{12}\boldsymbol{b}_1\\ \qquad\vdots\\ \boldsymbol{e}_{1v_1}\triangleq\boldsymbol{b}_1\end{cases} \tag{3.6.6}$$

and

$$\boldsymbol{A}^{v_2}\boldsymbol{b}_2=-\sum_{j=0}^{v_2-1}\alpha_{2j}\boldsymbol{A}^j\boldsymbol{b}_2+\sum_{i=1}^{1}\sum_{j=1}^{v_i}\gamma_{2ji}\boldsymbol{e}_{ij} \tag{3.6.7}$$

where, the linear combination of $\{\boldsymbol{b}_1,\boldsymbol{Ab}_1,\cdots,\boldsymbol{A}^{v_1-1}\boldsymbol{b}_1\}$ is equivalent to the linear combination relation of $\{\boldsymbol{e}_{11},\boldsymbol{e}_{12},\cdots,\boldsymbol{e}_{1v_1}\}$. Based on eq. (3.6.7), define base set

$$\begin{cases}\boldsymbol{e}_{21}\triangleq\boldsymbol{A}^{v_2-1}\boldsymbol{b}_2+\alpha_{2,v_2-1}\boldsymbol{A}^{v_2-2}\boldsymbol{b}_2+\cdots+\alpha_{21}\boldsymbol{b}_2\\ \boldsymbol{e}_{22}\triangleq\boldsymbol{A}^{v_2-2}\boldsymbol{b}_2+\alpha_{2,v_2-1}\boldsymbol{A}^{v_2-3}\boldsymbol{b}_2+\cdots+\alpha_{22}\boldsymbol{b}_2\\ \qquad\vdots\\ \boldsymbol{e}_{2v_2}\triangleq\boldsymbol{b}_2\end{cases} \tag{3.6.8}$$

Let

$$A^{v_3}b_3 = -\sum_{j=0}^{v_3-1}\alpha_{3j}A^jb_3 + \sum_{i=1}^{2}\sum_{j=1}^{v_i}\gamma_{3ji}e_{ij} \tag{3.6.9}$$

where, the linear combination of $\{b_1, Ab_1, \cdots, A^{v_1-1}b_1; b_2, Ab_2, \cdots, A^{v_2-1}b_2\}$ is equivalent to the linear combination relation of $\{e_{11}, e_{12}, \cdots, e_{1v_1}; e_{21}, e_{22}, \cdots, e_{2v_2}\}$. Based on eq. (3.6.9), define base set

$$\begin{cases} e_{31} \triangleq A^{v_3-1}b_3 + \alpha_{3,v_3-1}A^{v_3-2}b_3 + \cdots + \alpha_{31}b_3 \\ e_{32} \triangleq A^{v_3-2}b_3 + \alpha_{3,v_3-1}A^{v_3-3}b_3 + \cdots + \alpha_{32}b_3 \\ \vdots \\ e_{3v_3} \triangleq b_3 \end{cases} \tag{3.6.10}$$

Let

$$A^{v_l}b_l = -\sum_{j=0}^{v_l-1}\alpha_{lj}A^jb_l + \sum_{i=1}^{l-1}\sum_{j=1}^{v_i}\gamma_{lji}e_{ij} \tag{3.6.11}$$

where, the linear combination of $\{b_1, Ab_1, \cdots, A^{v_1-1}b_1; \cdots; b_{l-1}, Ab_{l-1}, \cdots, A^{v_{l-1}-1}b_{l-1}\}$ is equivalent to the linear combination relation of $\{e_{11}, e_{12}, \cdots, e_{1v_1}; \cdots; e_{l-1,1}, e_{l-1,2}, \cdots, e_{l-1,v_{l-1}}\}$. Based on eq. (3.6.11), define base set

$$\begin{cases} e_{l1} \triangleq A^{v_l-1}b_l + \alpha_{l,v_l-1}A^{v_l-2}b_l + \cdots + \alpha_{l1}b_l \\ e_{l2} \triangleq A^{v_l-2}b_l + \alpha_{l,v_l-1}A^{v_l-3}b_l + \cdots + \alpha_{l2}b_l \\ \vdots \\ e_{lv_l} \triangleq b_l \end{cases} \tag{3.6.12}$$

We construct nonsingular transformation matrix T based above base sets as

$$T = [e_{11}, e_{12}, \cdots, e_{1v_1}; \cdots; e_{l1}, e_{l2}, \cdots, e_{lv_l}] \tag{3.6.13}$$

The conclusionis as follow:

Conclusion 3.6.1 Consider an n-dimensional controllable MIMO LTI system eq. (3.6.1). The equivalence transformation

$$\bar{x} = T^{-1}x$$

will transform eq. (3.6.1) into the following Wonham controllable form:

$$\begin{cases} \dot{\bar{x}} = \bar{A}_c\bar{x} + \bar{B}_c u \\ y = \bar{C}_c\bar{x} \end{cases} \tag{3.6.14}$$

where

$$\bar{A}_c = T^{-1}AT = \begin{bmatrix} \bar{A}_{11} & \bar{A}_{12} & \cdots & \bar{A}_{1l} \\ & \bar{A}_{22} & \cdots & \bar{A}_{2l} \\ & & \ddots & \\ & & & \bar{A}_{ll} \end{bmatrix} \tag{3.6.15}$$

$$\underset{(v_i\times v_i)}{\bar{A}_{ii}} = \left[\begin{array}{c:ccc} 0 & 1 & & \\ \vdots & & \ddots & \\ 0 & & & 1 \\ \hdashline -a_{i0} & a_{i1} & \cdots & a_{i,v_i-1} \end{array}\right], i=1,2,\cdots,l \tag{3.6.16}$$

$$\underset{(v_i\times v_j)}{\bar{A}_{ij}} = \begin{bmatrix} \gamma_{j1i} & 0 & \cdots & 0 \\ \vdots & \vdots & & \vdots \\ \gamma_{jv_ii} & 0 & \cdots & 0 \end{bmatrix}, j=i+1,\cdots,l \tag{3.6.17}$$

$$\underset{(n\times p)}{\bar{B}_{c}} = T^{-1}B = \left[\begin{array}{c:c:c:ccc} 0 & & & * & \cdots & * \\ \vdots & & & & & \\ 0 & & & & & \\ 1 & & & \vdots & & \vdots \\ \hdashline & \ddots & & \vdots & & \vdots \\ \hdashline & & 0 & & & \\ & & \vdots & & & \\ & & 0 & & & \\ & & 1 & * & \cdots & * \end{array}\right] \begin{matrix} \left.\right\} v_1 \\ \vdots \\ \left.\right\} v_l \end{matrix} \tag{3.6.18}$$

(first l columns; last $p-l$ columns)

$$\underset{(q\times n)}{\bar{C}_{c}} = CT \quad (\text{don't have special form}) \tag{3.6.19}$$

The element denoted by $*$ is maybe not zero.

Proof omitted.

Example 3.6.1 For a LTI system

$$\dot{x} = \begin{bmatrix} -1 & -4 & -2 \\ 0 & 6 & -1 \\ 1 & 7 & -1 \end{bmatrix} x + \begin{bmatrix} 2 & 0 \\ 0 & 0 \\ 1 & 1 \end{bmatrix} u$$

determine its Wonham controllable form.

Answer First, using column-first searching scheme, find 3 independent columns in

$$Q_c = [B \vdots AB \vdots A^2B] = \begin{bmatrix} 2 & 0 & -4 & -2 & 6 & 8 \\ 0 & 0 & -1 & -1 & -7 & -5 \\ 1 & 1 & 1 & -1 & -12 & -8 \end{bmatrix}$$

We get

$$b_1 = \begin{bmatrix} 2 \\ 0 \\ 1 \end{bmatrix}, \quad Ab_1 = \begin{bmatrix} -4 \\ -1 \\ 1 \end{bmatrix}, \quad A^2b_1 = \begin{bmatrix} 6 \\ -7 \\ -12 \end{bmatrix}$$

Then let

$$\begin{bmatrix} 46 \\ -30 \\ -31 \end{bmatrix} = A^3b_1 = -(a_{12}A^2b_1 + a_{11}Ab_1 + a_{10}b_1) = -a_{12}\begin{bmatrix} 6 \\ -7 \\ -12 \end{bmatrix} - a_{11}\begin{bmatrix} -4 \\ -1 \\ 1 \end{bmatrix} - a_{10}\begin{bmatrix} 2 \\ 0 \\ 1 \end{bmatrix}$$

We get

$$\begin{bmatrix} 6 & -4 & 2 \\ -7 & -1 & 0 \\ -12 & 1 & 1 \end{bmatrix}\begin{bmatrix} a_{12} \\ a_{11} \\ a_{10} \end{bmatrix} = -\begin{bmatrix} 46 \\ -30 \\ -31 \end{bmatrix}$$

Solve the above equation set, we get

$$\begin{bmatrix} a_{12} \\ a_{11} \\ a_{10} \end{bmatrix} = -\begin{bmatrix} 6 & -4 & 2 \\ -7 & -1 & 0 \\ -12 & 1 & 1 \end{bmatrix}^{-1}\begin{bmatrix} 46 \\ -30 \\ -31 \end{bmatrix} = -\left(-\frac{1}{72}\right)\begin{bmatrix} -1 & 6 & 2 \\ 7 & 30 & -14 \\ -19 & 42 & -34 \end{bmatrix}\begin{bmatrix} 46 \\ -30 \\ -31 \end{bmatrix} = \begin{bmatrix} -4 \\ -2 \\ -15 \end{bmatrix}$$

let

$$\boldsymbol{e}_{11} = \boldsymbol{A}^2\boldsymbol{b}_1 + a_{12}\boldsymbol{A}\boldsymbol{b}_1 + a_{11}\boldsymbol{b}_1 = \begin{bmatrix} 18 \\ -3 \\ -18 \end{bmatrix}$$

$$\boldsymbol{e}_{12} = \boldsymbol{A}\boldsymbol{b}_1 + a_{12}\boldsymbol{b}_1 = \begin{bmatrix} -12 \\ -1 \\ -18 \end{bmatrix}$$

$$\boldsymbol{e}_{13} = \boldsymbol{b}_1 = \begin{bmatrix} 2 \\ 0 \\ 1 \end{bmatrix}$$

and

$$\boldsymbol{T} = [\boldsymbol{e}_{11} \quad \boldsymbol{e}_{12} \quad \boldsymbol{e}_{13}] = \begin{bmatrix} 18 & -12 & 2 \\ -3 & -1 & 0 \\ -18 & -3 & 1 \end{bmatrix}, \quad \boldsymbol{T}^{-1} = \left(-\frac{1}{72}\right)\begin{bmatrix} -1 & 6 & 2 \\ 3 & 54 & -6 \\ -9 & 270 & -54 \end{bmatrix}$$

Thus, using conclusion 3.7.1, yields

$$\overline{\boldsymbol{A}}_c = \boldsymbol{T}^{-1}\boldsymbol{A}\boldsymbol{T} = \begin{bmatrix} 0 & 1 & 0 \\ 0 & 0 & 1 \\ 15 & 2 & 4 \end{bmatrix}, \quad \overline{\boldsymbol{B}}_c = \boldsymbol{T}^{-1}\boldsymbol{B} = \begin{bmatrix} 0 & -\frac{1}{36} \\ 0 & \frac{1}{12} \\ 1 & \frac{3}{4} \end{bmatrix}$$

Its Wonham controllable form is

$$\dot{\overline{\boldsymbol{x}}} = \begin{bmatrix} 0 & 1 & 0 \\ 0 & 0 & 1 \\ 15 & 2 & 4 \end{bmatrix}\overline{\boldsymbol{x}} + \begin{bmatrix} 0 & -\frac{1}{36} \\ 0 & \frac{1}{12} \\ 1 & \frac{3}{4} \end{bmatrix}\boldsymbol{u}$$

3.6.3 Wonham observable form

Considering the dual relation between controllability and observability, the corresponding conclusion of Wonham observable form can be deduced from the Wonham controllable form by using the duality principle.

Conclusion 3.6.2 Considering the observable MIMO LTI system eq. (3.6.1), by using duality principle, we get its Wonham observable form:

$$\begin{cases} \dot{\tilde{\boldsymbol{x}}} = \tilde{\boldsymbol{A}}_o \tilde{\boldsymbol{x}} + \tilde{\boldsymbol{B}}_o \boldsymbol{u} \\ \boldsymbol{y} = \tilde{\boldsymbol{C}}_o \tilde{\boldsymbol{x}} \end{cases} \tag{3.6.20}$$

where

$$\tilde{\boldsymbol{A}}_o = \begin{bmatrix} \tilde{\boldsymbol{A}}_{11} & & & \\ \tilde{\boldsymbol{A}}_{21} & \tilde{\boldsymbol{A}}_{22} & & \\ \vdots & & \ddots & \\ \tilde{\boldsymbol{A}}_{m1} & \tilde{\boldsymbol{A}}_{m2} & \cdots & \tilde{\boldsymbol{A}}_{mm} \end{bmatrix} \tag{3.6.21}$$

$$\tilde{\boldsymbol{A}}_{ii} = \left[\begin{array}{ccc|c} 0 & \cdots & 0 & -\beta_{i0} \\ \hline 1 & & & -\beta_{i1} \\ & \ddots & & \vdots \\ & & 1 & -\beta_{i,\zeta_i-1} \end{array}\right], i = 1,2,\cdots,m \tag{3.6.22}$$

$$\tilde{\boldsymbol{A}}_{ij} = \begin{bmatrix} \rho_{i1j} & \cdots & \rho_{i\zeta_i j} \\ 0 & \cdots & 0 \\ \vdots & & \vdots \\ 0 & \cdots & 0 \end{bmatrix}, j = 1,2,\cdots,i-1 \tag{3.6.23}$$

$$\tilde{\boldsymbol{C}}_o = \left[\begin{array}{cccc|c|cccc} 0 & \cdots & 0 & 1 & & & & & \\ \hline & & & & \ddots & & & & \\ \hline & & & & & 0 & \cdots & 0 & 1 \\ \hline * & & & \cdots & \cdots & & & & * \\ \vdots & & & & & & & & \vdots \\ * & & & \cdots & \cdots & & & & * \end{array}\right] \tag{3.6.24}$$

$$\tilde{\boldsymbol{B}}_o \text{ don't have special form} \tag{3.6.25}$$

The " * " in eq. (3.6.24) denotes the element is maybe not zero.

3.6.4 Luenberger controllable form

Luenberger controllable form is widely used to design state feedback controllers. Considering a controllable MIMO LTI system, first find n linearly independent column vectors in $\boldsymbol{Q}_c = [\boldsymbol{B} \vdots \boldsymbol{AB} \vdots \cdots \vdots \boldsymbol{A}^{n-1}\boldsymbol{B}]$. Let $\boldsymbol{B} = [\boldsymbol{b}_1, \boldsymbol{b}_2, \cdots, \boldsymbol{b}_p]$ and $\boldsymbol{B} = r$. By using the row-first searching scheme, we find the n linearly independent column vectors which are used to construct the following matrix.

$$P^{-1}=[b_1,Ab_1,\cdots,A^{\mu_1-1}b_1;b_2,Ab_2,\cdots,A^{\mu_2-1}b_2;\cdots;b_r,Ab_r,\cdots,A^{\mu_r-1}b_r] \quad (3.6.26)$$

where

$$\mu_1+\mu_2+\cdots+\mu_r=n$$

So we have

$$P=(P^{-1})^{-1}=\begin{bmatrix} e_{11}^{\mathrm{T}} \\ \vdots \\ e_{1\mu_1}^{\mathrm{T}} \\ \vdots \\ e_{r1}^{\mathrm{T}} \\ \vdots \\ e_{r\mu_r}^{\mathrm{T}} \end{bmatrix} \quad (3.6.27)$$

Where, the number of rows in each block matrix of P is $\mu_i\ (i=1,2,\cdots,r)$. And then construct transformation matrix S by last row of each block matrix, $e_{1\mu_1}^{\mathrm{T}},e_{2\mu_2}^{\mathrm{T}},\cdots,e_{r\mu_r}^{\mathrm{T}}$, as

$$S^{-1}=\begin{bmatrix} e_{1\mu_1}^{\mathrm{T}} \\ e_{1\mu_1}^{\mathrm{T}}A \\ \vdots \\ e_{1\mu_1}^{\mathrm{T}}A^{\mu_1-1} \\ \hline \vdots \\ \hline e_{r\mu_r}^{\mathrm{T}} \\ e_{r\mu_r}^{\mathrm{T}}A \\ \vdots \\ e_{r\mu_r}^{\mathrm{T}}A^{\mu_r-1} \end{bmatrix} \quad (3.6.28)$$

Conclusion 3.6.3 Considering the controllable MIMO LTI System eq. (3.6.1), by using the linearly non-singular transformation $\hat{x}=S^{-1}x$, we get the its Luenberger controllable form:

$$\begin{cases} \dot{\hat{x}}=\hat{A}_c\hat{x}+\hat{B}_c u \\ y=\hat{C}_c\hat{x} \end{cases} \quad (3.6.29)$$

where

$$\underset{(n\times n)}{\hat{A}_c}=\begin{bmatrix} \hat{A}_{11} & \cdots & \hat{A}_{1r} \\ \vdots & & \vdots \\ \hat{A}_{r1} & \cdots & \hat{A}_{rr} \end{bmatrix} \quad (3.6.30)$$

$$\underset{(\mu_i\times\mu_i)}{\hat{A}_{ii}}=\left[\begin{array}{c|ccc} 0 & 1 & & \\ \vdots & & \ddots & \\ 0 & & & 1 \\ \hline * & * & \cdots & * \end{array}\right],i=1,2,\cdots,r \quad (3.6.31)$$

$$\underset{(\mu_i \times \mu_j)}{\hat{A}_{ij}} = \begin{bmatrix} 0 & \cdots & 0 \\ \vdots & & \vdots \\ 0 & \cdots & 0 \\ * & \cdots & * \end{bmatrix}, i \neq j \tag{3.6.32}$$

$$\underset{(n \times p)}{\hat{B}_c} = S^{-1}B = \left[\begin{array}{c:c:c:ccc} 0 & & & * & \cdots & * \\ \vdots & & & & & \\ 0 & & & & & \\ 1 & * & & \vdots & & \vdots \\ \hdashline & \ddots & & \vdots & & \vdots \\ \hdashline & & 0 & \vdots & & \vdots \\ & & \vdots & & & \\ & & 0 & & & \\ & & 1 & * & \cdots & * \end{array}\right] \tag{3.6.33}$$

$$\underset{(q \times n)}{\hat{C}_c} = CS \text{ (don't have special form)} \tag{3.6.34}$$

The " $*$ " inabove equations denotes the element is maybe not zero.

Proof Omitted

Example 3.6.2 Considering controllable LTI system described by

$$\dot{x} = \begin{bmatrix} -1 & -4 & -2 \\ 0 & 6 & -1 \\ 1 & 7 & -1 \end{bmatrix} x + \begin{bmatrix} 2 & 0 \\ 0 & 0 \\ 1 & 1 \end{bmatrix} u$$

determine its Luenberger controllable form.

Answer First, by using row-first searching scheme, the 3 linearly independent column vectors of controllability matrix

$$Q_c = [B \vdots AB \vdots A^2B] = \begin{bmatrix} 2 & 0 & -4 & -2 & 6 & 8 \\ 0 & 0 & -1 & -1 & -7 & -5 \\ 1 & 1 & 1 & -1 & -12 & -8 \end{bmatrix}$$

are

$$b_1 = \begin{bmatrix} 2 \\ 0 \\ 1 \end{bmatrix}, \quad b_2 = \begin{bmatrix} 0 \\ 0 \\ 1 \end{bmatrix}, \quad Ab_1 = \begin{bmatrix} -4 \\ -1 \\ 1 \end{bmatrix}$$

Then compose P^{-1} as

$$P^{-1} = [b_1 \quad Ab_1 \quad b_2] = \begin{bmatrix} 2 & -4 & 0 \\ 0 & -1 & 0 \\ 1 & 1 & 1 \end{bmatrix}$$

and

$$P = (P^{-1})^{-1} = \begin{bmatrix} 2 & -4 & 0 \\ 0 & -1 & 0 \\ 1 & 1 & 1 \end{bmatrix}^{-1} = \begin{bmatrix} 0.5 & -2 & 0 \\ 0 & -1 & 0 \\ -0.5 & 3 & 1 \end{bmatrix}$$

Then divide $\boldsymbol{P}$ into two block matrices which have two rows and one row, respectively. Form eq. (3.6.28), we have

$$\boldsymbol{S}^{-1}=\begin{bmatrix}\boldsymbol{e}_{12}^{\mathrm{T}}\\ \boldsymbol{e}_{12}^{\mathrm{T}}\boldsymbol{A}\\ \boldsymbol{e}_{21}^{\mathrm{T}}\end{bmatrix}=\begin{bmatrix}0 & -1 & 0\\ 0 & -6 & 1\\ -0.5 & 3 & 1\end{bmatrix}$$

$$\boldsymbol{S}=(\boldsymbol{S}^{-1})^{-1}=\begin{bmatrix}0 & -1 & 0\\ 0 & -6 & 1\\ -0.5 & 3 & 1\end{bmatrix}^{-1}=\begin{bmatrix}-18 & 2 & -2\\ -1 & 0 & 0\\ -6 & 1 & 0\end{bmatrix}$$

Hence, from conclusion 3.6.3, we get

$$\hat{\boldsymbol{A}}_{\mathrm{c}}=\boldsymbol{S}^{-1}\boldsymbol{A}\boldsymbol{S}=\begin{bmatrix}0 & -1 & 0\\ 0 & -6 & 1\\ -0.5 & 3 & 1\end{bmatrix}\begin{bmatrix}-1 & -4 & -2\\ 0 & 6 & -1\\ 1 & 7 & -1\end{bmatrix}\begin{bmatrix}-18 & 2 & -2\\ -1 & 0 & 0\\ -6 & 1 & 0\end{bmatrix}=\begin{bmatrix}0 & 1 & 0\\ -19 & 7 & -2\\ -36 & 0 & -3\end{bmatrix}$$

$$\hat{\boldsymbol{B}}_{\mathrm{c}}=\boldsymbol{S}^{-1}\boldsymbol{B}=\begin{bmatrix}0 & -1 & 0\\ 0 & -6 & 1\\ -0.5 & 3 & 1\end{bmatrix}\begin{bmatrix}2 & 0\\ 0 & 0\\ 1 & 1\end{bmatrix}=\begin{bmatrix}0 & 0\\ 1 & 1\\ 0 & 1\end{bmatrix}$$

And the Luenberger controllable form of the system is

$$\dot{\hat{\boldsymbol{x}}}=\begin{bmatrix}0 & 1 & 0\\ -19 & 7 & -2\\ -36 & 0 & -3\end{bmatrix}\hat{\boldsymbol{x}}+\begin{bmatrix}0 & 0\\ 1 & 1\\ 0 & 1\end{bmatrix}\boldsymbol{u}$$

3.6.5 Luenberger observable form

The Luenberger observable form and Luenberger controllable form have dual relationship. Below, based on the duality principle, we give the corresponding conclusions of the Luenberger observable form directly.

Conclusion 3.6.4 For a controllable MIMO LTI system eq. (3.6.1) and rank $\boldsymbol{C}=k$, its Luenberger observable form

$$\begin{cases}\dot{\hat{\boldsymbol{x}}}=\hat{\boldsymbol{A}}_{\mathrm{o}}\hat{\boldsymbol{x}}+\hat{\boldsymbol{B}}_{\mathrm{o}}\boldsymbol{u}\\ \boldsymbol{y}=\hat{\boldsymbol{C}}_{\mathrm{o}}\hat{\boldsymbol{x}}\end{cases}\tag{3.6.35}$$

where

$$\hat{\boldsymbol{A}}_{\mathrm{o}}=\begin{bmatrix}\hat{\boldsymbol{A}}_{11} & \cdots & \hat{\boldsymbol{A}}_{1k}\\ \vdots & & \vdots\\ \hat{\boldsymbol{A}}_{k1} & \cdots & \hat{\boldsymbol{A}}_{kk}\end{bmatrix}\tag{3.6.36}$$

$$\hat{\boldsymbol{A}}_{ii}=\left[\begin{array}{ccc:c}0 & \cdots & 0 & *\\ \hdashline 1 & & & *\\ & \ddots & & \vdots\\ & & 1 & *\end{array}\right],i=1,2,\cdots,k\tag{3.6.37}$$

$$\hat{A}_{ij} = \begin{bmatrix} 0 & \cdots & 0 & * \\ \vdots & & \vdots & \vdots \\ 0 & \cdots & 0 & * \end{bmatrix}, i \neq j \tag{3.6.38}$$

$$\hat{C}_o = \left[\begin{array}{cccc|c|cccc} 0 & \cdots & 0 & 1 & & & & & \\ \hline & & & * & \ddots & & & & \\ \hline & & & & & 0 & \cdots & 0 & 1 \\ \hline * & & & & \cdots & \cdots & & & * \\ \vdots & & & & & & & & \vdots \\ * & & & & \cdots & \cdots & & & * \end{array}\right] \tag{3.6.39}$$

$$\hat{B}_o \text{ don't have special form} \tag{3.6.40}$$

The " * " in above equations denotes the element is maybe not zero.

3.7 Canonical decomposition of linear systems

The controllability and observability of a linear system are preserved and the degree of incomplete controllability and observability remain unchanged under a linear nonsingular transformation no matter time-invariant or time-varying. The canonical decomposition of a linear system is to introduce a linear non-singular transformation, so that the state variables which cannot be controlled and observed can be revealed. Meanwhile, for the system which is not completely controllable and not completely observable, the system can be divided into four subsystems in general: controllable and observable subsystem, controllable but not observable subsystem, uncontrollable but observable subsystem, uncontrollable but not observable subsystem as shown in Figure 3.7.1. Furthermore, an important conclusion about the essential difference between the state-space description and the input-output description can be obtained by the canonical decomposition of systems.

3.7.1 Controllability and observability under nonsingular transformation

The canonical decomposition of linear systems is realized by introducing linear nonsingular transformation. We first discuss the characteristics of controllability and observability under linear nonsingular transformation.

Conclusion 3.7.1 Linear nonsingular transformation does not change the controllability and observability of LTI systems.

Proof Let $(\overline{A}, \overline{B}, \overline{C})$ is result using a nonsingular transformation to (A, B, C). From eq. (1.5.15), we have

$$\overline{A} = P^{-1}AP, \quad \overline{B} = P^{-1}B, \quad \overline{C} = CP \tag{3.7.1}$$

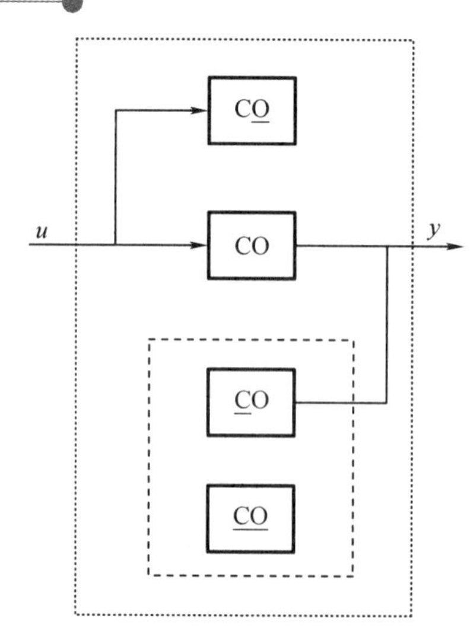

Figure 3.7.1 The four subsystems of a not completely controllable and not completely observable system

Hence

$$\overline{Q}_c = [\overline{B} \vdots \overline{A}\,\overline{B} \vdots \cdots \vdots \overline{A}^{n-1}\overline{B}] = [PB \vdots PAB \vdots \cdots \vdots PA^{n-1}B] = PQ_c \qquad (3.7.2)$$

where, Q_c and $\overline{Q}_c$ are controllabilty matrices, respectively. From $P = n$ and $Q_c \leqslant n$, we have

$$\text{rank } \overline{Q}_c \leqslant \min\{\text{rank } P, \text{rank } Q_c\} = \text{rank } Q_c \qquad (3.7.3)$$

Because P is non-sigular, from eq. (3.8.2), we have

$$Q_c = P^{-1}\overline{Q}_c \qquad (3.7.4)$$

Similarly we get

$$\text{rank } Q_c \leqslant \min\{\text{rank } P^{-1},\ \text{rank } \overline{Q}_c\} = \text{rank } \overline{Q}_c \qquad (3.7.5)$$

From eq. (3.8.3) and eq. (3.8.5), we have

$$\text{rank } Q_c = \text{rank } \overline{Q}_c \qquad (3.7.6)$$

Thus the controllablity is preserved. In a similar way it can be proved that the observability is preserved also.

Conclusion 3.7.2 For a LTV system, $\dot{x} = A(t)x + b(t)u, y = C(t)x, t \in J$, the properties of controllability and observability are preserved under differential nonsingular transformation $\hat{x} = R^{-1}(t)x$, where the elements of $R(t)$ are time continuous.

Proof Let the system after the transformation be described as

$$\begin{cases} \dot{\hat{x}} = \hat{A}(t)\hat{x} + \hat{B}(t)u \\ y = \hat{C}(t)\hat{x} \end{cases} \qquad (3.7.7)$$

From eq. (1.5.19), we have

$$\begin{gathered} \hat{A}(t) = -R^{-1}(t)\dot{R}(t) + R^{-1}(t)A(t)R(t) \\ \hat{B}(t) = R^{-1}(t)B(t), \hat{C} = C(t)R(t) \end{gathered} \qquad (3.7.8)$$

Considering $\hat{x} = R^{-1}(t)x$ and using state motion equation yields

$$\begin{aligned}\hat{\boldsymbol{x}}(t) &= \boldsymbol{R}^{-1}(t)\boldsymbol{x}\\ &= \boldsymbol{R}^{-1}(t)\boldsymbol{\Phi}(t,t_0)\boldsymbol{x}(t_0) + \int_{t_0}^{T} \boldsymbol{R}^{-1}(t)\boldsymbol{\Phi}(t,\tau)\boldsymbol{B}(\tau)\boldsymbol{u}(\tau)\mathrm{d}\tau\\ &= \boldsymbol{R}^{-1}(t)\boldsymbol{\Phi}(t,t_0)\boldsymbol{R}(t_0)\hat{\boldsymbol{x}}(t_0) + \int_{t_0}^{T} \boldsymbol{R}^{-1}(t)\boldsymbol{\Phi}(t,\tau)\boldsymbol{R}(\tau)\boldsymbol{B}(\tau)\boldsymbol{u}(\tau)\mathrm{d}\tau \end{aligned} \tag{3.7.9}$$

Thus we get

$$\hat{\boldsymbol{\Phi}}(t,\tau) = \boldsymbol{R}^{-1}(t)\boldsymbol{\Phi}(t,\tau)\boldsymbol{R}(\tau) \tag{3.7.10}$$

And the Grammian matrix $\hat{\boldsymbol{W}}_c[t_0,t_1]$ is

$$\begin{aligned}\hat{\boldsymbol{W}}_c[t_0,t_1] &= \int_{t_0}^{t_1} \hat{\boldsymbol{\Phi}}(t_0,\tau)\,\hat{\boldsymbol{B}}(\tau)\,\hat{\boldsymbol{B}}^{\mathrm{T}}(\tau)\,\hat{\boldsymbol{\Phi}}^{\mathrm{T}}(t_0,\tau)\mathrm{d}\tau\\ &= \int_{t_0}^{t_1} \boldsymbol{R}^{-1}(t_0)\boldsymbol{\Phi}(t_0,\tau)\boldsymbol{R}(\tau)\,\boldsymbol{R}^{-1}(\tau)\boldsymbol{B}(\tau)\,\boldsymbol{B}^{\mathrm{T}}(\tau)[\boldsymbol{R}^{-1}(\tau)]^{\mathrm{T}}[\boldsymbol{R}(\tau)]^{\mathrm{T}}\\ &\quad [\boldsymbol{\Phi}(t_0,\tau)]^{\mathrm{T}}[\boldsymbol{R}^{-1}(t_0)]^{\mathrm{T}}\mathrm{d}\tau\\ &= \int_{t_0}^{t_1} \boldsymbol{R}^{-1}(t_0)\boldsymbol{\Phi}(t_0,\tau)\boldsymbol{B}(\tau)\,\boldsymbol{B}^{\mathrm{T}}(\tau)[\boldsymbol{\Phi}(t_0,\tau)]^{\mathrm{T}}[\boldsymbol{R}^{-1}(t_0)]^{\mathrm{T}}\mathrm{d}\tau\\ &= \boldsymbol{R}^{-1}(t_0)\,\boldsymbol{W}_c[t_0,t_1][\boldsymbol{R}^{-1}(t_0)]^{\mathrm{T}} \end{aligned} \tag{3.7.11}$$

Because rank $\boldsymbol{R}(t_0) = n$ and rank $\boldsymbol{W}_c[t_0,t_1] \leqslant n$, so

$$\text{rank}\,\hat{\boldsymbol{W}}_c[t_0,t_1] \leqslant \min\{\text{rank}\,\boldsymbol{R}^{-1}(t_0), \text{rank}\,\boldsymbol{W}_c[t_0,t_1]\} = \text{rank}\,\boldsymbol{W}_c[t_0,t_1] \tag{3.7.12}$$

From eq. (3.7.11), we have

$$\boldsymbol{W}_c[t_0,t_1] = \boldsymbol{R}(t_0)\hat{\boldsymbol{W}}_c[t_0,t_1][\boldsymbol{R}(t_0)]^{\mathrm{T}}$$

and

$$\text{rank}\,\boldsymbol{W}_c[t_0,t_1] \leqslant \text{rank}\,\hat{\boldsymbol{W}}_c[t_0,t_1] \tag{3.7.13}$$

From eq. (3.7.12) and eq. (3.7.13), we have

$$\text{rank}\,\boldsymbol{W}_c[t_0,t_1] = \text{rank}\,\hat{\boldsymbol{W}}_c[t_0,t_1] \tag{3.7.14}$$

The above equation shows that linear non-singular transformation does not affect the controllability of the system. Similarly, it can be proved that linear non-singular transformation does not affect the observability of the system.

3.7.2 Canonical decomposition of LTI systems according to the controllability

Consider a MIMO LTI system described by

$$\begin{cases}\dot{\boldsymbol{x}} = \boldsymbol{A}\boldsymbol{x} + \boldsymbol{B}\boldsymbol{u}\\ \boldsymbol{y} = \boldsymbol{C}\boldsymbol{x}\end{cases} \tag{3.7.15}$$

where, $\boldsymbol{x}$ is an $n \times 1$ state vector, $\boldsymbol{u}$ is a $p \times 1$ input vector, $\boldsymbol{y}$ is a $q \times 1$ output vector. Suppose the system is not completely controllable and let

$$\text{rank}\,\boldsymbol{Q}_c = \text{rank}[\boldsymbol{B} \quad \boldsymbol{AB} \quad \cdots \quad \boldsymbol{A}^{n-1}\boldsymbol{B}] = k < n \tag{3.7.16}$$

We construct the $n \times n$ matrix

$$\boldsymbol{P}^{-1} \triangleq \boldsymbol{Q} = [\boldsymbol{q}_1, \cdots, \boldsymbol{q}_k \,\vdots\, \boldsymbol{q}_{k+1}, \cdots, \boldsymbol{q}_n] \tag{3.7.17}$$

where the first k columns are any linearly independent columns of $\boldsymbol{Q}_c$, and the remaining columns can arbitrarily be chosen as long as P is nonsingular. Then we have the following conclusion .

Conclusion 3.7.3 The equivalence transformation $\bar{\boldsymbol{x}} = \boldsymbol{P}\boldsymbol{x}$ will transform not complete controllable system eq. (3.7.15) into

$$\begin{cases} \begin{bmatrix} \dot{\bar{\boldsymbol{x}}}_c \\ \dot{\bar{\boldsymbol{x}}}_{\bar{c}} \end{bmatrix} = \begin{bmatrix} \bar{\boldsymbol{A}}_c & \bar{\boldsymbol{A}}_{12} \\ \boldsymbol{0} & \bar{\boldsymbol{A}}_{\bar{c}} \end{bmatrix} \begin{bmatrix} \bar{\boldsymbol{x}}_c \\ \bar{\boldsymbol{x}}_{\bar{c}} \end{bmatrix} + \begin{bmatrix} \bar{\boldsymbol{B}}_c \\ \boldsymbol{0} \end{bmatrix} \boldsymbol{u} \\ \boldsymbol{y} = [\bar{\boldsymbol{C}}_c \quad \bar{\boldsymbol{C}}_{\bar{c}}] \begin{bmatrix} \bar{\boldsymbol{x}}_c \\ \bar{\boldsymbol{x}}_{\bar{c}} \end{bmatrix} \end{cases} \tag{3.7.18}$$

where, $\bar{\boldsymbol{x}}_c$ is a $k \times 1$ controllable state, $\bar{\boldsymbol{x}}_{\bar{c}}$ is $(n-k) \times 1$ uncontrollable state, $k = \text{rank } \boldsymbol{Q}_c$, $\bar{\boldsymbol{A}}_c$ is $k \times k$, $\bar{\boldsymbol{A}}_{\bar{c}}$ is $(n-k) \times (n-k)$.

Proof Let

$$\boldsymbol{P} = \boldsymbol{Q}^{-1} = \begin{bmatrix} \boldsymbol{p}_1^{\mathrm{T}} \\ \vdots \\ \boldsymbol{p}_n^{\mathrm{T}} \end{bmatrix}, \ \boldsymbol{Q} = [\boldsymbol{q}_1 \quad \boldsymbol{q}_2 \quad \cdots \quad \boldsymbol{q}_n] \tag{3.7.19}$$

From $\boldsymbol{PQ} = \boldsymbol{I}$, we have

$$\boldsymbol{p}_i^{\mathrm{T}} \boldsymbol{q}_j = 0, \ \forall i \neq j, i, j = 1, 2, \cdots, n \tag{3.7.20}$$

If $\boldsymbol{q}_i$ belongs to the vector space spanned by $\boldsymbol{q}_1, \boldsymbol{q}_2, \cdots, \boldsymbol{q}_k$, then $n \times 1$ $\boldsymbol{A}\boldsymbol{q}_i$ is also belongs to the vector space. Thus we have

$$\begin{aligned} \boldsymbol{A}\boldsymbol{q}_1 &= a_1^1 \boldsymbol{q}_1 + \cdots + a_k^1 \boldsymbol{q}_k \\ &\vdots \\ \boldsymbol{A}\boldsymbol{q}_k &= a_1^k \boldsymbol{q}_1 + \cdots + a_k^k \boldsymbol{q}_k \end{aligned} \tag{3.7.21}$$

From eq. (3.7.20) and eq. (3.7.21), we get

$$\boldsymbol{p}_i^{\mathrm{T}} \boldsymbol{A} \boldsymbol{q}_j = 0, \quad i = k+1, \cdots, n, \quad j = 1, 2, \cdots, k \tag{3.7.22}$$

Thus, we have

$$\bar{\boldsymbol{A}} = \boldsymbol{PA}\boldsymbol{P}^{-1} = \left[\begin{array}{ccc|ccc} \boldsymbol{p}_1^{\mathrm{T}}\boldsymbol{A}\boldsymbol{q}_1 & \cdots & \boldsymbol{p}_1^{\mathrm{T}}\boldsymbol{A}\boldsymbol{q}_k & \boldsymbol{p}_1^{\mathrm{T}}\boldsymbol{A}\boldsymbol{q}_{k+1} & \cdots & \boldsymbol{p}_1^{\mathrm{T}}\boldsymbol{A}\boldsymbol{q}_n \\ \vdots & & \vdots & \vdots & & \vdots \\ \boldsymbol{p}_k^{\mathrm{T}}\boldsymbol{A}\boldsymbol{q}_1 & \cdots & \boldsymbol{p}_k^{\mathrm{T}}\boldsymbol{A}\boldsymbol{q}_k & \boldsymbol{p}_k^{\mathrm{T}}\boldsymbol{A}\boldsymbol{q}_{k+1} & \cdots & \boldsymbol{p}_k^{\mathrm{T}}\boldsymbol{A}\boldsymbol{q}_n \\ \boldsymbol{p}_{k+1}^{\mathrm{T}}\boldsymbol{A}\boldsymbol{q}_1 & \cdots & \boldsymbol{p}_{k+1}^{\mathrm{T}}\boldsymbol{A}\boldsymbol{q}_k & \boldsymbol{p}_{k+1}^{\mathrm{T}}\boldsymbol{A}\boldsymbol{q}_{k+1} & \cdots & \boldsymbol{p}_{k+1}^{\mathrm{T}}\boldsymbol{A}\boldsymbol{q}_n \\ \vdots & & \vdots & \vdots & & \vdots \\ \boldsymbol{p}_n^{\mathrm{T}}\boldsymbol{A}\boldsymbol{q}_1 & \cdots & \boldsymbol{p}_n^{\mathrm{T}}\boldsymbol{A}\boldsymbol{q}_k & \boldsymbol{p}_n^{\mathrm{T}}\boldsymbol{A}\boldsymbol{q}_{k+1} & \cdots & \boldsymbol{p}_n^{\mathrm{T}}\boldsymbol{A}\boldsymbol{q}_n \end{array}\right] = \begin{bmatrix} \bar{\boldsymbol{A}}_c & \bar{\boldsymbol{A}}_{12} \\ \boldsymbol{0} & \bar{\boldsymbol{A}}_{\bar{c}} \end{bmatrix} \tag{3.7.23}$$

Similarly, the columns of $\boldsymbol{B}$ belong to the vector space spanned by $\boldsymbol{q}_1$, $\boldsymbol{q}_2$, $\cdots$, $\boldsymbol{q}_k$, then i^{th} column of $\boldsymbol{B}$ can be expressed by

$$\boldsymbol{b}_i = a_1^i \boldsymbol{q}_1 + \cdots + a_k^i \boldsymbol{q}_k, \quad i = 1, \cdots, p \tag{3.7.24}$$

From eq. (3.8.20) and eq. (3.8.24), we have

$$\overline{\boldsymbol{B}} = \boldsymbol{P}\boldsymbol{B} = \begin{bmatrix} \boldsymbol{p}_1^{\mathrm{T}}\boldsymbol{B} \\ \vdots \\ \boldsymbol{p}_k^{\mathrm{T}}\boldsymbol{B} \\ \cdots \\ \boldsymbol{p}_{k+1}^{\mathrm{T}}\boldsymbol{B} \\ \vdots \\ \boldsymbol{p}_n^{\mathrm{T}}\boldsymbol{B} \end{bmatrix} = \begin{bmatrix} \overline{\boldsymbol{B}}_{\mathrm{c}} \\ \boldsymbol{0} \end{bmatrix} \tag{3.7.25}$$

Moreover, there is

$$\begin{aligned} k &= \operatorname{rank} \boldsymbol{Q}_{\mathrm{c}} \\ &= \operatorname{rank} \overline{\boldsymbol{Q}}_{\mathrm{c}} = \operatorname{rank}[\overline{\boldsymbol{B}} \vdots \overline{\boldsymbol{A}\boldsymbol{B}} \vdots \cdots \vdots \overline{\boldsymbol{A}}^{n-1}\overline{\boldsymbol{B}}] \\ &= \operatorname{rank}\begin{bmatrix} \overline{\boldsymbol{B}}_{\mathrm{c}} \vdots \overline{\boldsymbol{A}}_{\mathrm{c}}\overline{\boldsymbol{B}}_{\mathrm{c}} \vdots \cdots \vdots \overline{\boldsymbol{A}}_{\mathrm{c}}^{n-1}\overline{\boldsymbol{B}}_{\mathrm{c}} \\ \boldsymbol{0} \vdots \boldsymbol{0} \vdots \cdots \vdots \boldsymbol{0} \end{bmatrix} \\ &= \operatorname{rank}[\overline{\boldsymbol{B}}_{\mathrm{c}} \vdots \overline{\boldsymbol{A}}_{\mathrm{c}}\overline{\boldsymbol{B}}_{\mathrm{c}} \vdots \cdots \vdots \overline{\boldsymbol{A}}_{\mathrm{c}}^{n-1}\overline{\boldsymbol{B}}_{\mathrm{c}}] \end{aligned} \tag{3.7.26}$$

Because $\overline{\boldsymbol{A}}_{\mathrm{c}}$ is $k \times k$, based on Hamilton-Cayley theorem, $\overline{\boldsymbol{A}}_{\mathrm{c}}^k \overline{\boldsymbol{B}}_{\mathrm{c}}, \cdots, \overline{\boldsymbol{A}}_{\mathrm{c}}^{n-1} \overline{\boldsymbol{B}}_{\mathrm{c}}$ can be represented by the linear combination of $\overline{\boldsymbol{B}}_{\mathrm{c}}, \overline{\boldsymbol{A}}_{\mathrm{c}} \overline{\boldsymbol{B}}_{\mathrm{c}}, \cdots, \overline{\boldsymbol{A}}_{\mathrm{c}}^{k-1} \overline{\boldsymbol{B}}_{\mathrm{c}}$. So, from eq. (3.7.26), we have

$$\operatorname{rank}[\overline{\boldsymbol{B}}_{\mathrm{c}} \vdots \overline{\boldsymbol{A}}_{\mathrm{c}}\overline{\boldsymbol{B}}_{\mathrm{c}} \vdots \cdots \vdots \overline{\boldsymbol{A}}_{\mathrm{c}}^{k-1}\overline{\boldsymbol{B}}_{\mathrm{c}}] = k \tag{3.7.27}$$

It indicates $(\overline{\boldsymbol{A}}_{\mathrm{c}}, \overline{\boldsymbol{B}}_{\mathrm{c}})$ is controllable, i. e. $\overline{\boldsymbol{x}}_{\mathrm{c}}$ is controllable.

The canonical decomposition of LTI systems according to controllability is discussed as follows:

(1) The system is obviously divided into two subsystems: the controllable subsystem and the uncontrollable subsystem.

The k-dimensional controllable subsystem is described by

$$\begin{cases} \dot{\overline{\boldsymbol{x}}}_{\mathrm{c}} = \overline{\boldsymbol{A}}_{\mathrm{c}} \overline{\boldsymbol{x}}_{\mathrm{c}} + \overline{\boldsymbol{A}}_{12} \overline{\boldsymbol{x}}_{\bar{\mathrm{c}}} + \overline{\boldsymbol{B}}_{\mathrm{c}} \boldsymbol{u} \\ \boldsymbol{y}_1 = \overline{\boldsymbol{C}}_{\mathrm{c}} \overline{\boldsymbol{x}}_{\mathrm{c}} \end{cases} \tag{3.7.28}$$

The $(n-k)$-dimensional uncontrollable subsystem is described by

$$\begin{cases} \dot{\overline{\boldsymbol{x}}}_{\bar{\mathrm{c}}} = \overline{\boldsymbol{A}}_{\bar{\mathrm{c}}} \overline{\boldsymbol{x}}_{\bar{\mathrm{c}}} \\ \boldsymbol{y}_2 = \overline{\boldsymbol{C}}_{\bar{\mathrm{c}}} \overline{\boldsymbol{x}}_{\bar{\mathrm{c}}} \end{cases} \tag{3.2.29}$$

and $\boldsymbol{y} = \boldsymbol{y}_1 + \boldsymbol{y}_2$.

(2) After the decomposition, the diagram of the system is shown in Figure 3.7.2.

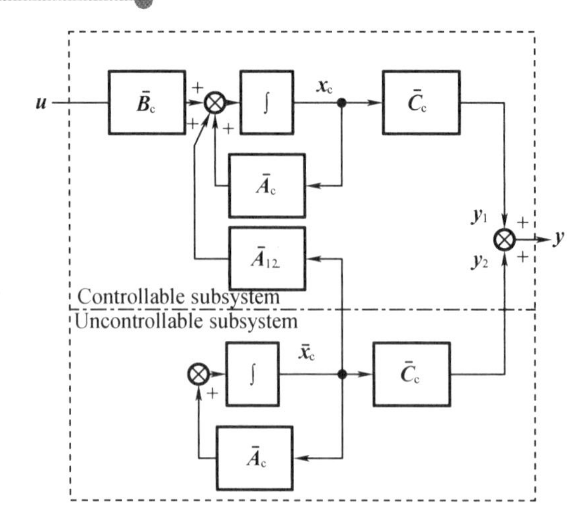

Figure 3.7.2 The diagram of the system after controllability decomposition

(3) The Figure 3.7.2 shows the state $\bar{x}_{\bar{c}}$ of uncontrollable subsystem is not effected by $\boldsymbol{u}$.

(4) With the different selection of $\boldsymbol{q}_1, \cdots, \boldsymbol{q}_k$ and $\boldsymbol{q}_{k+1}, \cdots, \boldsymbol{q}_n$, the form of eq. (3.7.18) remains unchanged, but the values of elements in each matrix may be different.

Example 3.7.1 Take canonical decomposition of the following LTI system according to controllability.

$$\dot{\boldsymbol{x}} = \begin{bmatrix} 1 & 1 & 1 \\ 0 & 1 & 0 \\ 1 & 1 & 1 \end{bmatrix} \boldsymbol{x} + \begin{bmatrix} 0 & 1 \\ 1 & 0 \\ 0 & 1 \end{bmatrix} \boldsymbol{u}$$

$$\boldsymbol{y} = [1 \quad 0 \quad 1] \boldsymbol{x}$$

Answer Because

$$\text{rank}\, [\boldsymbol{B} \ \vdots\ \boldsymbol{AB} \ \vdots\ \boldsymbol{A}^2\boldsymbol{B}] = \text{rank} \begin{bmatrix} 0 & 1 & 1 & 2 & \\ 1 & 0 & 1 & 0 & \cdots \\ 0 & 1 & 1 & 2 & \end{bmatrix} = 2 < n = 3,$$ so the system is not completely controllable.

Construct $\boldsymbol{P}^{-1} = \boldsymbol{Q} = \begin{bmatrix} 0 & 1 & 1 \\ 1 & 0 & 0 \\ 0 & 1 & 0 \end{bmatrix}$ according to conclusion 3.7.3. Then $\boldsymbol{P} = \begin{bmatrix} 0 & 1 & 0 \\ 0 & 0 & 1 \\ 1 & 0 & -1 \end{bmatrix}$.

$$\bar{\boldsymbol{A}} = \boldsymbol{PAP}^{-1} = \begin{bmatrix} 0 & 1 & 0 \\ 0 & 0 & 1 \\ 1 & 0 & -1 \end{bmatrix} \begin{bmatrix} 1 & 1 & 1 \\ 0 & 1 & 0 \\ 1 & 1 & 1 \end{bmatrix} \begin{bmatrix} 0 & 1 & 1 \\ 1 & 0 & 0 \\ 0 & 1 & 0 \end{bmatrix} = \left[\begin{array}{cc:c} 1 & 0 & 0 \\ 1 & 2 & 1 \\ \hdashline 0 & 0 & 0 \end{array}\right]$$

$$\bar{\boldsymbol{B}} = \boldsymbol{PB} = \begin{bmatrix} 0 & 1 & 0 \\ 0 & 0 & 1 \\ 1 & 0 & -1 \end{bmatrix} \begin{bmatrix} 0 & 1 \\ 1 & 0 \\ 0 & 1 \end{bmatrix} = \left[\begin{array}{cc} 1 & 0 \\ 0 & 1 \\ \hdashline 0 & 0 \end{array}\right]$$

$$\overline{C} = CP^{-1} = [1 \quad 0 \quad 1]\begin{bmatrix} 0 & 1 & 1 \\ 1 & 0 & 0 \\ 0 & 1 & 0 \end{bmatrix} = [0 \quad 2 \quad 1]$$

Thus the decomposition result is

$$\begin{bmatrix} \dot{\overline{x}}_c \\ \dot{\overline{x}}_{\bar{c}} \end{bmatrix} = \left[\begin{array}{cc:c} 1 & 0 & 0 \\ 1 & 2 & 1 \\ \hdashline 0 & 0 & 0 \end{array}\right] \begin{bmatrix} \overline{x}_c \\ \overline{x}_{\bar{c}} \end{bmatrix} + \left[\begin{array}{cc} 1 & 0 \\ 0 & 1 \\ \hdashline 0 & 0 \end{array}\right] u$$

$$y = [0 \quad 2 \quad 1] \begin{bmatrix} \overline{x}_c \\ \overline{x}_{\bar{c}} \end{bmatrix}$$

3.7.3 Canonical decomposition of LTI systems according to the observability

Conclusion 3.7.4 Consider a not completely observable MIMO LTI system described by

$$\begin{cases} \dot{x} = Ax + Bu \\ y = Cx \end{cases} \tag{3.7.30}$$

where x is an $n \times 1$ state vector, u is a $p \times 1$ input vector, y is a $q \times 1$ output vector. Let

$$\operatorname{rank} Q_o = \begin{bmatrix} C \\ CA \\ \vdots \\ CA^{n-1} \end{bmatrix} = m < n \tag{3.7.31}$$

Construct the $n \times n$ matrix

$$F = \begin{bmatrix} h_1 \\ \vdots \\ h_m \\ \cdots \\ h_{m+1} \\ \vdots \\ h_n \end{bmatrix} \tag{3.7.32}$$

where the first m rows are any linearly independent rows of Q_o, and the remaining rows can arbitrarily be chosen as long as F is nonsingular. Then the equivalence transformation $\hat{x} = Fx$ will transform eq. (3.7.30) into

$$\begin{cases} \begin{bmatrix} \dot{\hat{x}}_o \\ \dot{\hat{x}}_{\bar{o}} \end{bmatrix} = \begin{bmatrix} \hat{A}_o & 0 \\ \hat{A}_{21} & \hat{A}_{\bar{o}} \end{bmatrix} \begin{bmatrix} \hat{x}_o \\ \hat{x}_{\bar{o}} \end{bmatrix} + \begin{bmatrix} \hat{B}_o \\ \hat{B}_{\bar{o}} \end{bmatrix} u \\ y = [\hat{C}_o \quad 0] \begin{bmatrix} \hat{x}_o \\ \hat{x}_{\bar{o}} \end{bmatrix} \end{cases} \tag{3.7.33}$$

where $\hat{x}_o$ is $m \times 1$ observable state vector, $\hat{x}_o$ is $(n-m) \times 1$ unobservable state vector.

From eq. (3.7.33), we get the m dimensional observable subsystem described by

$$\begin{cases} \dot{\hat{\boldsymbol{x}}}_{o} = \hat{\boldsymbol{A}}_{o} \hat{\boldsymbol{x}}_{o} + \hat{\boldsymbol{B}}_{o} \boldsymbol{u} \\ \boldsymbol{y}_1 = \hat{\boldsymbol{C}}_{o} \hat{\boldsymbol{x}}_{o} \end{cases}$$

and $n-m$ dimensional unobservable subsystem described by

$$\begin{cases} \dot{\hat{\boldsymbol{x}}}_{\bar{o}} = \hat{\boldsymbol{A}}_{21} \hat{\boldsymbol{x}}_{o} + \hat{\boldsymbol{A}}_{\bar{o}} \hat{\boldsymbol{x}}_{\bar{o}} + \hat{\boldsymbol{B}}_{\bar{o}} \boldsymbol{u} \\ \boldsymbol{y}_2 = \boldsymbol{0} \end{cases}$$

The decomposed diagram of the system is shown in Figure 3.7.3.

Example 3.7.2 Take canonical decomposition of the following LTI system according to observability.

$$\dot{\boldsymbol{x}} = \begin{bmatrix} 0 & 0 & -1 \\ 1 & 0 & -3 \\ 0 & 1 & -3 \end{bmatrix} \boldsymbol{x} + \begin{bmatrix} 1 \\ 1 \\ 0 \end{bmatrix} \boldsymbol{u}$$

$$\boldsymbol{y} = [0 \quad 1 \quad -2] \boldsymbol{x}$$

Answer Because

Construct $\boldsymbol{F} = \begin{bmatrix} 0 & 1 & -2 \\ 1 & -2 & 3 \\ 0 & 0 & 1 \end{bmatrix}$. Then $\boldsymbol{F}^{-1} = \begin{bmatrix} 2 & 1 & 1 \\ 1 & 0 & 2 \\ 0 & 0 & 1 \end{bmatrix}$, we have

$$\hat{\boldsymbol{A}}_{o} = \boldsymbol{F}\boldsymbol{A}\boldsymbol{F}^{-1} = \left[\begin{array}{cc:c} 0 & 1 & 0 \\ -1 & -2 & 0 \\ \hdashline 1 & 0 & 1 \end{array}\right], \quad \hat{\boldsymbol{B}}_{o} = \boldsymbol{F}\boldsymbol{B} = \left[\begin{array}{c} 1 \\ -1 \\ \hdashline 0 \end{array}\right] = \begin{bmatrix} \hat{\boldsymbol{B}}_{o} \\ \hat{\boldsymbol{B}}_{\bar{o}} \end{bmatrix}$$

$$\hat{\boldsymbol{C}}_{o} = \boldsymbol{C}\boldsymbol{F}^{-1} = [1 \quad 0 \quad 0] = [\hat{\boldsymbol{C}}_{o} \quad \vdots \quad 0]$$

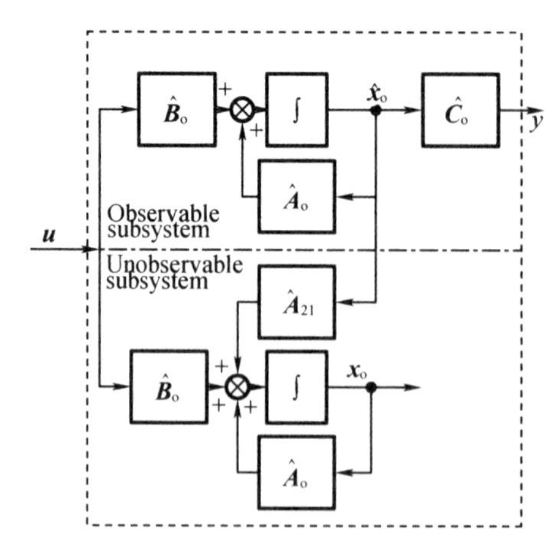

Figure 3.7.3 The diagram of the system after observable decomposition

Thus after observable decomposition, the system is described by

$$\begin{cases}\begin{bmatrix}\dot{\hat{x}}_o \\ \dot{\hat{x}}_{\bar{o}}\end{bmatrix}=\left[\begin{array}{cc:c}0 & 1 & 0 \\ -1 & -2 & 0 \\ \hdashline 1 & 0 & 1\end{array}\right]\begin{bmatrix}\hat{x}_o \\ \hat{x}_{\bar{o}}\end{bmatrix}+\left[\begin{array}{c}1 \\ -1 \\ \hdashline 0\end{array}\right]u \\ y=[1 \quad 0 \quad 0]\begin{bmatrix}\hat{x}_o \\ \hat{x}_{\bar{o}}\end{bmatrix}\end{cases}$$

3.7.4 Canonical decomposition of LTI systems

Consider a not completely controllable and not completely observable MIMO LTI system described by

$$\begin{cases}\dot{x}=Ax+Bu \\ y=Cx\end{cases} \tag{3.7.34}$$

where, x is an $n\times 1$ state vector, u is a $p\times 1$ input vector, y is a $q\times 1$ output vector.

By introducing up to three linear nonsingular transformations to the states, the state of the system is divided into four groups: controllable and observable, controllable but not observable, uncontrollable and observable, and uncontrollable but not observable. There are two ways to do this.

Method 1 First using conclusion 3.7.3, decompose eq. (3.7.34) into controllable and uncontrollable sub-equations. Then decompose each sub-equation, using conclusion 3.7.4, into observable and unobservable part. So we get the four subsystems shown in Figure 3.7.1.

Method 2 First decompose eq. (3.7.34), using conclusion 3.7.4, into observable and unobservable sub-equations. Then decompose each sub-equation, using conclusion 3.7.3, into controllable and uncontrollable part. The following, taking method 1 as an example, discusses the canonical decomposition method of systems in detail.

1. Taking canonical decomposition according to controllability to eq (3.7.34). Linearly nonsingular transformation $\bar{x}=Px$ transforms eq. (3.7.34) into

$$\begin{cases}\begin{bmatrix}\dot{\bar{x}}_c \\ \dot{\bar{x}}_{\bar{c}}\end{bmatrix}=\begin{bmatrix}\bar{A}_c & \bar{A}_{12} \\ 0 & \bar{A}_{\bar{c}}\end{bmatrix}\begin{bmatrix}\bar{x}_c \\ \bar{x}_{\bar{c}}\end{bmatrix}+\begin{bmatrix}\bar{B}_c \\ 0\end{bmatrix}u \\ y=[\bar{C}_c \quad \bar{C}_{\bar{c}}]\begin{bmatrix}\bar{x}_c \\ \bar{x}_{\bar{c}}\end{bmatrix}\end{cases} \tag{3.7.35}$$

From eq. (3.8.35) we get controllable subsystem described by

$$\begin{cases}\dot{\bar{x}}_c=\bar{A}_c\bar{x}_c+\bar{A}_{12}\bar{x}_{\bar{c}}+\bar{B}_c u \\ y_1=\bar{C}_c\bar{x}_c\end{cases} \tag{3.7.36}$$

and uncontrollable subsystem described by

$$\begin{cases} \dot{\bar{x}}_{\bar{c}} = \bar{A}_{\bar{c}} \bar{x}_{\bar{c}} \\ y_2 = \bar{C}_{\bar{c}} \bar{x}_{\bar{c}} \end{cases} \tag{3.7.37}$$

The output equation is $y = y_1 + y_2$.

2. Taking canonical decomposition according to observability to obtained subsystems

(1) Suppose controllable subsystem eq. (3.7.36) is not completely observable. Based on conclusion 3.7.4, decompose eq. (3.7.36), using the equivalence transformation $\tilde{x}_c = F_c \bar{x}_c$, into

$$\begin{cases} \dot{\tilde{x}}_c = \begin{bmatrix} \dot{\tilde{x}}_{co} \\ \dot{\tilde{x}}_{c\bar{o}} \end{bmatrix} = \begin{bmatrix} \tilde{A}_{co} & 0 \\ \tilde{A}_{21} & \tilde{A}_{c\bar{o}} \end{bmatrix} \begin{bmatrix} \tilde{x}_{co} \\ \tilde{x}_{c\bar{o}} \end{bmatrix} + \begin{bmatrix} \tilde{B}_{co} \\ \tilde{B}_{c\bar{o}} \end{bmatrix} u + F_c \bar{A}_{12} \bar{x}_{\bar{c}} \\ y_c = [\tilde{C}_{co} \quad 0] \begin{bmatrix} \tilde{x}_{co} \\ \tilde{x}_{c\bar{o}} \end{bmatrix} \end{cases} \tag{3.7.38}$$

(2) Suppose uncontrollable subsystem (3.7.37) is not completely observable. Based on conclusion 3.7.4, decompose eq. (3.7.37), using the equivalence transformation $\tilde{x}_{\bar{c}} = F_{\bar{c}} \bar{x}_{\bar{c}}$, into

$$\begin{cases} \dot{\tilde{x}}_{\bar{c}} = \begin{bmatrix} \dot{\tilde{x}}_{\bar{c}o} \\ \dot{\tilde{x}}_{\bar{c}\bar{o}} \end{bmatrix} = \begin{bmatrix} \tilde{A}_{\bar{c}o} & 0 \\ \tilde{A}_{43} & \tilde{A}_{\bar{c}\bar{o}} \end{bmatrix} \begin{bmatrix} \tilde{x}_{\bar{c}o} \\ \tilde{x}_{\bar{c}\bar{o}} \end{bmatrix} \\ y_{\bar{c}} = [\tilde{C}_{\bar{c}o} \quad 0] \begin{bmatrix} \tilde{x}_{\bar{c}o} \\ \tilde{x}_{\bar{c}\bar{o}} \end{bmatrix} \end{cases} \tag{3.7.39}$$

Substituting $\bar{x}_{\bar{c}} = F_{\bar{c}}^{-1} \tilde{x}_{\bar{c}}$ into eq. (3.7.38) and using eq. (3.7.39), yields

$$\begin{cases} \begin{bmatrix} \dot{\tilde{x}}_{co} \\ \dot{\tilde{x}}_{c\bar{o}} \\ \dot{\tilde{x}}_{\bar{c}o} \\ \dot{\tilde{x}}_{\bar{c}\bar{o}} \end{bmatrix} = \left[\begin{array}{cc|cc} \tilde{A}_{co} & 0 & \tilde{A}_{13} & 0 \\ \tilde{A}_{21} & \tilde{A}_{c\bar{o}} & \tilde{A}_{23} & \tilde{A}_{24} \\ \hline 0 & 0 & \tilde{A}_{\bar{c}o} & 0 \\ 0 & 0 & \tilde{A}_{43} & \tilde{A}_{\bar{c}\bar{o}} \end{array}\right] \begin{bmatrix} \tilde{x}_{co} \\ \tilde{x}_{c\bar{o}} \\ \tilde{x}_{\bar{c}o} \\ \tilde{x}_{\bar{c}\bar{o}} \end{bmatrix} + \left[\begin{array}{c} \tilde{B}_{co} \\ \tilde{B}_{c\bar{o}} \\ \hline 0 \\ 0 \end{array}\right] u \\ y = [\tilde{C}_{co} \quad 0 \quad \tilde{C}_{\bar{c}o} \quad 0] \tilde{x} \end{cases} \tag{3.7.40}$$

According to the definition of controllability and observability, eq. (3.7.40) can also be verified. For example, if the submatrix in the first row and fourth column of $\tilde{A}$ is not the zero matrix, then the uncontrollable and unobservable state $\tilde{x}_{\bar{c}\bar{o}}$ will affect $\dot{\tilde{x}}_{co}$, and then $\tilde{x}_{co}$. From the output equation, $\tilde{\boldsymbol{x}}_{co}$ will affect the output $\boldsymbol{y}$, which indicates that $\tilde{\boldsymbol{x}}_{c\bar{o}}$ is related to the output y, which contradicts the fact that $\tilde{\boldsymbol{x}}_{c\bar{o}}$ is the unobservable state. Thus the submatrix in the first row and fourth column of $\tilde{\boldsymbol{A}}$ must be zero matrix.

Example 3.7.3 Take canonical decomposition to the following LTI system.

$$\dot{\boldsymbol{x}} = \begin{bmatrix} 1 & 0 & 1 \\ 0 & 2 & 0 \\ 0 & 0 & 3 \end{bmatrix} \boldsymbol{x} + \begin{bmatrix} 1 \\ 1 \\ 0 \end{bmatrix} \boldsymbol{u}$$

$$y = [1 \quad 0 \quad 1] \boldsymbol{x}$$

Answer (1) Taking canonical decomposition according to controllability. Because

rank $[\boldsymbol{b} \ \vdots \ \boldsymbol{Ab} \ \vdots \ \boldsymbol{A}^2 \boldsymbol{b}] = \text{rank} \begin{bmatrix} 1 & 1 & 1 \\ 1 & 2 & 4 \\ 0 & 0 & 0 \end{bmatrix} = 2 < n = 3$, the system is not completely controllable. Introduce linearly nonsingular transformation $\bar{\boldsymbol{x}} = \boldsymbol{Px}$. Using conclusion 3.7.3, we chose

$$\boldsymbol{P}^{-1} = \boldsymbol{Q} = \begin{bmatrix} 1 & 1 & 0 \\ 1 & 2 & 0 \\ 0 & 0 & 1 \end{bmatrix}, \text{ then } \boldsymbol{P} = \begin{bmatrix} 2 & -1 & 0 \\ -1 & 1 & 0 \\ 0 & 0 & 1 \end{bmatrix}$$

$$\bar{\boldsymbol{A}} = \boldsymbol{PAP}^{-1} = \begin{bmatrix} 2 & -1 & 0 \\ -1 & 1 & 0 \\ 0 & 0 & 1 \end{bmatrix} \begin{bmatrix} 1 & 0 & 1 \\ 0 & 2 & 0 \\ 0 & 0 & 3 \end{bmatrix} \begin{bmatrix} 1 & 1 & 0 \\ 1 & 2 & 0 \\ 0 & 0 & 1 \end{bmatrix} = \begin{bmatrix} 0 & -2 & 2 \\ 1 & 3 & -1 \\ 0 & 0 & 3 \end{bmatrix}$$

$$\bar{\boldsymbol{B}} = \boldsymbol{PB} = \begin{bmatrix} 2 & -1 & 0 \\ -1 & 1 & 0 \\ 0 & 0 & 1 \end{bmatrix} \begin{bmatrix} 1 \\ 1 \\ 0 \end{bmatrix} = \begin{bmatrix} 1 \\ 0 \\ 0 \end{bmatrix}$$

$$\bar{\boldsymbol{c}} = \boldsymbol{cP}^{-1} = [1 \quad 0 \quad 1] \begin{bmatrix} 1 & 1 & 0 \\ 1 & 2 & 0 \\ 0 & 0 & 1 \end{bmatrix} = [1 \quad 1 \quad 1]$$

Obtained controllable subsystem is

$$\dot{\bar{\boldsymbol{x}}}_c = \begin{bmatrix} 0 & -2 \\ 1 & 3 \end{bmatrix} \bar{\boldsymbol{x}}_c + \begin{bmatrix} 2 \\ -1 \end{bmatrix} \bar{\boldsymbol{x}}_{\bar{c}} + \begin{bmatrix} 1 \\ 0 \end{bmatrix} \boldsymbol{u}$$

$$y_c = [1 \quad 1] \bar{\boldsymbol{x}}_c \tag{3.7.41}$$

where

$$\bar{\boldsymbol{A}}_c = \begin{bmatrix} 0 & -2 \\ 1 & 3 \end{bmatrix}, \quad \bar{\boldsymbol{B}}_c = \begin{bmatrix} 1 \\ 0 \end{bmatrix}, \quad \bar{\boldsymbol{A}}_{12} = \begin{bmatrix} 2 \\ -1 \end{bmatrix}, \quad \bar{\boldsymbol{C}}_c = [1 \quad 1]$$

Obtained uncontrollable subsystem is

$$\dot{\bar{x}}_{\bar{c}} = 3\bar{x}_{\bar{c}}$$

$$\bar{y}_{\bar{c}} = \bar{x}_{\bar{c}} \tag{3.7.42}$$

(2) Decompose subsystem eq. (3.7.41) according to controllbility. Because

$$\text{rank}\begin{bmatrix} \bar{c}_c \\ \bar{c}_c \bar{A}_c \end{bmatrix} = \text{rank}\begin{bmatrix} 1 & 1 \\ 1 & 1 \end{bmatrix} = 1 < 2$$

Subsystem eq. (3. 7. 41) is not completely controllable. Introduce linearly nonsingular transformation $\tilde{x}_c = F\bar{x}_c$. Using conclusion 3.7.4, we chose

$$F = \begin{bmatrix} 1 & 1 \\ 0 & 1 \end{bmatrix}, \text{ then } F^{-1} = \begin{bmatrix} 1 & -1 \\ 0 & 1 \end{bmatrix}$$

$$\tilde{A}_c = F\bar{A}_c F^{-1} = \begin{bmatrix} 1 & 1 \\ 0 & 1 \end{bmatrix}\begin{bmatrix} 0 & -2 \\ 1 & 3 \end{bmatrix}\begin{bmatrix} 1 & -1 \\ 0 & 1 \end{bmatrix} = \begin{bmatrix} 1 & 0 \\ 1 & 2 \end{bmatrix}$$

$$\tilde{B}_c = F\bar{B}_c = \begin{bmatrix} 1 & 1 \\ 0 & 1 \end{bmatrix}\begin{bmatrix} 1 \\ 0 \end{bmatrix} = \begin{bmatrix} 1 \\ 0 \end{bmatrix}$$

$$\tilde{C}_c = \bar{C}_c F^{-1} = [1 \quad 1]\begin{bmatrix} 1 & -1 \\ 0 & 1 \end{bmatrix} = [1 \quad 0]$$

$$F\bar{A}_{12} = \begin{bmatrix} 1 & 1 \\ 0 & 1 \end{bmatrix}\begin{bmatrix} 2 \\ -1 \end{bmatrix} = \begin{bmatrix} 1 \\ -1 \end{bmatrix}$$

Rewrite the result as

$$\dot{\tilde{x}}_c = \begin{bmatrix} 1 & 0 \\ 1 & 2 \end{bmatrix}\tilde{x}_c + \begin{bmatrix} 1 \\ -1 \end{bmatrix}\bar{x}_{\bar{c}} + \begin{bmatrix} 1 \\ 0 \end{bmatrix}u$$

$$\tilde{y}_c = [1 \quad 0]\tilde{x}_c \tag{3.7.43}$$

From eq. (3.7.42) and eq. (3.7.43), the final decomposed equation is described by

$$\begin{cases} \begin{bmatrix} \dot{\tilde{x}}_{co} \\ \dot{\tilde{x}}_{c\bar{o}} \\ \dot{\tilde{x}}_{\bar{c}o} \end{bmatrix} = \left[\begin{array}{cc:c} 1 & 0 & 1 \\ 1 & 2 & -1 \\ \hdashline 0 & 0 & 3 \end{array}\right]\begin{bmatrix} \tilde{x}_{co} \\ \tilde{x}_{c\bar{o}} \\ \tilde{x}_{\bar{c}o} \end{bmatrix} + \left[\begin{array}{c} 1 \\ 0 \\ \hdashline 0 \end{array}\right]u \\ y = \left[\begin{array}{cc:c} 1 & 0 & 1 \end{array}\right]\begin{bmatrix} \tilde{x}_{co} \\ \tilde{x}_{c\bar{o}} \\ \tilde{x}_{\bar{c}o} \end{bmatrix} \end{cases}$$

From the above discussion about the canonical decomposition of systems, an important conclusion about the difference between the internal and external descriptions of systems can be obtained.

Conclusion 3. 7. 5 Considering the not completely controllable and not completely observable LTI system eq. (3. 7. 34), its input-output description (such as transfer function matrix) can only reflect the controllable and observable part of the system, i. e.

$$G(s) = C(sI - A)^{-1}B = \widetilde{C}_{co}(sI - \widetilde{A}_{co})^{-1}\widetilde{B}_{co} \tag{3.7.44}$$

Proof According to the discussion in Section 1.5, linear non-singular transformation does not change the transfer function matrix of systems, so the transfer function matrix $G(s)$ of the system eq. (3.7.34) can be obtained by eq. (3.7.40), i. e.

$$G(s) = [\widetilde{C}_{co} \quad 0 \quad \widetilde{C}_{\bar{c}o} \quad 0]\,[sI - \widetilde{A}]^{-1}\begin{bmatrix} \widetilde{B}_{co} \\ \widetilde{B}_{c\bar{o}} \\ 0 \\ 0 \end{bmatrix} \tag{3.7.45}$$

The inverse of a partitioned matrix has the following formula:

$$\begin{bmatrix} A_1 & 0 \\ A_2 & A_3 \end{bmatrix}^{-1} = \begin{bmatrix} A_1^{-1} & 0 \\ -A_3^{-1}A_2A_1^{-1} & A_3^{-1} \end{bmatrix}, \quad \begin{bmatrix} A_1 & A_2 \\ 0 & A_3 \end{bmatrix}^{-1} = \begin{bmatrix} A_1^{-1} & A_1^{-1}A_2A_3^{-1} \\ 0 & A_3^{-1} \end{bmatrix} \tag{3.7.46}$$

From eq. (3.7.40) and eq. (3.7.46), we have

$$[sI - \widetilde{A}]^{-1} = \left[\begin{array}{cc:cc} sI - \widetilde{A}_{co} & 0 & -\widetilde{A}_{13} & 0 \\ -\widetilde{A}_{21} & sI - \widetilde{A}_{c\bar{o}} & -\widetilde{A}_{23} & -\widetilde{A}_{24} \\ \hdashline 0 & 0 & sI - \widetilde{A}_{\bar{c}o} & 0 \\ 0 & 0 & -\widetilde{A}_{43} & sI - \widetilde{A}_{\bar{c}\bar{o}} \end{array}\right]^{-1}$$

$$= \left[\begin{array}{c:c} \begin{bmatrix} sI - \widetilde{A}_{co} & 0 \\ -\widetilde{A}_{21} & sI - \widetilde{A}_{co} \end{bmatrix}^{-1} & * \\ \hline 0 & * \end{array}\right]$$

$$= \left[\begin{array}{cc:c} (sI - \widetilde{A}_{co})^{-1} & 0 & * \\ * & * & \\ \hdashline \multicolumn{2}{c:}{0} & * \end{array}\right]$$

Substituting the above equation into eq. (3.7.45), yields

$$G(s) = [\widetilde{C}_{co} \quad 0 \quad \widetilde{C}_{\bar{c}o} \quad 0]\begin{bmatrix} (sI - \widetilde{A}_{co})^{-1}\widetilde{B}_{co} \\ * \\ 0 \\ 0 \end{bmatrix} = \widetilde{C}_{co}(sI - \widetilde{A}_{co})^{-1}\widetilde{B}_{co}$$

It indicates that the input-output description of systems, such as the transfer function matrix, is an incomplete description of systems, which only describes the part of the system that is controllable and observable. If the system is controllable and observable, the input-output description is equivalent to the state-space description.

Problems

3.1 Check the controllability of the following LTI systems

(1) $\dot{x} = \begin{bmatrix} 0 & 1 & 0 \\ 0 & 0 & 1 \\ -2 & -4 & -3 \end{bmatrix} x + \begin{bmatrix} 1 & 0 \\ 0 & 1 \\ -1 & 1 \end{bmatrix} u$

(2) $\dot{x} = \begin{bmatrix} 0 & 4 & 3 \\ 0 & 20 & 21 \\ 0 & -25 & -20 \end{bmatrix} x + \begin{bmatrix} -1 \\ 3 \\ 0 \end{bmatrix} u$

(3) $\dot{x} = \begin{bmatrix} 2 & 0 & 0 & 0 \\ 0 & 3 & 0 & 0 \\ 0 & 0 & 4 & 1 \\ 0 & 0 & 0 & 4 \end{bmatrix} x + \begin{bmatrix} 2 & 0 \\ 4 & 1 \\ 0 & 0 \\ 1 & 0 \end{bmatrix} u$

(4) $\dot{x} = \begin{bmatrix} 4 & 1 & 0 & 0 \\ 0 & 4 & 0 & 0 \\ 0 & 0 & 4 & 1 \\ 0 & 0 & 0 & 4 \end{bmatrix} x + \begin{bmatrix} 0 & 0 \\ 1 & 2 \\ 0 & 0 \\ 2 & 1 \end{bmatrix} u$

(5) $\dot{x} = \begin{bmatrix} -1 & 0 & 0 \\ 0 & -1 & 0 \\ 0 & 0 & -1 \end{bmatrix} x + \begin{bmatrix} 1 & 1 & 1 \\ 2 & 2 & 2 \\ 1 & 2 & 3 \end{bmatrix} u$

3.2 Determine the range of value of unknown parameters such that the following systems controllable.

(1) $\dot{x} = \begin{bmatrix} -2 & 0 & 0 \\ 0 & -2 & 0 \\ 0 & 0 & -2 \end{bmatrix} x + \begin{bmatrix} a & 1 \\ 2 & 4 \\ b & 1 \end{bmatrix} u$

(2) $\dot{x} = \begin{bmatrix} 0 & a \\ b & c \end{bmatrix} x + \begin{bmatrix} 1 \\ 0 \end{bmatrix} u$

3.3 Check the observability of the following LTI systems:

(1) $\dot{x} = \begin{bmatrix} 0 & 1 & 0 \\ 0 & 0 & 1 \\ -2 & -4 & -3 \end{bmatrix} x, y = [1 \quad 4 \quad 2] x$

(2) $\dot{x} = \begin{bmatrix} -2 & 1 & 0 \\ 0 & -2 & 0 \\ 0 & 0 & -2 \end{bmatrix} x, y = \begin{bmatrix} 1 & 0 & 4 \\ 2 & 0 & 8 \end{bmatrix} x$

(3) $\dot{\boldsymbol{x}} = \begin{bmatrix} 1 & 3 & 2 \\ 1 & 4 & 6 \\ 2 & 1 & 7 \end{bmatrix}\boldsymbol{x}, \boldsymbol{y} = \begin{bmatrix} 1 & 0 & 0 \\ 2 & 1 & 0 \end{bmatrix}\boldsymbol{x}$

3.4 Determine the range of value of unknown parameters such that the following systems observable.

(1) $\dot{\boldsymbol{x}} = \begin{bmatrix} a & b \\ c & 0 \end{bmatrix}\boldsymbol{x}, \boldsymbol{y} = [1 \quad 0]\boldsymbol{x}$

(2) $\dot{\boldsymbol{x}} = \begin{bmatrix} -2 & 0 & 0 \\ 1 & -2 & 0 \\ 0 & 0 & -2 \end{bmatrix}\boldsymbol{x}, \boldsymbol{y} = \begin{bmatrix} 1 & a & b \\ 4 & 0 & 4 \end{bmatrix}\boldsymbol{x}$

3.5 Determine the range of value of unknown parameters such that the following systems controllable and observable.

(1) $\dot{\boldsymbol{x}} = \begin{bmatrix} -1 & 1 & a \\ 0 & -2 & 1 \\ 0 & 0 & -3 \end{bmatrix}\boldsymbol{x} + \begin{bmatrix} 0 \\ 0 \\ 1 \end{bmatrix}\boldsymbol{u}$

$\boldsymbol{y} = [0 \quad 0 \quad 1]\boldsymbol{x}$

(2) $\dot{\boldsymbol{x}} = \begin{bmatrix} 0 & 0 & 1 \\ 0 & 1 & 0 \\ -2 & -3 & -5 \end{bmatrix}\boldsymbol{x} + \begin{bmatrix} 0 \\ 1 \\ a \end{bmatrix}\boldsymbol{u}$

$\boldsymbol{y} = [0 \quad 1 \quad b]\boldsymbol{x}$

3.6 Compute the controllability index and the observability index.

$\dot{\boldsymbol{x}} = \begin{bmatrix} 0 & 1 & 0 \\ 0 & 0 & 1 \\ 0 & 3 & -1 \end{bmatrix}\boldsymbol{x} + \begin{bmatrix} 0 & 1 \\ 1 & 0 \\ 0 & 0 \end{bmatrix}\boldsymbol{u}$

$\boldsymbol{y} = \begin{bmatrix} 1 & 0 & 1 \\ 0 & 1 & 0 \end{bmatrix}\boldsymbol{x}$

3.7 For a LTV system $\dot{\boldsymbol{x}} = \boldsymbol{A}(t)\boldsymbol{x} + \boldsymbol{B}(t)\boldsymbol{u}$, if it is controllable at t_0, check whether the system is controllable at t_1 or t_2, where $t_1 > t_0$ and $t_2 < t_0$.

3.8 Check the controllability of the following LTV systems.

(1) $\dot{\boldsymbol{x}} = \begin{bmatrix} 0 & 1 \\ 0 & t \end{bmatrix}\boldsymbol{x} + \begin{bmatrix} 0 \\ 1 \end{bmatrix}\boldsymbol{u}, t \geqslant 0$

(2) $\dot{\boldsymbol{x}} = \begin{bmatrix} 0 & 0 \\ 0 & 1 \end{bmatrix}\boldsymbol{x} + \begin{bmatrix} 1 \\ \mathrm{e}^{-2t} \end{bmatrix}\boldsymbol{u}, t \geqslant 0$

(3) $\dot{\boldsymbol{x}} = \begin{bmatrix} t & 1 & 0 \\ 0 & t & 0 \\ 0 & 0 & t^2 \end{bmatrix}\boldsymbol{x} + \begin{bmatrix} 0 \\ 1 \\ 1 \end{bmatrix}\boldsymbol{u}, t \in [0 \quad 2]$

3.9 Consider the parallel connection system shown in Figure 3. 1. Try to prove the necessary condition for the parallel connection system to be controllable (observable) is that both subsystems $\Sigma 1$ and $\Sigma 2$ are controllable (observable).

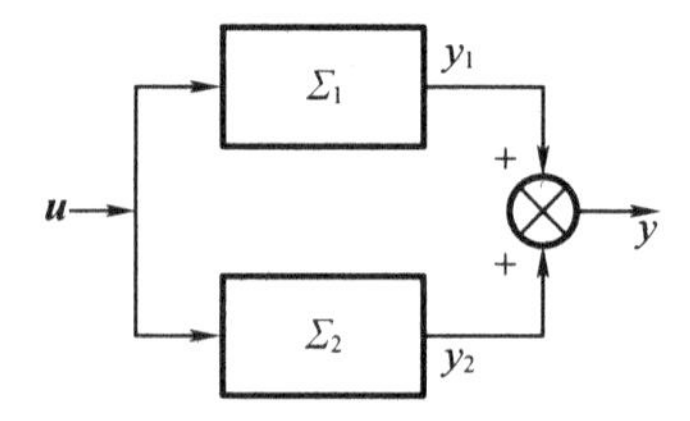

Figure P3.1

3.10 Consider the following LTI system

$$\dot{\boldsymbol{x}} = \begin{bmatrix} -1 & -2 & -2 \\ 0 & -1 & 1 \\ 1 & 0 & 1 \end{bmatrix} \boldsymbol{x} + \begin{bmatrix} 2 \\ 0 \\ 1 \end{bmatrix} \boldsymbol{u}$$

$$y = [1 \quad 1 \quad 0] \boldsymbol{x}$$

(1) Determine its controllable canonical form and corresponding transformation matrix.

(2) Determine its observable canonical form and corresponding transformation matrix.

3.11 Consider a single input LTI system described by

$$\dot{\boldsymbol{x}} = \boldsymbol{A}\boldsymbol{x} + \boldsymbol{b}\boldsymbol{u}$$

where $\boldsymbol{A}$ is $n \times n$ constant matrix, $\boldsymbol{b}$ is $n \times 1$ constant matrix. Now take nonsingular transformation $\bar{\boldsymbol{x}} = \boldsymbol{P}\boldsymbol{x}$, where $\boldsymbol{P}^{-1} = [\boldsymbol{b} \quad \boldsymbol{A}\boldsymbol{b} \quad \cdots \quad \boldsymbol{A}^{n-1}\boldsymbol{b}]$. Try to determine the state equation after transformation, and demonstrate whether the system is still completely controllable after transformation.

3.12 Given a controllable LTI system.

$$\dot{\boldsymbol{x}} = \begin{bmatrix} 1 & 0 & 1 \\ 0 & 1 & 0 \\ 1 & 1 & 0 \end{bmatrix} \boldsymbol{x} + \begin{bmatrix} 1 & 0 \\ 0 & 1 \\ 1 & 0 \end{bmatrix} \boldsymbol{u}$$

Compute its controllable canonical form.

3.13 Take canonical decomposition of the following system according to the controllability.

$$\dot{\boldsymbol{x}} = \begin{bmatrix} -1 & 1 \\ 0 & 0 \end{bmatrix} \boldsymbol{x} + \begin{bmatrix} 1 \\ 1 \end{bmatrix} \boldsymbol{u}$$

3.14 Determine the controllable and observable subsystem of

$$\dot{x} = \begin{bmatrix} \lambda_1 & 1 & 0 & 0 & 0 \\ 0 & \lambda_1 & 1 & 0 & 0 \\ 0 & 0 & \lambda_1 & 0 & 0 \\ 0 & 0 & 0 & \lambda_2 & 1 \\ 0 & 0 & 0 & 0 & \lambda_2 \end{bmatrix} \boldsymbol{x} + \begin{bmatrix} 0 \\ 1 \\ 0 \\ 0 \\ 1 \end{bmatrix} \boldsymbol{u}$$

$$y = [0 \quad 1 \quad 1 \quad 0 \quad 1]\boldsymbol{x}$$

where, $\lambda_1 \neq \lambda_2$.

3.15 Consider a SISO LTI system described by

$$\dot{x} = \boldsymbol{A}\boldsymbol{x} + \boldsymbol{b}\boldsymbol{u}, \quad y = \boldsymbol{c}\boldsymbol{x} + \mathrm{d}u$$

where, $\boldsymbol{A}$, $\boldsymbol{b}$ and $\boldsymbol{c}$ are nonzero constant matrix, $\dim(\boldsymbol{A}) = n$, and there exist

$$\boldsymbol{c}\boldsymbol{b} = 0, \quad \boldsymbol{c}\boldsymbol{A}\boldsymbol{b} = 0, \quad \cdots, \quad \boldsymbol{c}\boldsymbol{A}^{n-1}\boldsymbol{b} = 0$$

Try to check whether the system is controllable and observable.

3.16 Compute transfer function $g(s)$ of the system in problem 3.15.

3.17 Consider a SISO LTI system described by

$$\dot{x} = \boldsymbol{A}\boldsymbol{x} + \boldsymbol{b}\boldsymbol{u}, \quad y = \boldsymbol{c}\boldsymbol{x}$$

If $\{\boldsymbol{A}, \boldsymbol{b}\}$ is controllable, check whether there exists $\boldsymbol{c}$ that makes the $\{\boldsymbol{A}, \boldsymbol{c}\}$ observable. Please demonstrate and give an example to support your argument.

Chapter 4 Stability

Stability is an important characteristic of systems. In general, stability is first analyzed. An unstable system is useless in practice and asymptotic properties are the basic requirement for all practical systems. There are two ways to describe the stability of systems. One is input-output stability, which describes the stability of a system through its input-output relationship. The other method is internal stability, which describes the stability of systems by the state motion response under zero-input. In this chapter, we mainly study the internal stability of linear systems and the Lyapunov stability criterion which is widely used. In 1892, a Russian mathematician, Lyapunov gave a precise definition of the motion stability of systems through the stability of equilibrium points in his doctoral thesis "The General Problem of Stability of Motion", and proposed a general theory to solve the stability problem. Due to the difficulties in calculation and application, the backwardness of computing tools, and the language barrier, Lyapunov stability theory was not paid attention to by the west and even the whole world until about half a century later. In the following half-century, the influence of Lyapunov theory on applied mathematics, mechanics, systems theory, and many other disciplines proved that it is the basis of modern stability theory. In this chapter we will discuss the stability of linear and nonlinear systems, time-invariant and time-varying systems, continuous and discrete systems.

4.1 Input-output stability and internal stability

4.1.1 Input-output stability

Definition For a causal system, under zero initial conditions, if every bounded input p-dimensional $\boldsymbol{u}(t)$, i. e.

$$\|\boldsymbol{u}(t)\| \leqslant k_1 < \infty \quad \forall t \in [t_0, \infty) \tag{4.1.1}$$

excites a bounded output q-dimensional output $\boldsymbol{y}(t)$, i. e.

$$\|\boldsymbol{y}(t)\| \leqslant k_2 < \infty \quad \forall t \in [t_0, \infty) \tag{4.1.2}$$

then the system is said to be input-output stable (i. e. BIBO stable-bounded input bounded output stable).

It should be noted that the initial condition of systems must be assumed to be zero when discussing input-output stability, because only under this assumption the input-output relationship of the system can be unique and meaningful.

Conclusion 4.1.1 Input-output stability criterion of LTV systems

A MIMO causal LTV system is BIBO stable if and only if each element, $g_{ij}(t,\tau)$ $(i=1,2,\cdots,q;j=1,2,\cdots,p)$, of its impulse response matrix $\boldsymbol{G}(t,\tau)$, is absolutely integrable in $t\in[t_0,t]$, or

$$\int_{t_0}^{T}|g_{ij}(t,\tau)|\mathrm{d}\tau \leqslant k < \infty \tag{4.1.3}$$

Proof Use two steps to prove it.

Step 1. Consider the situation when $p=q=1$, i.e. it has single input and single output.

First prove sufficiency. If

$$\int_{t_0}^{T}|g(t,\tau)|\mathrm{d}\tau \leqslant k < \infty \tag{4.1.4}$$

and any input $u(t)$ which satisfies $|u(t)|\leqslant k_1<\infty$, $t\in[t_0,\infty)$, then from eq. (1.1.12), we get

$$\begin{aligned}|y(t)| &= \left|\int_{-\infty}^{+\infty}g(t,\tau)u(\tau)\mathrm{d}\tau\right| \\ &\leqslant \int_{-\infty}^{+\infty}|g(t,\tau)||u(\tau)|\mathrm{d}\tau \\ &\leqslant k_1\int_{-\infty}^{+\infty}|g(t,\tau)|\mathrm{d}\tau \leqslant k_1 k \leqslant \infty\end{aligned}$$

Thus according to the definition of input-output stability, the system is BIBO stable.

Next prove necessity. If the system is BIBO stable, prove eq. (4.1.4) is correct.

Adopt proof by contradiction to prove it. We suppose there exists a $t_1\in[t_0,\infty)$ such that

$$\int_{t_0}^{t_1}|g(t,\tau)|\mathrm{d}\tau = \infty$$

Chose a bounded input $u(t)$ as

$$u(t)=\operatorname{sgn} g(t_1,t)=\begin{cases}1, \text{ when } g(t_1,t)>0\\ 0, \text{ when } g(t_1,t)=0\\ -1, \text{ when } g(t_1,t)<0\end{cases} \tag{4.1.5}$$

Then the response excited by the $u(t)$ at t_1 is

$$y(t_1) = \int_{t_0}^{t_1}g(t,\tau)u(\tau)\mathrm{d}\tau = \int_{t_0}^{t_1}|g(t,\tau)|\mathrm{d}\tau = \infty \tag{4.1.6}$$

The eq. (4.1.6) indicates $y(t)$ is unlimited that contradicts the system is BIBO stable. Thus the assumption $\int_{t_0}^{t_1}|g(t,\tau)|\mathrm{d}\tau = \infty$ is not correct. So if the system is BIBO stable, eq. (4.14) holds.

Step 2. Consider the multiple-input and multiple-output situation. Each element $y_i(t)$ of output $y(t)$ satisfies the following equation

$$\begin{aligned}|y_i(t)| &= \left|\int_{t_0}^{t}g_{i1}(t,\tau)u_1(\tau)\mathrm{d}\tau + \int_{t_0}^{t}g_{i2}(t,\tau)u_2(\tau)\mathrm{d}\tau + \cdots + \int_{t_0}^{t}g_{ip}(t,\tau)u_p(\tau)\mathrm{d}\tau\right| \\ &\leqslant \left|\int_{t_0}^{t}g_{i1}(t,\tau)u_p(\tau)\mathrm{d}\tau\right| + \left|\int_{t_0}^{t}g_{i2}(t,\tau)u_p(\tau)\mathrm{d}\tau\right| + \cdots +\end{aligned}$$

$$\left|\int_{t_0}^{t} g_{ip}(t,\tau)u_p(\tau)\mathrm{d}\tau\right| \quad i=1,2,\cdots,q$$

Using the conclusion for single-input and single output system, we can prove the situation of the MIMO system.

From the conclusion 4.1.1, we get input-output stability criterion of LTI systems.

Conclusion 4.1.2 Input-output stability criterion of LTI systems.

A MIMO LTI system is BIBO stable if and only if each element, $g_{ij}(t)(i=1,2,\cdots,q;j=1,2,\cdots,p)$, of its impulse response matrix $\boldsymbol{G}(t)$ satisfies the following equation:

$$\int_0^{\infty} |g_{ij}(t)|\mathrm{d}t \leqslant k < \infty \tag{4.1.7}$$

or all poles of every non-zero element $\hat{g}_{ij}(s)$ of $\hat{\boldsymbol{G}}(s)$ have negative real part, where $\hat{\boldsymbol{G}}(s)$ is proper or strictly proper.

Proof From conclusion 4.1.1, we get the first part of the conclusion. Next we prove the second part. When $\hat{g}_{ij}(s)$ is a proper or strictly proper rational fraction, it can be expanded to the sum of partial fractions as follows:

$$\hat{g}_{ij}(s) = \sum_{l=1}^{m} \frac{\beta_l}{(s-\lambda_l)^{\alpha_l}} + \mathrm{d}_{ij} \tag{4.1.8}$$

where α_l, β_l and d_{ij} are zero or nonzero constant, λ_l is the lth pole of $\hat{g}_{ij}(s)$. Applying inverse Laplace transform to $\hat{g}_{ij}(s)$, yields

$$g_{ij}(t) = L^{-1}(\hat{g}_{ij}(s)) = \sum_{l=1}^{m} \frac{\beta_l}{(\alpha_l-1)!} t^{\alpha_l-1}\mathrm{e}^{\lambda_l t} + \mathrm{d}_{ij}\delta(t) \tag{4.1.9}$$

If λ_l has negative real part, let $\lambda_l = -a_l + b_l j, a_l > 0$, then substitute $\mathrm{e}^{\lambda_l t} = \mathrm{e}^{(-a_l+b_l j)t}$ into eq. (4.1.9), we have

$$\begin{aligned} g_{i,j}(t) &= \sum_{l=1}^{m} \frac{\beta_l}{(\alpha_l-1)!} t^{\alpha_l-1}\mathrm{e}^{(-a_l+b_l j)t} \\ &= \sum_{l=1}^{m} \frac{\beta_l}{(\alpha_l-1)!}\frac{t^{\alpha_l-1}}{\mathrm{e}^{a_l t}}\mathrm{e}^{b_l j t} + \mathrm{d}_{ij}\delta(t) \\ &= \sum_{l=1}^{m} \frac{\beta_l}{(\alpha_l-1)!}\frac{t^{\alpha_l-1}}{\mathrm{e}^{a_l t}}(\cos b_l t + j\sin b_l t) + \mathrm{d}_{ij}\delta(t) \end{aligned} \tag{4.1.10}$$

where, $(-\sin b_l t + j\cos b_l t)$ is limited, and $\lim_{t\to\infty}\frac{t^{\alpha_l-1}}{\mathrm{e}^{a_l t}}=0$. Thus we get,

$$\lim_{t\to\infty}\sum_{l=1}^{m} \frac{\beta_l}{(\alpha_l-1)!}\frac{t^{\alpha_l-1}}{\mathrm{e}^{a_l t}}(\cos b_l t + j\sin b_l t) = 0 \tag{4.1.11}$$

From eq. (4.1.11), we get eq. (4.1.7).

4.1.2 Internal stability

Definition Consider a linear system described by

$$\begin{cases} \dot{\boldsymbol{x}} = \boldsymbol{A}(t)\boldsymbol{x} + \boldsymbol{B}(t)\boldsymbol{u}, \ \boldsymbol{x}(t_0) = \boldsymbol{x}_0, \ t \in [t_0, t_\alpha] \\ \boldsymbol{y} = \boldsymbol{C}(t)\boldsymbol{x} + \boldsymbol{D}(t)\boldsymbol{u} \end{cases}$$

If every finite initial state $\boldsymbol{x}_0$ excites zero-input response $\boldsymbol{\Phi}(t;0,\boldsymbol{x}_0,\boldsymbol{0})$ which satisfies

$$\lim_{t\to\infty}\boldsymbol{\Phi}(t;0,\boldsymbol{x}_0,\boldsymbol{0})=\boldsymbol{0} \tag{4.1.12}$$

then the system is said to be internal stable or a symptotically stable.

Internal stability refers to the stability of the free movement of the system state, namely the asymptotic stability in the sense of Lyapunov, which is determined by the structure and parameters of systems. The relevant content of stability in the sense of Lyapunov will be discussed in the following sections.

4.1.3 The relationship between input-output stability and internal stability of LTI systems

Conclusion 4.1.3 If a LTI system

$$\begin{cases}\dot{\boldsymbol{x}}=\boldsymbol{A}\boldsymbol{x}+\boldsymbol{B}\boldsymbol{u},\ \boldsymbol{x}(0)=\boldsymbol{x}_0,\ t\geqslant 0\\ \boldsymbol{y}=\boldsymbol{C}\boldsymbol{x}+\boldsymbol{D}\boldsymbol{u}\end{cases} \tag{4.1.13}$$

is internal stable, then it must be BIBO stable.

Proof The transfer matrix of the system eq. (4.1.13) is

$$\boldsymbol{G}(s)=\boldsymbol{C}(s\boldsymbol{I}-\boldsymbol{A})^{-1}\boldsymbol{B}+\boldsymbol{D}$$

Applying inverse Laplace transform to the above equation, we get the impulse response matrix of the system as

$$\begin{aligned}\boldsymbol{G}(t)&=L^{-1}(\boldsymbol{G}(s))=L^{-1}(\boldsymbol{C}(s\boldsymbol{I}-\boldsymbol{A})-1\boldsymbol{B}+\boldsymbol{D})\\&=\boldsymbol{C}L^{-1}((s\boldsymbol{I}-\boldsymbol{A})^{-1})\boldsymbol{B}+L^{-1}(\boldsymbol{D})\\&=\boldsymbol{C}\mathrm{e}^{\boldsymbol{A}t}\boldsymbol{B}+\boldsymbol{D}\delta(t)\end{aligned} \tag{4.1.14}$$

If the system is internal stable, then $\lim\limits_{t\to\infty}\mathrm{e}^{\boldsymbol{A}t}\boldsymbol{x}_0=\boldsymbol{0}$ for any initial state $\boldsymbol{x}_0$. Thus, we have

$$\lim_{t\to\infty}\mathrm{e}^{\boldsymbol{A}t}=\boldsymbol{0} \tag{4.1.15}$$

So, from eq. (4.1.14) and eq. (4.1.15), each element $g_{ij}(t)$, $(i=1,2,\cdots,q;j=1,2,\cdots,p)$ of $\boldsymbol{G}(t)$, satisfies

$$\int_0^{\infty}|g_{ij}(t)|\mathrm{d}t\leqslant k<\infty$$

Thus the system is BIBO stable.

Conclusion 4.1.4 If a LTI system eq. (4.1.13) is BIBO stable, it doesn't ensure the system is internal stable.

Proof Stability is the inherent property of systems and linearly nonsingular transformation doesn't change the system's stability. Through linearly nonsingular transformation, the system is decomposed to controllable and observable, controllable but not observable, observable but not controllable, unobservable and uncontrollable parts. The BIBO stable is equivalent to that the controllable and observable subsystem is stable, but it doesn't ensure the other three subsystems are stable also. Hence the BIBO stable doesn't ensure the system is internal stable.

Conclusion 4.1.5 For a completely controllable and completely observable LTI system, the BIBO stable is equivalent to the internal stable.

Proof From the conclusion 4.1.3, if a system is internal stable, it must be BIBO stable. And from the proof of conclusion 4.1.4, for a completely controllable and completely observable LTI system, if the system is BIBO stable, it must be internal stable.

4.2 Several concepts about stability of Lyapunov

Before discussing Lyapunov stability criterion, we first discuss some basic concepts related to it.

4.2.1 Equilibrium

When discussing the internal stability of the system $\dot{\boldsymbol{x}} = \boldsymbol{f}(\boldsymbol{x},t) + \boldsymbol{u}(\boldsymbol{x},t)$, we always let $\boldsymbol{u}(\boldsymbol{x},t) = \boldsymbol{0}$. So its state equation becomes

$$\dot{\boldsymbol{x}} = \boldsymbol{f}(\boldsymbol{x},t) \tag{4.2.1}$$

where, $\boldsymbol{x}$ is an n-dimensional state vector, $\boldsymbol{f}(\cdot,\cdot)$ is an n-dimensional vector function of state vector $\boldsymbol{x}$ and time t, which can be linear or nonlinear. The system (4.2.1) is usually referred to as an autonomous system. A point $\boldsymbol{x}_e \in R^n$ is called an equilibrium point of (4.2.1), or simply an equilibrium of (4.2.1), if

$$\dot{\boldsymbol{x}}_e = \boldsymbol{f}(\boldsymbol{x}_e,t) = \boldsymbol{0}, \quad \forall t \geqslant t_0 \tag{4.2.2}$$

Because $\dot{\boldsymbol{x}}_e = \boldsymbol{0}$, if system (4.2.1) arrives at the equilibrium $\boldsymbol{x}_e$, then the system will stay at the equilibrium forever, unless there exists some external force to make the system leave from the equilibrium.

In most cases, the origin of the state space is an equilibrium point of a system. In addition, the system can also have non-zero equilibrium points. For example, for a LTI system $\dot{\boldsymbol{x}} = \boldsymbol{A}\boldsymbol{x}$, when $\boldsymbol{A}$ is singular, the system has an infinite number of non-zero equilibrium points. The equilibrium point of a system is called isolated equilibrium if it is an isolated point in the state space. The isolated equilibrium can be shifted to the origin of the state space by moving the coordinate system. So, in the following discussions, we often assumed the equilibrium point to be the origin.

Example 4.2.1 Consider a angular control system described by

$$\begin{cases} \dot{x}_1(t) = x_2(t) \\ \dot{x}_2(t) = -k\sin x_1(t) - ax_2(t) \end{cases}$$

Compute its equilibria.

Answer Let

$$\dot{x}_{1e} = x_{2e} = 0$$

$$\dot{x}_{2e} = -k\sin x_{1e} = 0$$

We get the equilibria of the nonlinear time-varying system as

$$\begin{bmatrix} x_{1e} \\ x_{2e} \end{bmatrix} = \begin{bmatrix} k\pi \\ 0 \end{bmatrix}, \quad k = 0, \pm 1, \pm 2, \cdots$$

It can be seen that the system has an infinite number of equilibria, and these equilibria are isolated and discontinuous. The state trajectory of the system is shown in Figure 4.2.1.

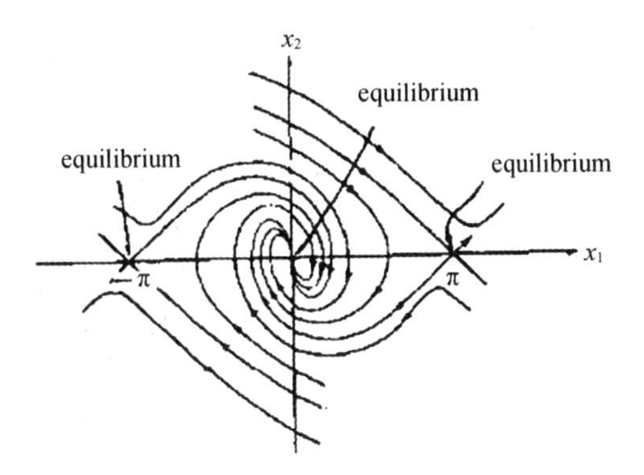

Figure 4.2.1 The state trajectory of the angular control system

4.2.2 Stability in the sense of Lyapunov

The equilibrium $\boldsymbol{x}_e$ of the system eq. (4.2.1) is said to be stable in the sense of Lyapunov (or Lyapunov stable) at t_0, if, for every $\varepsilon > 0$, there exists a $\delta(\varepsilon, t_0) > 0$, such that, if

$$\| \boldsymbol{x}_0 - \boldsymbol{x}_e \| \leqslant \delta(\varepsilon, t_0) \tag{4.2.3}$$

then zero-input response $\boldsymbol{\Phi}(t; t_0, \boldsymbol{x}_0, \boldsymbol{0})$ excited by $\boldsymbol{x}_0$ satisfies the following formula

$$\| \boldsymbol{\Phi}(t; t_0, \boldsymbol{x}_0, \boldsymbol{0}) - \boldsymbol{x}_e \| \leqslant \varepsilon, \quad \forall t \geqslant t_0 \tag{4.2.4}$$

We use $S(\varepsilon)$ to denote an n-dimensional hypersphere with the equilibrium $\boldsymbol{x}_e$ as its center and ε as its radius. $S(\delta)$ denotes an n-dimensional hypersphere with the equilibrium $\boldsymbol{x}_e$ as its center and $\delta(\varepsilon, t_0)$ as its radius. When $\boldsymbol{x}_e$ is Lyapunov stable, all trajectories start from $\boldsymbol{x}_0$ which lies inside $S(\delta)$, will lie inside $S(\varepsilon)$. Figure 4.2.2(a) and Figure 4.2.4(a) show $\boldsymbol{x}_e = 0$ is Lyapnov stable for a two-dimensional system.

In other word, Lyapunov stability of an equilibrium means that solutions starting "close enough" to the equilibrium (within a distance δ) remain "close enough" forever (within a distance ε). The stability in the sense of Lyapunov is the critical stable in engineering practice.

Below we give the definition of uniformly stability in the sense of Lyapunov.

The equilibrium $\boldsymbol{x}_e$ of the system eq. (4.2.1) is said to be uniformly stable in the sense of Lyapunov at t_0, if for every $\varepsilon > 0$, there exists a $\delta(\varepsilon) > 0$, such that, if

$$\| \boldsymbol{x}_0 - \boldsymbol{x}_e \| \leqslant \delta(\varepsilon) \tag{4.2.5}$$

then zero-input response $\boldsymbol{\Phi}(t; t_0, \boldsymbol{x}_0, \boldsymbol{0})$ excited by $\boldsymbol{x}_0$ satisfies the following formula

$$\| \boldsymbol{\Phi}(t;t_1,\boldsymbol{x}_0,\boldsymbol{0}) - \boldsymbol{x}_e \| \leqslant \varepsilon, \quad \forall t \geqslant t_1 \text{ and } t_1 \geqslant t_0 \tag{4.2.6}$$

For time-invariant systems, the uniform stability is equivalent to the stability.

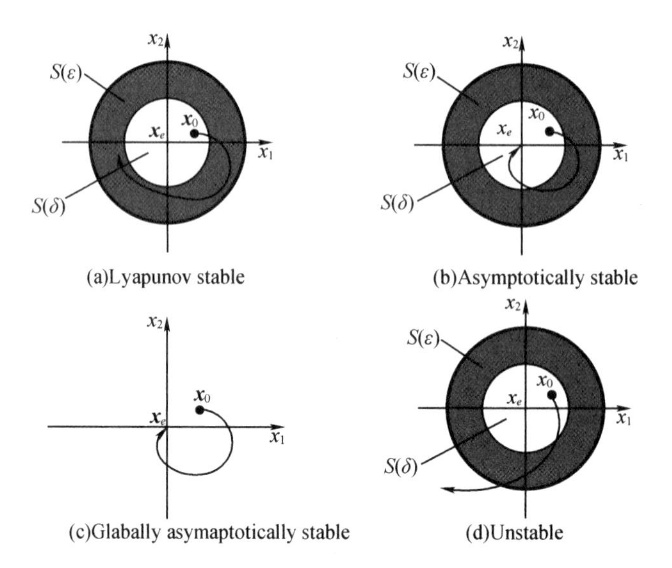

Figure 4.2.2 Geometric explanation of stability of the equilibrium

Example 4.2.2 Consider the simple gravity pendulum system shown in Figure 4.2.3. The pendulum length is l. The mass of the end ball is m. Without considering the mass and damping of the pendulum rod, determine the equilibrium and its stability.

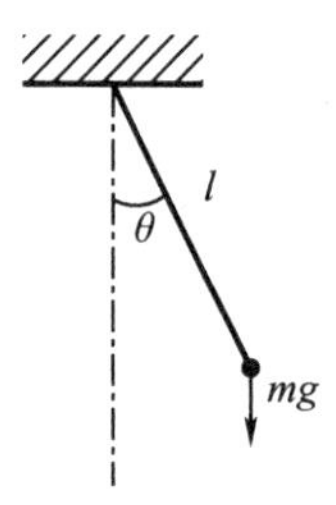

Figure 4.2.3 The simple gravity pendulum system

Answer The rod's kinetic equation of free motion without external force is

$$-mg\sin\theta = ml\ddot{\theta}$$

Let $x_1 = \theta$, $x_2 = \dot{\theta}$, the state equation is

$$\begin{cases} \dot{x}_1 = x_2 \\ \dot{x}_2 = (-g/l)\sin x_1 \end{cases}$$

Let $\dot{x}_1 = 0$ 和 $\dot{x}_2 = 0$, we get the equilibria

$$\begin{bmatrix} x_{1e} \\ x_{2e} \end{bmatrix} = \begin{bmatrix} k\pi \\ 0 \end{bmatrix}, \quad k = 0, \pm 1, \pm 2, \cdots$$

Linearizing the state equation at the equilibrium x_e, yields

$$\begin{cases}\dot{x}_1 = x_2 \\ \dot{x}_2 = (-g/l)\sin x_1 \big|_{x_1 = k\pi} + (-g/l)\dfrac{\partial \sin x_1}{\partial x_1}\Big|_{x_1 = k\pi}(x_1 - k\pi) = (-g/l)\cos x_1 \big|_{x_1 = k\pi}(x_1 - k\pi)\end{cases}$$

Let $\bar{x}_1 = x_1 - k\pi, \bar{x}_2 = x_2$, the linearized state equation of the single pendulum system is

$$\begin{cases}\dot{\bar{x}}_1 = \bar{x}_2 \\ \dot{\bar{x}}_2 = (-g/l)\bar{x}_1, \quad 当 k = 0, \pm 2, \pm 4, \cdots\end{cases} \tag{4.2.7}$$

and

$$\begin{cases}\dot{\bar{x}}_1 = \bar{x}_2 \\ \dot{\bar{x}}_2 = (g/l)\bar{x}_1, \quad 当 k = \pm 1, \pm 3, \cdots\end{cases} \tag{4.2.8}$$

Refer to the discussion in the Section 4.5, for the equilibria

$$x_e = \begin{bmatrix} k\pi \\ 0 \end{bmatrix}, \quad k = 0, \pm 2, \pm 4, \cdots$$

corresponding state equation is $\dot{\boldsymbol{x}} = \begin{bmatrix} 0 & 1 \\ -\dfrac{g}{l} & 0 \end{bmatrix}$, the eigenvalues are $\pm\sqrt{\dfrac{g}{l}}j$. So the equilibria are stable in the sense of Lyapunov.

Whereas, for the equilibria

$$\boldsymbol{x}_e = \begin{bmatrix} k\pi \\ 0 \end{bmatrix}, \quad k = \pm 1, \pm 3, \cdots$$

corresponding state equation is $\dot{\boldsymbol{x}} = \begin{bmatrix} 0 & 1 \\ \dfrac{g}{l} & 0 \end{bmatrix}$, the eigenvalues are $\pm\sqrt{\dfrac{g}{l}}$, So the equilibria are not stable in the sense of Lyapunov.

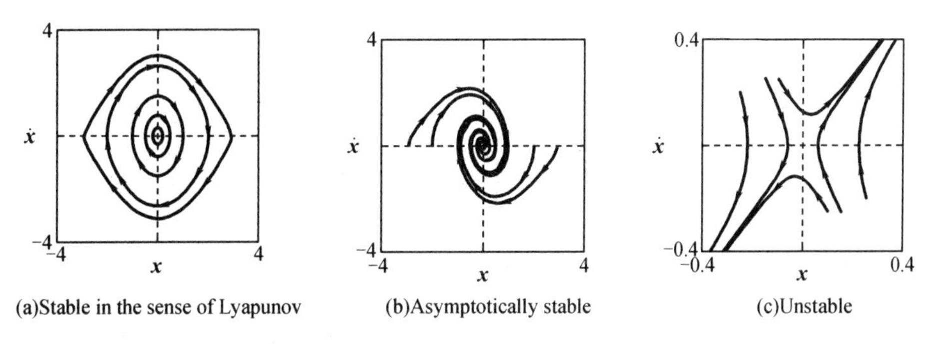

Figure 4.2.4 Phase portraits for stable and unstable equilibrium points

4.2.3 Asymptotically stability

The equilibrium $\boldsymbol{x}_e$ of the system eq. (4.2.1) is said to be asymptotically stable if

(1) It is stable in the sense of Lyapunov.

(2) For any initial point which sufficiently nears the equilibrium $\boldsymbol{x}_e$, the state response approaches $\boldsymbol{x}_e$ as $t \to \infty$.

Using mathematic language asymptotically stability can be expressed as:

There exists an $\gamma > 0$, any $\bar{\varepsilon} > 0$ and a positive T which depends on $\bar{\varepsilon}$, γ and t_0, such that, if $\| \boldsymbol{x}_0 - \boldsymbol{x}_e \| \leqslant \gamma$, then

$$\| \boldsymbol{\Phi}(t;t_0,\boldsymbol{x}_0,\boldsymbol{0}) - \boldsymbol{x}_e \| \leqslant \bar{\varepsilon}, \quad \forall t \geqslant t_0 + T(\bar{\varepsilon},\gamma,t_0) \tag{4.2.9}$$

The geometric explanation of asymptotically stability is given in Figure 4.2.2(b), Figure 4.2.4(b) and Figure 4.2.5.

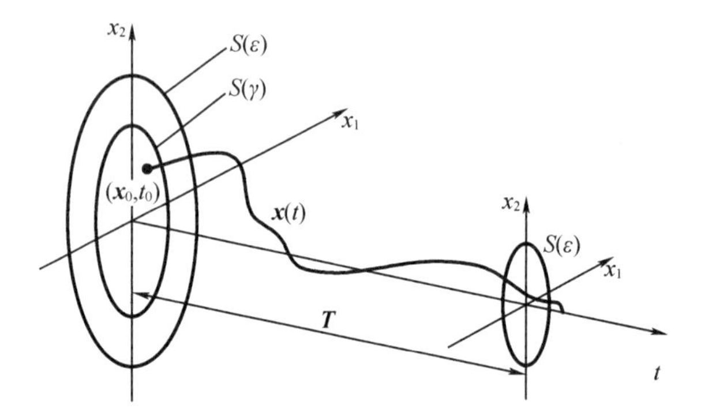

Figure 4.2.5 Explanation for the asymptotically stability

As in the previous definition, the asymptotically stability is defined at t_0. The equilibrium $\boldsymbol{x}_e$ is uniformly asymptotically stable if $\boldsymbol{x}_e$ is asymptotically stable and there exists T independent of t_0 such that eq. (4.2.9) holds for any $t \in [t_0, \infty)$.

Asymptotically stability is stability in the engineering sense, and stability in the sense of Lyapunov is critical stability in the engineering sense.

4.2.4 Globally asymptotically stability

The equilibrium $\boldsymbol{x}_e$ of the system eq. (4.2.1) is globally asymptotically stable if it is asymptotically stable for all initial conditions $\boldsymbol{x}_0 \in R^n$.

Conclusion 4.2.1 If a linear system is asymptotically stable, then it is globally asymptotically stable.

Proof If the equilibrium $\boldsymbol{x}_e = \boldsymbol{0}$ is asymptotically stable for a linear system, given any initial points $\boldsymbol{x}_0$ which sufficiently nears the equilibrium $\boldsymbol{x}_e$, $\boldsymbol{x}(t)$ satisfies:

$$\lim_{t\to\infty} \boldsymbol{x}(t) = \lim_{t\to\infty} \boldsymbol{\Phi}(t,t_0)\boldsymbol{x}_0 = \boldsymbol{x}_e = \boldsymbol{0}$$

Then the state response $\boldsymbol{x}(t)$ excited by other initial state $\boldsymbol{x}'_0(\boldsymbol{x}'_0 = a\,\boldsymbol{x}_0, a$ is non-zreo constant) satisfies

$$\lim_{t\to\infty} \boldsymbol{x}(t) = \lim_{t\to\infty} \boldsymbol{\Phi}(t,t_0)\boldsymbol{x}'_0 = \lim_{t\to\infty} \boldsymbol{\Phi}(t,t_0)a\,\boldsymbol{x}_0 = a \lim_{t\to\infty} \boldsymbol{\Phi}(t,t_0)\boldsymbol{x}_0 = \boldsymbol{0}$$

For non-zero equilibrium, we can transfer it to the origin by coordinate transformation. So If the equilibrium is asymptotically stable, then it is globally asymptotically stable for linear systems.

The geometric explanation of globally asymptotically stability is given in Figure 4.2.2(c).

4.2.5 Unstability

If for a $\varepsilon > 0$ and any $\delta > 0$, at least there exists a $\boldsymbol{x}_0$ in the region which constrained by $\| \boldsymbol{x}_0 - \boldsymbol{x}_e \| < \delta$, such that the zero-input response excited by it satisfies

$$\| \boldsymbol{x}(t) - \boldsymbol{x}_e \| > \varepsilon, \quad t \geqslant t_0 + \boldsymbol{T} \tag{4.2.10}$$

no matter δ is how small, then the system is unstable.

The geometric explanation ofunstability is given in Figure 4.2.2(d)and Figure 4.2.4(c).

4.3 Main theorems of Lyapunov's second method for stability

Lyapunov, in his original 1892 work, proposed two methods for demonstrating stability based on ordinary differential equations which are used to describe the system. The first method, also known as indirect method, through linearizing the system $\dot{\boldsymbol{x}} = \boldsymbol{f}(\boldsymbol{x},t)$ at the equilibrium point and analyzing the stability of the linearized equation, ascertains the stability of the original nonlinear system in a small range near the equilibrium point.

The second method, which is now referred to as the Lyapunov stability criterion or the Direct Method, uses a Lyapunov function $V(\boldsymbol{x})$ which has an analogy to the potential function of classical dynamics, and analyzes the positive definiteness of $V(\boldsymbol{x})$ and $\dot{V}(\boldsymbol{x})$ ($\dot{V}(\boldsymbol{x})$ is the derivative of $V(\boldsymbol{x})$ with respect to time (t)) to ascertain the stability of the system. In this section, we mainly discuss the Lyapunov's second method which has been widely used.

Consider a time-continuous nonlinear autonomous system

$$\dot{\boldsymbol{x}} = \boldsymbol{f}(\boldsymbol{x},t), \ t \geqslant t_0 \tag{4.3.1}$$

where $\boldsymbol{f}(\boldsymbol{0},t) = \boldsymbol{0}$, i.e. the origin of the state space is an equilibrium point of the system.

4.3.1 Criterion theorems of globally asymptotically stability

We first discuss the criterion theorems about globally asymptotically stability of the origin equilibrium.

Conclusion 4.3.1 The criterion theorem of globally uniformly asymptotically stability for time-varying systems. Consider the system eq. (4.3.1), $\boldsymbol{x}_e = \boldsymbol{0}$ is it's a equilibrium. If there exists a $V(\boldsymbol{x},t)$, $V(\boldsymbol{0},t) = 0$, for any non-zero solution of eq. (4.3.1), $\boldsymbol{x}(t)$, such that $V(\boldsymbol{x},t)$ is positive definite and limited, i. e. there exist nondecreasing time-continuous scalar function $\alpha(\|\boldsymbol{x}\|)$ and $\beta(\|\boldsymbol{x}\|)$, where $\alpha(0) = 0$and $\beta(0) = 0$, for any $t \geqslant t_0$, such that

$$\beta(\|\boldsymbol{x}\|) \geqslant V(\boldsymbol{x},t) \geqslant \alpha(\|\boldsymbol{x}\|) > 0 \tag{4.3.2}$$

$\dot{V}(\boldsymbol{x},t)$ is negative definite and limited, i. e. there exist nondecreasing time-continuous scalar function $\gamma(\|\boldsymbol{x}\|)$, where $\gamma(0) = 0$, for any $t \geqslant t_0$, such that

$$\dot{V}(\boldsymbol{x},t) \leqslant -\gamma(\|\boldsymbol{x}\|) < 0 \tag{4.3.3}$$

Where $\dot{V}(\boldsymbol{x},t)\big|_{\dot{x}=f(x,t)} = \dfrac{\partial V}{\partial t} + \dfrac{\partial V}{\partial x}\boldsymbol{f}(\boldsymbol{x},t)$.

(3) When $\|\boldsymbol{x}\| \to \infty$, $\alpha(\|\boldsymbol{x}\|) \to \infty$, i. e. $V(\boldsymbol{x},t) \to \infty$.

Then the origin of the system is globally uniformly asymptotically stable.

Note that conclusion 4.3.1 is a sufficient condition to ensure the equilibrium point is globally uniformly asymptotically stable, rather than a necessary condition. In other words, if the $V(\boldsymbol{x},t)$ that meets the condition in conclusion 4.3.1 cannot be found, it cannot mean the equilibrium point is not globally uniformly asymptotically stable.

Figure 4.3.1 can help us to intuitively understand conclusion 4.3.1. For a 2-order system, the state variables are x_1 and x_2. Since $V(\boldsymbol{x},t)$ is a positive definite function, the contour surfaces of $V(\boldsymbol{x},t) = c$ (c is a positive constant) form a family of closed surfaces centered on the origin. For example, $V(\boldsymbol{x},t) = x_1^2 + x_2^2 = c$. When $\dot{V}(\boldsymbol{x},t)$ is negative definition, the solution trajectory of the system (4.3.1) travels from the outside to the inside of the closed surfaces and converges to the origin when $t \to \infty$.

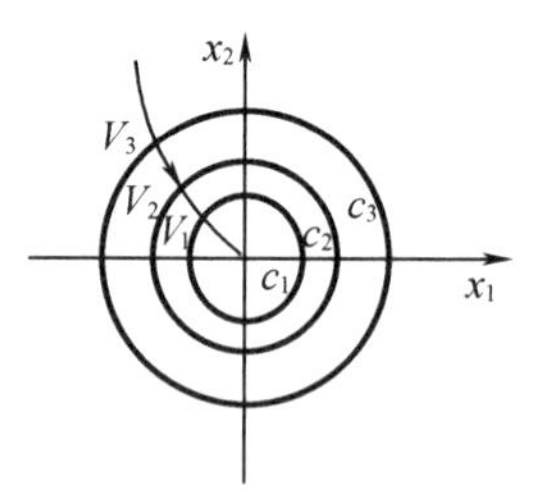

Figure 4.3.1 The intuitively understand to conclusion 4.3.1

Conclusion 4.3.2 The criterion theorem of globally uniformly asymptotically stability for time-invariant systems. Consider a time-invariant system $\dot{\boldsymbol{x}} = \boldsymbol{f}(\boldsymbol{x})$, $t \geqslant 0$ having a point of equilibrium at $\boldsymbol{x} = \boldsymbol{0}$. If there exists a scale function $V(\boldsymbol{x})$, and $V(\boldsymbol{0}) = 0$, for any non-zero solution of the system, $\boldsymbol{x}(t)$, such that

(1) $V(\boldsymbol{x})$ is negative definite, i. e. $V(\boldsymbol{x})>0$;

(2) $\dot{V}(\boldsymbol{x})$ is negative definite, i. e. $\dot{V}(\boldsymbol{x})<0$, where $\dot{V}(\boldsymbol{x}) \triangleq \frac{\mathrm{d}V(\boldsymbol{x})}{\mathrm{d}t}$;

(3) When $\|\boldsymbol{x}\| \to \infty$, $V(\boldsymbol{x}) \to \infty$.

Then $V(\boldsymbol{x})$ is called a Lyapunov function and the origin of the system is globally uniformly asymptotically stable. Note that if only the condition (1) and (2) are satisfied, then the origin of the system is uniformly asymptotically stable, not globally stable.

Example 4.3.1 Analyse the stability of the origin equilibrium of the following system

$$\begin{cases} \dot{x}_1 = x_2 - x_1(x_1^2 + x_2^2) \\ \dot{x}_2 = -x_1 - x_2(x_1^2 + x_2^2) \end{cases}$$

Answer Let $\dot{x}_1 = 0$ and $\dot{x}_2 = 0$, we get $x_1 = 0$ and $x_2 = 0$, i. e. the origin is uniquely equilibrium.

Chose $V(\boldsymbol{x}) = x_1^2 + x_2^2$, we have

(1) $V(\boldsymbol{x})>0$, $\forall \boldsymbol{x} \neq 0$;

(2) The first order derivative of $V(\boldsymbol{x})$ with respect to t is:

$$\begin{aligned} \dot{V}(x) &= \frac{\partial V(\boldsymbol{x})}{\partial x_1} \cdot \frac{\mathrm{d}x_1}{\mathrm{d}t} + \frac{\partial V(\boldsymbol{x})}{\partial x_2} \cdot \frac{\mathrm{d}x_2}{\mathrm{d}t} \\ &= 2x_1 \cdot \dot{x}_1 + 2x_2 \cdot \dot{x}_2 \\ &= -2(x_1^2 + x_2^2) \end{aligned}$$

So, $\dot{V}(\boldsymbol{x},t)<0$, $\forall \boldsymbol{x} \neq 0$.

(3) when $\|\boldsymbol{x}\| \to \infty$, $V(\boldsymbol{x}) \to \infty$.

According to the conclusion 4.3.2, the origin of the system is globally asymptotically stable.

Relaxing the restriction on the negative definition of $\dot{V}(\boldsymbol{x})$ in conclusion 4.3.2 leads to conclusion 4.3.3.

Conclusion 4.3.3 The criterion theorem of globally uniformly asymptotically stability for time-invariant systems. Considering the system $\dot{\boldsymbol{x}} = \boldsymbol{f}(\boldsymbol{x})$, if there exists a scalar function $V(\boldsymbol{x})$, *and* $\boldsymbol{V}(0)=0$, for any non-zero solution of the system, $\boldsymbol{x}(t)$, such that

(1) $\mathrm{V}(\boldsymbol{x})$ is positive definite, i. e. $V(\boldsymbol{x})>0$;

(2) $\dot{V}(\boldsymbol{x},t)$ is negative semi-definite, i. e. $\dot{V}(\boldsymbol{x},t) \leqslant 0$;

(3) For any non-zero initial state $\boldsymbol{x}_0$, $\dot{V}(\boldsymbol{\Phi}(t;0,\boldsymbol{x}_0,\boldsymbol{0}))$ is not identical to zero, where $\boldsymbol{\Phi}(t;0,\boldsymbol{x}_0,\boldsymbol{0})$ is zero-input response;

(4) When $\|\boldsymbol{x}\| \to \infty$, $V(\boldsymbol{x}) \to \infty$.

Then the origin of the system is globally uniformly asymptotically stable.

Example 4.3.2 Analyse the stability of the origin equilibrium of the following system

$$\begin{cases} \dot{x}_1 = x_2 \\ \dot{x}_2 = -x_1 - (1 + x_2)^2 x_2 \end{cases}$$

Answer Let $\dot{x}_1=0$ and $\dot{x}_2=0$, we get $x_1=0$ and $x_2=0$, i. e. the origin is uniquely equilibrium.

Chose $V(\boldsymbol{x})=x_1^2+x_2^2$, we have

(1) $V(\boldsymbol{x})>0$, $\forall \boldsymbol{x}\neq 0$;

(2) The first order derivative of with respect t to is:

$$\begin{aligned}\dot{V}(\boldsymbol{x})&=\frac{\partial V(\boldsymbol{x})}{\partial x_1}\cdot\frac{\mathrm{d}x_1}{\mathrm{d}t}+\frac{\partial V(\boldsymbol{x})}{\partial x_2}\cdot\frac{\mathrm{d}x_2}{\mathrm{d}t}\\&=2x_1\cdot\dot{x}_1+2x_2\cdot\dot{x}_2\\&=2x_1x_2+2x_2(-x_1-(1+x_2)2x_2)\\&=-2x_2^2(1+x_2)2\end{aligned}$$

Thus, when

(a) $x_2=0$;

(b) $x_2=-1$

$\dot{V}(\boldsymbol{x},t)=0$. In addition to the above two cases, $\dot{V}(\boldsymbol{x},t)<0$. So $\dot{V}(\boldsymbol{x},t)$ is negative semi-definite;

(3) Check whether $\dot{V}(\boldsymbol{\Phi}(t;0,\boldsymbol{x}_0,\boldsymbol{0}))$ is identical to zero for cases (a) and (b).

Case (a): when $x_2=0$, the state equation becomes

$$\begin{cases}\dot{x}_1(t)=x_2(t)=0\\\dot{x}_2=-x_1(t)\end{cases}$$

The above state equation does not have any solutions. It indicates $[x_1(t)\quad 0]^T$ is not solution of the system apart from $(x_1=0, x_2=0)$.

Case (b): when $x_2=-1$, the state equation becomes

$$\begin{cases}\dot{x}_1(t)=x_2(t)=-1\\\dot{x}_2=-x_1(t)\end{cases}$$

The above state equation does not have any solutions. It indicates $[x_1(t)\quad -1]^T$ is not solution of the system apart from $(x_1=0, x_2=0)$.

(4) When $\|\boldsymbol{x}\|\to\infty$, $V(\boldsymbol{x})\to\infty$.

According to conclusion 4.3.3, the origin of the system is globally asymptotically stable.

4.3.2 Criterion theorem of stability in the sense of Lyapunov

Conclusion 4.3.4 The criterion theorem of uniformly stability in the sense of Lyapunov for time-varying systems. Consider a time-varying system eq. (4.3.1) having a point of equilibrium at $\boldsymbol{x}=\boldsymbol{0}$. If there exists a scale function $V(\boldsymbol{x},t)$, and $V(\boldsymbol{0},t)=0$, a region around the origin $\boldsymbol{\Omega}$, for any non-zero solution of the system, $\boldsymbol{x}(t)\in\boldsymbol{\Omega}$, and any $t\geqslant t_0$, such that

(1) $V(\boldsymbol{x},t)$ is positive definite and limited;

(2) $\dot{V}(\boldsymbol{x},t)$ is negative semi-definite and limited.

Then the origin of the system is uniformly stable in the sense of Lyapunov in the region Ω.

Conclusion 4.3.5 The criterion theorem of uniformly stability in the sense of Lyapunov for time-invariant systems. Consider a time-invariant system $\dot{\boldsymbol{x}} = \boldsymbol{f}(\boldsymbol{x})$ having a point of equilibrium at $\boldsymbol{x} = \boldsymbol{0}$. If there exists a scale function $V(\boldsymbol{x})$, and $V(\boldsymbol{0}) = 0$, a region around the origin Ω, for any non-zero solution of the system, $\boldsymbol{x}(t) \in \Omega$, and any $t \geqslant t_0$, such that

(1) $V(\boldsymbol{x})$ is positive definite;

(2) $\dot{V}(\boldsymbol{x})$ is negative semi-definite.

Then the origin of the system is stable in the sense of Lyapunov in the region Ω.

4.3.3 Unstable criterion

The above stability criterion theorems can only judge the stability of systems, but not the unstability of systems. When we chose $V(\boldsymbol{x},t)$ several times and do not find $V(\boldsymbol{x},t)$ satisfying the stable conditions, we should consider the system may be unstable. The following is the criterion for unstability.

Conclusion 4.3.6 Consider a time-varying system $\dot{\boldsymbol{x}} = \boldsymbol{f}(\boldsymbol{x},t)$ or a time-invariant system $\dot{\boldsymbol{x}} = \boldsymbol{f}(\boldsymbol{x})$ having a point of equilibrium at $\boldsymbol{x} = \boldsymbol{0}$. If there exists a scale function $V(\boldsymbol{x},t)$ or $V(\boldsymbol{x})$, and $V(\boldsymbol{0},t) = 0$ or $V(\boldsymbol{0}) = 0$, a region around the origin Ω, for any non-zero solution of the system, $\boldsymbol{x}(t) \in \Omega$, and any $t \geqslant t_0$, such that

(1) $V(\boldsymbol{x},t)$ is positive definite and limited, or $V(\boldsymbol{x})$ is positive definite;

(2) $\dot{V}(\boldsymbol{x},t)$ is positive definite and limited, or $\dot{V}(\boldsymbol{x})$ is positive definite.

Then the origin of the system isunstable in the sense of Lyapunov.

4.4 Common construction methods of Lyapunov function

Lyapunov's second method for stability is an effective method to determine whether the equilibrium points of systems are stable or not, which is suitable for nonlinear time-varying systems. This method is widely used in fields like designing neural network adaptive stable controllers. The main difficulty of applying this method is that there is no general method to construct Lyapunov functions for nonlinear systems. Now there are mature theories for LTI systems and some effective methods for nonlinear time-invariant systems. Next, we introduce two methods of constructing Lyapunov functions.

4.4.1 Variable gradient method

The variable gradient method is suitable for constructing Lyapunov functions for nonlinear time-invariant systems. Consider a nonlinear time-invariant system described by

$$\dot{\boldsymbol{x}} = \boldsymbol{f}(\boldsymbol{x}) \tag{4.4.1}$$

where $\boldsymbol{x}(0) = \boldsymbol{0}$, i. e. the origin is unique equilibrium point.

We chose

$$\nabla V(\boldsymbol{x}) = \begin{bmatrix} a_{11}x_1 + a_{12}x_2 + \cdots + a_{1n}x_n \\ \vdots \\ a_{n1}x_1 + a_{n2}x_2 + \cdots + a_{nn}x_n \end{bmatrix} \tag{4.4.2}$$

as the gradient of candidate Lyapunov function $V(\boldsymbol{x})$, where a_{ij} is constant or function of vector $\boldsymbol{x}, \boldsymbol{x} = [x_1\ x_2 \cdots\ x_n]^{\mathrm{T}}$. The gradient of $V(\boldsymbol{x})$ is defined as

$$\nabla V(\boldsymbol{x}) \triangleq \frac{\partial V(\boldsymbol{x})}{\partial \boldsymbol{x}} \triangleq \begin{bmatrix} \dfrac{\partial V(\boldsymbol{x})}{\partial x_1} \\ \vdots \\ \dfrac{\partial V(\boldsymbol{x})}{\partial x_n} \end{bmatrix} = \begin{bmatrix} \nabla V_1(\boldsymbol{x}) \\ \vdots \\ \nabla V_n(\boldsymbol{x}) \end{bmatrix} \tag{4.4.3}$$

We can get the constraints on $\nabla V(\boldsymbol{x})$ from the conditions given by the stability conclusion. From $\dot{V}(\boldsymbol{x}) < 0$, we get

$$0 > \frac{\mathrm{d}V(\boldsymbol{x})}{\mathrm{d}t} = \frac{\partial V(\boldsymbol{x})}{\partial x_1} \cdot \frac{\mathrm{d}x_1}{\mathrm{d}t} + \cdots + \frac{\partial V(\boldsymbol{x})}{\partial x_n} \cdot \frac{\mathrm{d}x_n}{\mathrm{d}t} = \begin{bmatrix} \dfrac{\partial V(\boldsymbol{x})}{\partial x_1} & \cdots & \dfrac{\partial V(\boldsymbol{x})}{\partial x_n} \end{bmatrix} \begin{bmatrix} \dot{x}_1 \\ \vdots \\ \dot{x}_n \end{bmatrix}$$

$$= [\nabla V(\boldsymbol{x})]^{\mathrm{T}} \dot{\boldsymbol{x}}$$

Thus

$$[\nabla V(\boldsymbol{x})]^{\mathrm{T}} \dot{\boldsymbol{x}} < 0 \tag{4.4.4}$$

Second, according to the field theory of vector analysis, there are

$$\frac{\partial \nabla V_j(\boldsymbol{x})}{\partial x_i} = \frac{\partial \nabla V_i(\boldsymbol{x})}{\partial x_j}, \quad \forall i \neq j \tag{4.4.5}$$

From eq. (4.4.4) and eq. (4.4.5), we can obtain the a_{ij} in eq. (4.4.2).

Based in the obtained $\nabla V(\boldsymbol{x})$, we have

$$\begin{aligned} V(\boldsymbol{x}) &= \int_0^{V(x)} \mathrm{d}V(\boldsymbol{x}) \\ &= \int_0^t \frac{\mathrm{d}V(\boldsymbol{x})}{\mathrm{d}t}\mathrm{d}t \\ &= \int_0^t [\nabla V(\boldsymbol{x})]^{\mathrm{T}} \dot{\boldsymbol{x}}\mathrm{d}t \\ &= \int_0^x [\nabla V(\boldsymbol{x})]^{\mathrm{T}}\mathrm{d}x \\ &= \int_0^x [\nabla V_1(\boldsymbol{x}) \ \cdots \ \nabla V_n(\boldsymbol{x})] \begin{bmatrix} \mathrm{d}x_1 \\ \vdots \\ \mathrm{d}x_n \end{bmatrix} \end{aligned} \tag{4.4.6}$$

According to the relevant characteristics of the potential field, the above integration have nothing to do with the integration path, so the integration path is selected as follows:

Let $x_2 = \cdots = x_n = 0$, select x_1: $0 \rightarrow x_1$;

Then fix x_1, let $x_3 = \cdots = x_n = 0$, select x_2: $0 \rightarrow x_2$;

$\vdots$

Fix $x_1, x_2, \cdots, x_{n-2}$, and let $x_n = 0$, select x_{n-1}: $0 \rightarrow x_{n-1}$;

Final fix $x_1, x_2, \cdots, x_{n-1}$, select x_n: $0 \rightarrow x_n$.

Thus, eq. (4.4.6) becomes

$$V(\boldsymbol{x}) = \int_0^{x_1(x_2 = \cdots = x_n = 0)} \nabla V_1(\boldsymbol{x}) \mathrm{d}x_1 + \int_0^{x_2(x_1 = x_1, x_3 = \cdots = x_n = 0)} \nabla V_2(\boldsymbol{x}) \mathrm{d}x_2 + \cdots + \int_0^{x_n(x_1 = x_1, \cdots, x_{n-1} = x_{n-1})} \nabla V_n(\boldsymbol{x}) \mathrm{d}x_n \tag{4.4.7}$$

Determine whether $V(\boldsymbol{x})$ is positive. When $V(\boldsymbol{x}) > 0$, the obtained $V(\boldsymbol{x})$ satisfies the conditions (1) and (2) in conclusion 4.3.2, the system is asymptotically stable. Otherwise, when $V(\boldsymbol{x}) < 0$, we need to choose other methods to construct the $V(\boldsymbol{x})$ to check the stability of systems.

Example 4.4.1 Given a system

$$\begin{cases} \dot{x}_1 = x_2 \\ \dot{x}_2 = -x_1^3 - x_2 \end{cases}$$

try to determine its internal stability usingthe variable gradient method.

Answer From $\dot{x}_1 = 0$ and $\dot{x}_2 = 0$, we get $x_1 = 0$ and $x_2 = 0$, i. e. the origin is uniquely equilibrium. The system order $n = 2$, the gradient of candidate Lyapunov function $V(\boldsymbol{x})$, $\nabla V(\boldsymbol{x})$, is

$$\nabla V(\boldsymbol{x}) = \begin{bmatrix} \nabla V_1(\boldsymbol{x}) \\ \nabla V_2(\boldsymbol{x}) \end{bmatrix} = \begin{bmatrix} a_{11}x_1 + a_{12}x_2 \\ a_{21}x_1 + a_{22}x_2 \end{bmatrix}$$

Let $a_{22} = 2$, we get

$$\frac{\partial \nabla V_1(\boldsymbol{x})}{\partial x_2} = \frac{\partial \nabla V_2(\boldsymbol{x})}{\partial x_1}$$

Thus

$$a_{12} = a_{21} \tag{4.4.8}$$

Then, from $[\nabla V(\boldsymbol{x})]^{\mathrm{T}} \dot{\boldsymbol{x}} < 0$ we have

$$[a_{11}x_1 + a_{12}x_2 \quad a_{21}x_1 + a_{22}x_2] \begin{bmatrix} x_2 \\ -x_1^3 - x_2 \end{bmatrix} = (a_{11} - a_{21} - 2x_1^2)x_1x_2 + (a_{12} - 2)x_2^2 - a_{21}x_1^4 < 0 \tag{4.4.9}$$

chose

$$\begin{cases} a_{12} = a_{21} \\ a_{11} = a_{21} + 2x_1^2 \\ 0 < a_{12} < 2 \end{cases}$$

Such that eq. (5.4.8) and eq. (5.4.9) hold. Hence, we have

$$\nabla V(\boldsymbol{x}) = \begin{bmatrix} (a_{21} + 2x_1^2)x_1 + a_{21}x_2 \\ a_{21}x_1 + 2x_2 \end{bmatrix} \tag{4.4.10}$$

From eq. (4.4.7) and eq. (4.4.10), we have

$$\begin{aligned} V(x) &= \int_0^{x_1(x_2=0)} (a_{21} + 2x_1^2)x_1 \mathrm{d}x_1 + \int_0^{x_2(x_1=x)} (a_{21}x_1 + 2x_2)\mathrm{d}x_2 \\ &= \frac{1}{2}x_1^4 + \frac{1}{2}a_{21}x_1^2 + a_{21}x_1x_2 + x_2^2 \\ &= \frac{1}{2}x_1^4 + [x_1 \quad x_2]\begin{bmatrix} \frac{a_{21}}{2} & \frac{a_{21}}{2} \\ \frac{a_{21}}{2} & 1 \end{bmatrix}\begin{bmatrix} x_1 \\ x_2 \end{bmatrix} \end{aligned}$$

Now check if the obtained $V(\boldsymbol{x})$ is positive definite:

From $0 < a_{12} < 2$, we get

$$\det\left(\frac{a_{21}}{2}\right) > 0 \quad \text{and} \quad \det\begin{bmatrix} \frac{a_{21}}{2} & \frac{a_{21}}{2} \\ \frac{a_{21}}{2} & 1 \end{bmatrix} = \frac{a_{21}}{2}\left(1 - \frac{a_{21}}{2}\right) > 0$$

Thus, $\forall x \neq \boldsymbol{0}$, $V(\boldsymbol{x}) > 0$, i.e. $V(\boldsymbol{x})$ is positive definition. And when $\|\boldsymbol{x}\| \to \infty$, $V(\boldsymbol{x}) \to \infty$. So the all conditions in conclusion 4.3.2 are satisfied, $\boldsymbol{x}_e = \boldsymbol{0}$ is globally asymptotically stable.

4.4.2 Krasovskii method

Consider a nonlinear time-invariant system described by

$$\dot{\boldsymbol{x}} = \boldsymbol{f}(\boldsymbol{x}), \quad t \geqslant 0 \tag{4.4.11}$$

Where, $\boldsymbol{x} = [x_1 \quad \cdots \quad x_n]^{\mathrm{T}}$ and $\boldsymbol{f}(\boldsymbol{0}) = \boldsymbol{0}$, i.e. $\boldsymbol{x}_e = \boldsymbol{0}$ is uniquely equilibrium. Let

$$\boldsymbol{f}(\boldsymbol{x}) = \begin{bmatrix} f_1(\boldsymbol{x}) \\ \vdots \\ f_n(\boldsymbol{x}) \end{bmatrix} \tag{4.4.12}$$

The Jacobian matrix of the system eq. (4.4.11), $\boldsymbol{F}(\boldsymbol{x})$, is

$$\boldsymbol{F}(\boldsymbol{x}) \triangleq \frac{\partial \boldsymbol{f}(\boldsymbol{x})}{\partial \boldsymbol{x}^T} = \begin{bmatrix} \frac{\partial f_1(\boldsymbol{x})}{\partial x_1} & \cdots & \frac{\partial f_1(\boldsymbol{x})}{\partial x_n} \\ \vdots & & \vdots \\ \frac{\partial f_n(\boldsymbol{x})}{\partial x_1} & \cdots & \frac{\partial f_n(\boldsymbol{x})}{\partial x_n} \end{bmatrix} \tag{4.4.13}$$

Proposition 4.4.1 Consider a non-linear time-invariant system (4.4.11) and a region around the origin equilibrium $\Omega \subset \mathbf{R}^n$. If

$$\boldsymbol{F}^{\mathrm{T}}(\boldsymbol{x}) + \boldsymbol{F}(\boldsymbol{x}) < 0, \quad \forall \boldsymbol{x} \neq \boldsymbol{0} \tag{4.4.14}$$

i. e. $\boldsymbol{F}^{\mathrm{T}}(\boldsymbol{x})+\boldsymbol{F}(\boldsymbol{x})$ is negative definite, then

$$\dot{V}(\boldsymbol{x})<0 \tag{4.4.15}$$

i. e. $\dot{V}(\boldsymbol{x})$ is negative definite, where $V(\boldsymbol{x})=\boldsymbol{f}^{\mathrm{T}}(\boldsymbol{x})\boldsymbol{f}(\boldsymbol{x})$.

Proof Because

$$\begin{aligned}\frac{\mathrm{d}V(\boldsymbol{x})}{\mathrm{d}t}&=\frac{\mathrm{d}}{\mathrm{d}t}[\boldsymbol{f}^{\mathrm{T}}(x)\boldsymbol{f}(x)]\\&=\frac{\mathrm{d}\boldsymbol{f}^{\mathrm{T}}(\boldsymbol{x})}{\mathrm{d}t}\cdot\boldsymbol{f}(\boldsymbol{x})+\boldsymbol{f}^{\mathrm{T}}(\boldsymbol{x})\cdot\frac{\mathrm{d}\boldsymbol{f}(\boldsymbol{x})}{\mathrm{d}t}\\&=[\frac{\partial\boldsymbol{f}(\boldsymbol{x})}{\partial\boldsymbol{x}^{\mathrm{T}}}\cdot\frac{\mathrm{d}\boldsymbol{x}}{\mathrm{d}t}]^{\mathrm{T}}\boldsymbol{f}(\boldsymbol{x})+\boldsymbol{f}^{\mathrm{T}}(\boldsymbol{x})[\frac{\partial\boldsymbol{f}(\boldsymbol{x})}{\partial\boldsymbol{x}^{\mathrm{T}}}\cdot\frac{\mathrm{d}\boldsymbol{x}}{\mathrm{d}t}]\\&=\boldsymbol{f}^{\mathrm{T}}(\boldsymbol{x})[\boldsymbol{F}^{\mathrm{T}}(\boldsymbol{x})+\boldsymbol{F}(\boldsymbol{x})]\boldsymbol{f}(\boldsymbol{x})\end{aligned}$$

From eq. (4.4.14), we get $\frac{\mathrm{d}V(\boldsymbol{x})}{\mathrm{d}t}\leqslant 0,\ \forall\boldsymbol{x}\neq\boldsymbol{0}$.

Conclusion 4.4.1 Krasovskii theorem. Consider a non-linear time-invariant system (4.4.11), $\Omega\subset R^n$ is a region around unique origin equilibrium. If

$$\boldsymbol{F}^{\mathrm{T}}(\boldsymbol{x})+\boldsymbol{F}(\boldsymbol{x})<0,\quad\forall\boldsymbol{x}\neq 0 \tag{4.4.16}$$

Then the origin equilibrium is asymptotically stable in the region Ω and $V(\boldsymbol{x})=\boldsymbol{f}^{\mathrm{T}}(\boldsymbol{x}),\boldsymbol{f}(\boldsymbol{x})$ is its a Lyapunov function. Moreover, if when $\|\boldsymbol{x}\|\to\infty$, $\boldsymbol{f}^{\mathrm{T}}(\boldsymbol{x})\boldsymbol{f}(\boldsymbol{x})\to\infty$, then origin equilibrium is globally asymptotically stable.

Proof Using Proposition 4.4.1 and conclusion 4.3.2, we can prove the conclusion.

Conclusion 4.4.2 Consider a LTI system $\dot{\boldsymbol{x}}=\boldsymbol{A}\boldsymbol{x}$, $\boldsymbol{A}$ is nonsingular. If

$$\boldsymbol{A}+\boldsymbol{A}^{\mathrm{T}}<0 \tag{4.4.17}$$

then $\boldsymbol{x}=0$ is globally asymptotically stable.

Proof Note $\boldsymbol{F}(\boldsymbol{x})=\boldsymbol{A}$, using conclusion 4.4.1, we can prove the conclusion 4.4.2.

Example 4.4.2 Given a system

$$\begin{cases}\dot{x}_1=-3x_1+x_2\\\dot{x}_2=2x_1-x_2-x_2^3\end{cases}$$

Try to determine its internal stability using Krasovskii theorem.

Answer From $\dot{x}_1=0$ and $\dot{x}_2=0$, we get $x_1=0$ and $x_2=0$, i. e. the origin is uniquely equilibrium. And

$$\boldsymbol{F}(\boldsymbol{x})=\frac{\partial\boldsymbol{f}(\boldsymbol{x})}{\partial\boldsymbol{x}^{\mathrm{T}}}=\begin{bmatrix}\frac{\partial f_1(\boldsymbol{x})}{\partial x_1}&\frac{\partial f_1(\boldsymbol{x})}{\partial x_2}\\\frac{\partial f_2(\boldsymbol{x})}{\partial x_1}&\frac{\partial f_2(\boldsymbol{x})}{\partial x_2}\end{bmatrix}=\begin{bmatrix}-3&1\\2&-1-3x_2^2\end{bmatrix}$$

$$\boldsymbol{F}^{\mathrm{T}}(\boldsymbol{x})+\boldsymbol{F}(\boldsymbol{x})=-\begin{bmatrix}6&-3\\-3&2+6x_2^2\end{bmatrix}$$

Because $6>0, \det\begin{bmatrix} 6 & -3 \\ -3 & 2+6x_2^2 \end{bmatrix} = 36x_2^2+3>0$, $F^{\mathrm{T}}(x)+F(x)<0$, $\forall x \neq 0$.

When $\|x\| \to \infty$, $V(x) \to \infty$. So, according to conclusion 4. 4. 1, $x_e = 0$ is globally asymptotically stable.

4.5 State motion stability criteria of linear systems

By using the related concepts of Lyapunov stability and main theorems of Lyapunov's second method, we can obtain the criteria of stability of state motion of linear systems. In this section, we discuss the state motion stability criteria of linear time-varying and time-invariant systems, respectively.

4.5.1 State motion stability criterion of LTI systems

Consider an n-order LTI system described by

$$\dot{x} = Ax, \ x(0) = x_0, \ t \geqslant 0 \tag{4.5.1}$$

$x_e = 0$ is its an equilibrium. The stability of the origin equilibrium is completely determined by the system matrix A. The stability of the system can be judged by the distribution of the eigenvalues of A.

Conclusion 4. 5. 1 Eigenvalue criterion. For the LTI system (4. 5. 1), we have the following conclusions:

(1) Every equilibrium point of the system is stable in the sense of Lyapunov, if and only if all eigenvalues of A have nonpositive real parts, and every eigenvalue with zero-real part is a single root of the minimal polynomial of A.

(2) The unique equilibrium $x_e = 0$ of eq. (4. 5. 1) is global asymptotically stable, if and only if all eigenvalues of A have negative real parts.

Proof (1) We prove the first part of the conclusion based on the Lyapunov stability definition. x_e is a equilibrium of the system, so the state motion excited by initial state x_e is $x(t) = x_e = e^{At}x_e$. For every $\varepsilon > 0$, if $\|x_0 - x_e\| < \delta(\varepsilon)$, then

$$\begin{aligned} \|\Phi(t;0,x_0,0) - x_e\| = \|e^{At}x_0 - x_e\| &= \|e^{At}x_0 - e^{At}x_e\| \\ &= \|e^{At}(x_0 - x_e)\| \\ &\leqslant \|e^{At}\| \cdot \|(x_0 - x_e)\| \end{aligned}$$

So, as long as $\|e^{At}\| < k < \infty$, let $\delta(\varepsilon) = \dfrac{\varepsilon}{k}$, we have

$$\|\Phi(t;0,x_0,0) - x_e\| \leqslant k \cdot \delta(\varepsilon) = k \cdot \frac{\varepsilon}{k} = \varepsilon$$

It means the system is stable in the sense of Lyapunov.

Thus as long as $\| e^{At} \|$ is limited, then the system is Lyapnov stable. Introducing a nonsingular matrix $\boldsymbol{P}$, transforms $\boldsymbol{A}$ into Jordan form, i. e. $\hat{\boldsymbol{A}} = \boldsymbol{P}^{-1}\boldsymbol{A}\boldsymbol{P}$. Then there is

$$\| e^{At} \| \leqslant \| \boldsymbol{P} \| \cdot \| e^{\hat{A}t} \| \cdot \| \boldsymbol{P}^{-1} \|$$

So, $\| e^{At} \|$ is limited is equivalent to $\| e^{\hat{A}t} \|$ is limited. And

$$\hat{\boldsymbol{A}} = \begin{bmatrix} \boldsymbol{J}_1 & & \\ & \ddots & \\ & & \boldsymbol{J}_m \end{bmatrix} = \begin{bmatrix} \boldsymbol{J}_{11} & & & & & & \\ & \ddots & & & & & \\ & & \boldsymbol{J}_{1\alpha_1} & & & & \\ & & & \ddots & & & \\ & & & & \boldsymbol{J}_{m1} & & \\ & & & & & \ddots & \\ & & & & & & \boldsymbol{J}_{m\alpha_m} \end{bmatrix}$$

where each small Jordan block has same form like $\boldsymbol{J}_{11}$:

$$J_{11} = \begin{bmatrix} \lambda_1 & 1 & & & \\ & \lambda_1 & 1 & & \\ & & \ddots & & \\ & & & \lambda_1 & 1 \\ & & & & \lambda_1 \end{bmatrix}_{k \times k}$$

J_{11} is first small Jordan block corresponding to λ_1 and

$$e^{\hat{A}t} = \begin{bmatrix} e^{J_{11}t} & & & & & & \\ & \ddots & & & & & \\ & & e^{J_{1\alpha_1}t} & & & & \\ & & & \ddots & & & \\ & & & & e^{J_{m1}t} & & \\ & & & & & \ddots & \\ & & & & & & e^{J_{m\alpha_m}t} \end{bmatrix}$$

$$e^{J_{11}t} = \begin{bmatrix} e^{\lambda_1 t} & te^{\lambda_1 t} & \dfrac{t^2 e^{\lambda_1 t}}{2!} & \cdots & \dfrac{t^{k-1} e^{\lambda_1 t}}{(k-1)!} \\ 0 & e^{\lambda_1 t} & te^{\lambda_1 t} & \cdots & \dfrac{t^{k-2} e^{\lambda_1 t}}{(k-2)!} \\ & & \cdots & & \\ 0 & 0 & 0 & \cdots & e^{\lambda_1 t} \end{bmatrix}$$

Thus, as long as each element of J_{ij} is limited, $\| e^{\hat{A}t} \|$ is limited. The non-zero element of $e^{\hat{A}t}$ has the following form

$$\frac{1}{(p-1)!} t^{p-1} e^{\lambda_i t}, \ p = 1, \cdots, k_{im}$$

where, λ_i is i^{th} eigenvalue of $\boldsymbol{A}$, k_{im} is the order of the mth small block of the Jordan block respect with λ_i. Let $\lambda_i = \alpha_i + \omega_i j$, the element of $e^{\hat{A}t}$ can be written as

$$\frac{1}{(p-1)!}t^{p-1}e^{\alpha_i t} \cdot e^{\omega_i jt},\ p = 1, \cdots, k_{im}$$

If the real part of λ_i, α_i, is less than zero, then we have

$$\lim_{t \to +\infty} \frac{1}{(p-1)!} \cdot \frac{t^{p-1}}{e^{|\alpha_i|t}} \cdot e^{\omega_i jt} = 0$$

It indicates if the all eigenvalues of $\boldsymbol{A}$ have negative real parts, the system is asymptotically stable.

When the real part of λ_i, α_i, is zero, we have

$$\frac{1}{(p-1)!}t^{p-1}e^{\omega_i jt},\ p = 1, \cdots, k_{i,m}$$

In order to the above term has a limited value when $t \to \infty$, p must be one. That is, the maximum value of the orders of all small Jordan blocks corresponding to the eigenvalue with zero real-part is 1, i. e. the corresponding eigenvalues with zero real-part is the single root of the minimal polynomial of $\boldsymbol{A}$. So, we finished the proof of the first part.

(2) According to the proof process of the first part, it can be seen that the unique equilibrium $\boldsymbol{x}_e = \boldsymbol{0}$ of the system is asymptotically stable if all eigenvalues of $\boldsymbol{A}$ have negative real parts.

Example 4.5.1 Determine the internal stability of the following systems

$$(1)\ \dot{\boldsymbol{x}} = \begin{bmatrix} -2 & 0 & 0 \\ 0 & 0 & 0 \\ 0 & 0 & 0 \end{bmatrix}\boldsymbol{x}$$

$$(2)\ \dot{\boldsymbol{x}} = \begin{bmatrix} -2 & 0 & 0 \\ 0 & 0 & 1 \\ 0 & 0 & 0 \end{bmatrix}\boldsymbol{x}$$

Answer The eigenvalues of the system (1) and (2) are -2, 0 and 0. The minimal polynomial of $\boldsymbol{A}$ of the first system is $(s-2)(s-0)$, and the second system is $(s-2)(s-0)^2$. So 0 is a single root of the minimal polynomial of $\boldsymbol{A}$ of the first system but is not a single root of the minimal polynomial of $\boldsymbol{A}$ of the second system. According to conclusion 4.5.1, the first system is stable in the sense of Lyapunov, but the second system is not.

Conclusion 4.5.2 Routh criterion for asymptotic stability of LTI systems. From conclusion 4.5.1, for a LTI system (4.5.1), the equilibrium $\boldsymbol{x}_e = \boldsymbol{0}$ is asymptotic stable, if only if all eigenvalues of the system have negative real parts. We can use the Routh criterion to determine whether all eigenvalues of systems have negative real parts instead of calculating all eigenvalues. First, construct the Routh table based on the characteristic polynomial of the system matrix. If the coefficients of the first column of the Routh table are all greater than 0, then the real part of all eigenvalues of the system is less than zero, i. e. the system is asymptotically stable. Example 4.5.2 shows how to construct the Routh table.

Example 4.5.2 Given the characteristic polynomial of a system by

$$f(s)=s^4+a_1s^3+a_2s^2+a_3s+a_4$$

Try to construct the Routh table.

Answer Construct the Routh table as the following form

$$\begin{array}{lll}
1 & a_2 & a_4 \\
a_1 & a_3 & 0 \\
\dfrac{a_1a_2-a_3}{a_1} & \dfrac{a_1a_4-1\times 0}{a_1}=a_4 & \\
\dfrac{\dfrac{a_1a_2-a_3}{a_1}\times a_3-a_1a_4}{\dfrac{a_1a_2-a_3}{a_1}} & 0 & \\
\dfrac{\dfrac{\dfrac{a_1a_2-a_3}{a_1}\times a_3-a_1a_4}{\dfrac{a_1a_2-a_3}{a_1}}\times a_4-\dfrac{a_1a_2-a_3}{a_1}\times 0}{\dfrac{\dfrac{a_1a_2-a_3}{a_1}\times a_3-a_1a_4}{\dfrac{a_1a_2-a_3}{a_1}}}=a_4 & &
\end{array}$$

Conclusion 4.5.3 Lyapunov criterion.

For the LTI system eq. (4.5.1), the equilibrium $\boldsymbol{x}_e=\boldsymbol{0}$ is asymptotically stable, if only if the Lyapunov matrix equation

$$\boldsymbol{A}^{\mathrm{T}}\boldsymbol{P}+\boldsymbol{P}\boldsymbol{A}=-\boldsymbol{Q} \tag{4.5.2}$$

has a unique positive-definite and symmetric solution matrix $\boldsymbol{P}$, where $\boldsymbol{Q}^T=\boldsymbol{Q}$, $\boldsymbol{Q}>0$,$\boldsymbol{P}$ is $n\times n$.

Proof First prove sufficiency, i. e. prove the equilibrium $\boldsymbol{x}_e=\boldsymbol{0}$ is asymptotically stable, if $\boldsymbol{P}$ is the positive-definite and symmetric solution of eq. (4.5.2).

Chose $\boldsymbol{V}(\boldsymbol{x})=\boldsymbol{x}^{\mathrm{T}}\boldsymbol{P}\boldsymbol{x}$. Because $\boldsymbol{P}=\boldsymbol{P}^{\mathrm{T}}>0$, the $\boldsymbol{V}(\boldsymbol{x})$ is positive - definite. Then

$$\dot{\boldsymbol{V}}(\boldsymbol{x})=\dot{\boldsymbol{x}}^{\mathrm{T}}\boldsymbol{P}\boldsymbol{x}+x^{\mathrm{T}}\boldsymbol{P}\,\dot{\boldsymbol{x}}=(\boldsymbol{A}\boldsymbol{x})^{\mathrm{T}}\boldsymbol{P}\boldsymbol{x}+\boldsymbol{x}^{\mathrm{T}}\boldsymbol{P}\boldsymbol{A}\boldsymbol{x}=\boldsymbol{x}^{\mathrm{T}}(\boldsymbol{A}\boldsymbol{P}+\boldsymbol{P}\boldsymbol{A})\boldsymbol{x}=-\boldsymbol{x}^{\mathrm{T}}\boldsymbol{Q}\boldsymbol{x}$$

Because $\boldsymbol{Q}=\boldsymbol{Q}^{\mathrm{T}}>0$, $\dot{\boldsymbol{V}}(\boldsymbol{x})$ is negative definite. So, $\boldsymbol{x}_e$ is asymptotically stable.

Next prove necessity, i. e. prove $\boldsymbol{P}$ is the positive-definite and symmetric solution of (4.5.2), if $\boldsymbol{x}_e=\boldsymbol{0}$ is asymptotically stable.

Consider the following matrix equation

$$\dot{\boldsymbol{X}}=\boldsymbol{A}^{\mathrm{T}}\boldsymbol{X}+\boldsymbol{X}\boldsymbol{A},\ \boldsymbol{X}(0)=\boldsymbol{Q},\ t\geqslant 0 \tag{4.5.3}$$

where, $\boldsymbol{X}$ is $n\times n$ matrix.

Because

$$\frac{\mathrm{d}}{\mathrm{d}t}(\mathrm{e}^{A^{\mathrm{T}}t}\boldsymbol{Q}\mathrm{e}^{At}) = \boldsymbol{A}^{\mathrm{T}}\mathrm{e}^{A^{\mathrm{T}}t}\boldsymbol{Q}\mathrm{e}^{At} + \mathrm{e}^{A^{\mathrm{T}}t}\boldsymbol{Q}\mathrm{e}^{At}\boldsymbol{A} = \boldsymbol{A}^{\mathrm{T}}(\mathrm{e}^{A^{\mathrm{T}}t}\boldsymbol{Q}\mathrm{e}^{At}) + (\mathrm{e}^{A^{\mathrm{T}}t}\boldsymbol{Q}\mathrm{e}^{At})\boldsymbol{A}$$

The solution of the matrix equation (4.5.3) is

$$\boldsymbol{X}(t) = \mathrm{e}^{A^{\mathrm{T}}t}\boldsymbol{Q}\mathrm{e}^{At} \tag{4.5.4}$$

Integrating eq. (4.5.3) from $t=0$ to $t=\infty$, yields

$$\boldsymbol{X}(\infty) - \boldsymbol{X}(0) = \boldsymbol{A}^{\mathrm{T}}\left(\int_0^\infty \boldsymbol{X}(t)\mathrm{d}t\right) + \left(\int_0^\infty \boldsymbol{X}(t)\mathrm{d}t\right)\boldsymbol{A} \tag{4.5.5}$$

Because the system is asymptotically stable, when $t\to\infty$, $\mathrm{e}^{At}\to\boldsymbol{0}$. From eq. (4.5.4), we have

$$\boldsymbol{X}(\infty) = \lim_{t\to\infty}\mathrm{e}^{A^{\mathrm{T}}t}\boldsymbol{Q}\mathrm{e}^{At} = \boldsymbol{0}$$

Let $\boldsymbol{P} = \int_0^\infty \boldsymbol{X}(t)\mathrm{d}t$, we rewrite eq. (4.5.5) as

$$\boldsymbol{A}^{\mathrm{T}}\boldsymbol{P} + \boldsymbol{P}\boldsymbol{A} = -\boldsymbol{Q}$$

It indicates $\boldsymbol{P} = \int_0^\infty \boldsymbol{X}(t)\mathrm{d}t$ is a solution of eq. (4.5.3). Further, $\boldsymbol{X}(t)$ is existing and unique, and $\boldsymbol{X}(\infty)=\boldsymbol{0}$. Hence $\boldsymbol{P} = \int_0^\infty \boldsymbol{X}(t)\mathrm{d}t$ is existing and unique. And

$$\boldsymbol{P}^{\mathrm{T}} = \int_0^\infty \boldsymbol{X}^{\mathrm{T}}(t)\mathrm{d}t = \int_0^\infty (\mathrm{e}^{A^{\mathrm{T}}t}\boldsymbol{Q}\mathrm{e}^{At})^{\mathrm{T}}\mathrm{d}t = \int_0^\infty \mathrm{e}^{A^{\mathrm{T}}t}\boldsymbol{Q}\mathrm{e}^{At}\mathrm{d}t = \boldsymbol{P}$$

So, $\boldsymbol{P} = \int_0^\infty \boldsymbol{X}(t)\mathrm{d}t$ is symmetric.

For any non-zero $n\times 1$ vector $\boldsymbol{x}_0$, there is

$$\boldsymbol{x}_0^{\mathrm{T}}\boldsymbol{P}\boldsymbol{x}_0 = \boldsymbol{x}_0^{\mathrm{T}}\left(\int_0^\infty \boldsymbol{X}(t)\mathrm{d}t\right)\boldsymbol{x}_0 = \boldsymbol{x}_0^{\mathrm{T}}\int_0^\infty (\mathrm{e}^{A^{\mathrm{T}}t}\boldsymbol{Q}\mathrm{e}^{At}\mathrm{d}t)\boldsymbol{x}_0 = \int_0^\infty (\mathrm{e}^{At}\boldsymbol{x}_0)^{\mathrm{T}}\boldsymbol{Q}(\mathrm{e}^{At}\boldsymbol{x}_0)\mathrm{d}t$$

Because $\boldsymbol{Q}$ is positive-definite, let $\boldsymbol{Q}=\boldsymbol{N}^{\mathrm{T}}\boldsymbol{N}$, where $\boldsymbol{N}$ is a nonsingular matrix. We rewrite the above equation as

$$\begin{aligned}
\boldsymbol{x}_0^{\mathrm{T}}\boldsymbol{P}\boldsymbol{x}_0 &= \int_0^\infty (\mathrm{e}^{At}\boldsymbol{x}_0)^{\mathrm{T}}\boldsymbol{N}^{\mathrm{T}}\boldsymbol{N}(\mathrm{e}^{At}\boldsymbol{x}_0)\mathrm{d}t \\
&= \int_0^\infty (\boldsymbol{N}\mathrm{e}^{At}\boldsymbol{x}_0)^{\mathrm{T}}(\boldsymbol{N}\mathrm{e}^{At}\boldsymbol{x}_0)\mathrm{d}t \\
&= \int_0^\infty \|\boldsymbol{N}\mathrm{e}^{At}\boldsymbol{x}_0\|^2\mathrm{d}t > 0
\end{aligned}$$

Thus, $\boldsymbol{P}$ is positive definite.

Example 4.5.3 According to Lyapunov criterion, determine the internal stability of the following system:

$$\begin{cases} \dot{x}_1 = -x_1 + x_2 \\ \dot{x}_2 = x_1 + 5x_2 \end{cases}$$

Answer

$$A=\begin{bmatrix}-1 & 1\\ 1 & 5\end{bmatrix},\ \text{chose } \boldsymbol{Q}=\boldsymbol{I} \text{ and let } \boldsymbol{P}=\boldsymbol{P}^{\mathrm{T}}=\begin{bmatrix}p_{11} & p_{12}\\ p_{12} & p_{22}\end{bmatrix}$$

The Lyapunov equation is

$$\boldsymbol{A}^{\mathrm{T}}\boldsymbol{P}+\boldsymbol{P}\boldsymbol{A}=-\boldsymbol{Q}=-\boldsymbol{I}$$

$$\begin{bmatrix}-1 & 1\\ 1 & 5\end{bmatrix}\cdot\begin{bmatrix}p_{11} & p_{12}\\ p_{12} & p_{22}\end{bmatrix}+\begin{bmatrix}p_{11} & p_{12}\\ p_{12} & p_{22}\end{bmatrix}\cdot\begin{bmatrix}-1 & 1\\ 1 & 5\end{bmatrix}=\begin{bmatrix}-1 & 0\\ 0 & -1\end{bmatrix}$$

$$\begin{bmatrix}p_{12}-2p_{11}+p_{12} & p_{11}+4p_{12}+p_{22}\\ p_{11}+4p_{12}+p_{22} & 2p_{12}+10p_{22}\end{bmatrix}=\begin{bmatrix}-1 & 0\\ 0 & -1\end{bmatrix}$$

The elements of matrices on both sides of the above equation are equal to each other. So, we get

$$\begin{cases}p_{12}-2p_{11}+p_{12}=-1\\ p_{11}+4p_{12}+p_{22}=0\\ 2p_{21}+10p_{22}=-1\end{cases}$$

The solution of the above equation set is $\begin{cases}p_{11}=\dfrac{5}{12}\\ p_{12}=-\dfrac{1}{12}\\ p_{22}=-\dfrac{1}{12}\end{cases}$, $\boldsymbol{P}=\begin{bmatrix}\dfrac{5}{12} & -\dfrac{1}{12}\\ -\dfrac{1}{12} & -\dfrac{1}{12}\end{bmatrix}$.

We can use Matlab function lyap($\boldsymbol{A}^T$, [1 0; 0 1]) to solve the above Lyapunov equation to get $\boldsymbol{P}$. Because first-order and second-order principle minors of $\boldsymbol{P}$ are 5/12, −0.0417, respectively. So P is not positive-definite and the system is not asymptotically stable.

4.5.2 State motion stability criterion of linear time-varying systems

Consider a LTV autonomous system described by

$$\dot{\boldsymbol{x}}=\boldsymbol{A}(t)\boldsymbol{x},\ \boldsymbol{x}(t_0)=\boldsymbol{x}_0, t\geqslant t_0 \tag{4.5.6}$$

The criteria for stability of the system eq. (4.5.6) are discussed as the following.

Conclusion 4.5.4 State transition matrix criterion. For the LTV system eq. (4.5.6),

(1) The equilibria of the system are stable in the sense of Lyapunov at time t_0 if and only if

$$\|\boldsymbol{\Phi}(t,t_0)\|\leqslant k(t_0)<\infty,\ \forall t\geqslant t_0 \tag{4.5.7}$$

where $k(t_0)$ is a constant depends on t_0 and $\boldsymbol{\Phi}(t,t_0)$ is the state transition matrix of the system.

Moreover, if for any t_0, there exists a constant k that doesn't depend on t_0, such that

$$\|\boldsymbol{\Phi}(t,t_0)\|\leqslant k<\infty,\ \forall t\geqslant t_0$$

Then every equilibrium point of the system is uniformly stable in the sense of Lyapunov.

(2) The system is asymptotically stable at time t_0 if and only if

$$\begin{cases}\|\boldsymbol{\Phi}(t,t_0)\|\leqslant k(t_0)<\infty,\ \forall t\geqslant t_0\\ \lim\limits_{t\to\infty}\|\boldsymbol{\Phi}(t,t_0)\|=\boldsymbol{0}\end{cases} \tag{4.5.8}$$

The equilibrium $\boldsymbol{x}_e$ is uniformly asymptotically stable at $[0,\infty)$, if only if there exists the constants k_1 and k_2 that doesn't depend on t_0, for any $t \geqslant t_0$, such that

$$\|\boldsymbol{\Phi}(t,t_0)\| \leqslant k_1 \mathrm{e}^{-k_2(t-t_0)} \tag{4.5.9}$$

Proof omitted.

Conclusion 4.5.5 Lyapunov criterion. For the LTV system (4.5.6), $\boldsymbol{x}_e = \boldsymbol{0}$ is it's a unique equilibrium. The origin equilibrium is uniform asymptotically stable if and only if equation

$$-\dot{\boldsymbol{P}}(t) = \boldsymbol{P}(t)\boldsymbol{A}(t) + \boldsymbol{A}^{\mathrm{T}}(t)\boldsymbol{P}(t) + \boldsymbol{Q}(t),\ \forall t \geqslant t_0 \tag{4.5.10}$$

has a unique real symmetric bounded positive-definite solution $\boldsymbol{P}(t)$ for any given real symmetric bounded positive-definite matrix $\boldsymbol{Q}(t)$ which satisfies

$$0 < \beta_1 \boldsymbol{I} \leqslant \boldsymbol{Q}(t) \leqslant \beta_2 \boldsymbol{I},\ \forall t \geqslant t_0 \tag{4.5.11}$$

where $\beta_2 > \beta_1 > 0$. $\boldsymbol{P}(t)$ is bounded positive-definite means

$$0 < a_1 \boldsymbol{I} \leqslant \boldsymbol{P}(t) \leqslant a_2 \boldsymbol{I},\quad \forall t \geqslant t_0 \tag{4.5.12}$$

where $a_2 > a_1 > 0$.

Proof omitted.

Problems

4.1 Given a SISO LTI system

$$\dot{\boldsymbol{x}} = \begin{bmatrix} 0 & 1 & 0 \\ 0 & 0 & 1 \\ 250 & 0 & -5 \end{bmatrix} \boldsymbol{x} + \begin{bmatrix} 0 \\ 0 \\ 10 \end{bmatrix} \boldsymbol{u}$$

$$y = [-25 \quad 5 \quad 0]\boldsymbol{u}$$

(1) Determine whether the system is asymptotically stable or not;

(2) Determine whether the system is BIBO stable or not.

4.2 Given a time-continuous nonlinear time-invariant system

$$\begin{cases} \dot{x}_1 = x_2 \\ \dot{x}_2 = -x_1 - x_1^2 x_2 \end{cases}$$

Determine whether the origin equilibrium $\boldsymbol{x}_e = \boldsymbol{0}$ is global asymptotically stable or not.

4.3 Given a time-continuous nonlinear time-invariant system

$$\begin{cases} \dot{x}_1 = x_2 \\ \dot{x}_2 = -x_1^3 - x_2 \end{cases}$$

Determine whether the origin equilibrium $\boldsymbol{x}_e = \boldsymbol{0}$ is global asymptotically stable or not.

4.4 Given a time-continuous nonlinear time-varying system

$$\dot{\boldsymbol{x}} = \begin{bmatrix} 0 & 1 \\ -\dfrac{1}{t+1} & -10 \end{bmatrix} \boldsymbol{x},\quad t \geqslant 0$$

Determinewhether the origin equilibrium $\boldsymbol{x}_e = \boldsymbol{0}$ is global asymptotically stable or not.

(Here, chose $V(\boldsymbol{x},t) = \frac{1}{2}[x_1^2 + (t+1)x_2^2]$)

4.5 Use the Krasovskii method to determine whether the following system are asymptotically stable

$$\begin{cases} \dot{x}_1 = -3x_1 + x_2 \\ \dot{x}_2 = x_1 - x_2 - x_2^3 \end{cases}$$

4.6 Given a LTI system:

$$\dot{\boldsymbol{x}} = \begin{bmatrix} a_{11} & a_{12} \\ a_{21} & a_{22} \end{bmatrix} \boldsymbol{x} \triangleq \boldsymbol{A}\boldsymbol{x}$$

Using the Lyapunov criterion to prove the origin equilibrium $\boldsymbol{x}_e = \boldsymbol{0}$ is globally asymptotically stable if

$$\det \boldsymbol{A} > 0, \quad a_{11} + a_{22} < 0$$

(Tip: chose $\boldsymbol{Q} = \boldsymbol{I}$).

4.7 Using the Lyapunov criterion to check whether the following system is globally asymptotically stable or not.

$$\dot{\boldsymbol{x}} = \begin{bmatrix} -1 & 1 \\ 2 & -3 \end{bmatrix} \boldsymbol{x}, \quad \boldsymbol{Q} = \boldsymbol{I}$$

4.8 Consider a asymptotically stable SISO LTI system

$$\dot{\boldsymbol{x}} = \boldsymbol{A}\boldsymbol{x} + \boldsymbol{b}u, \quad y = \boldsymbol{c}\boldsymbol{x}, \quad \boldsymbol{x}(0) = \boldsymbol{x}_0$$

Where $u \equiv 0$. Let $\boldsymbol{P}$ is the positive definite symmetric solution of the following Lyapunov equation

$$\boldsymbol{P}\boldsymbol{A} + \boldsymbol{A}^{\mathrm{T}}\boldsymbol{P} = -\boldsymbol{c}^{\mathrm{T}}\boldsymbol{c}$$

Try to prove

$$\int_0^{\infty} y^2(t)\,dt = \boldsymbol{x}_0^{\mathrm{T}}\boldsymbol{p}\boldsymbol{x}_0$$

4.9 Consider a controllable LTI system

$$\dot{\boldsymbol{x}} = \boldsymbol{A}\boldsymbol{x} + \boldsymbol{B}\boldsymbol{u}$$

Chose $\boldsymbol{u} = -\boldsymbol{B}^{\mathrm{T}}\mathrm{e}^{-\boldsymbol{A}^{\mathrm{T}}t}\boldsymbol{W}^{-1}(0,T)\boldsymbol{x}_0$ and

$$W(0,T) = \int_0^{\mathrm{T}} \mathrm{e}^{-\boldsymbol{A}t}\boldsymbol{B}\boldsymbol{B}^{\mathrm{T}}\mathrm{e}^{-\boldsymbol{A}^{\mathrm{T}}t}\mathrm{d}t, \quad T > 0$$

Try to prove the closed loop system is asymptotically stable.

Chapter 5 Time-domain Synthesis of Linear Systems

In the previous chapters, we discuss both quantitative and qualitative analyses of systems. System quantitative analysis study on state motion. The qualitative analysis includes controllability and observability analysis, system stability analysis, etc. In this chapter, we mainly discuss system synthesis. Based on qualitative analysis, if the system is unstable, the system synthesis method such as state feedback can be used to place the eigenvalues of the system, so that the unstable system can become stable system, and the performance of the system can be improved. System synthesis is also called system design, refers to the design of a control law under the condition that the structure and parameters of the system are known or partially so that the behavior of the system under its action meets the expected performance specification.

The performance indexes of synthetic problems can be divided into non-optimal performance indexes and optimal performance indexes. If asymptotically stability is taken as the performance index, the corresponding synthesis problem is called the stabilization problem. If a set of desired poles of a closed-loop system is used as the performance index, the corresponding synthesis problem is called the pole assignment problem. If one input controls only one output in a multi-input-multiple output system as a performance index, the corresponding synthesis problem is called the decoupling control problem. If an external signal $\boldsymbol{r}(t)$ is followed by the output of the system, $\boldsymbol{y}$, without steady error as a performance indicator, the corresponding synthesis problem is called the tracking problem. All the above performance indexes are non-optimization performance indexes. If a quadratic integral with respect to the state $\boldsymbol{x}$ and control $\boldsymbol{u}$ is taken as the performance index, and a control law is determined to minimize the performance index, this synthesis problem is called optimal control. We can use the state feedback to realized all the above system synthesis problems. In addition to the state feedback, there are unit feedback, cascade compensation and other synthesis methods.

This chapter mainly includes the state feedback for single-input systems and multiple-input systems, the effect of state feedback on controllability, observability, and the transfer function matrix, stabilization and decoupling control through the state feedback, the state estimator, the state feedback using the state estimator, etc.

5.1 State feedback and output feedback

5.1.1 Definition of state feedback and output feedback

1. State feedback

Consider the LTI system described by

$$\begin{cases} \dot{\boldsymbol{x}} = \boldsymbol{A}\boldsymbol{x} + \boldsymbol{B}\boldsymbol{u} \\ \boldsymbol{y} = \boldsymbol{C}\boldsymbol{x} \end{cases} \tag{5.1.1}$$

Introduce the following state feedback control law:

$$\boldsymbol{u} = -\boldsymbol{K}\boldsymbol{x} + \boldsymbol{v} \tag{5.1.2}$$

where, $\boldsymbol{v}$ is reference input. For the regulator problem, $\boldsymbol{v} = \boldsymbol{0}$. For the tracking problem, $\boldsymbol{v}$ is nonzero given function vector. $\boldsymbol{K}$ is a $p \times n$ constant gain matrix. We call the system after introducing the state feedback the closed-loop system. Substitute eq. (5.1.2) into eq. (5.1.1), we get the state space description of the closed-loop system as

$$\begin{cases} \dot{\boldsymbol{x}} = (\boldsymbol{A} - \boldsymbol{B}\boldsymbol{K})\boldsymbol{x} + \boldsymbol{B}\boldsymbol{v} \\ \boldsymbol{y} = \boldsymbol{C}\boldsymbol{x} \end{cases} \tag{5.1.3}$$

Figure 5.1.1 shows the diagram of the closed-loop system. The transfer matrix of the closed-loop system is

$$\boldsymbol{G}_K(s) = \boldsymbol{C}(s\boldsymbol{I} - \boldsymbol{A} + \boldsymbol{B}\boldsymbol{K})^{-1}\boldsymbol{B} \tag{5.1.4}$$

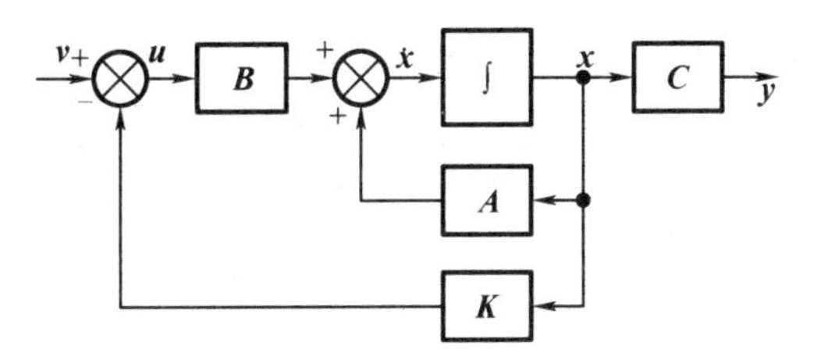

Figure 5.1.1 State feedback

2. Output feedback

Given the LTI system (5.1.1), introduce the following output feedback control law:

$$\boldsymbol{u} = -\boldsymbol{F}\boldsymbol{y} + \boldsymbol{v} \tag{5.1.5}$$

where, $\boldsymbol{v}$ is a reference input, $\boldsymbol{F}$ is a $p \times q$ constant gain matrix. We call the system after introducing the output feedback the closed-loop system. Substitute eq. (5.1.5) into eq. (5.1.1), we get the state space description of the closed-loop system as

$$\begin{cases} \dot{\boldsymbol{x}} = (\boldsymbol{A} - \boldsymbol{B}\boldsymbol{F}\boldsymbol{C})\boldsymbol{x} + \boldsymbol{B}\boldsymbol{v} \\ \boldsymbol{y} = \boldsymbol{C}\boldsymbol{x} \end{cases} \tag{5.1.6}$$

Figure 5.1.2 shows the diagram of the closed-loop system. The transfer matrix of the closed-loop system is

$$G_F(s) = C(sI - A + BFC)^{-1}B \tag{5.1.7}$$

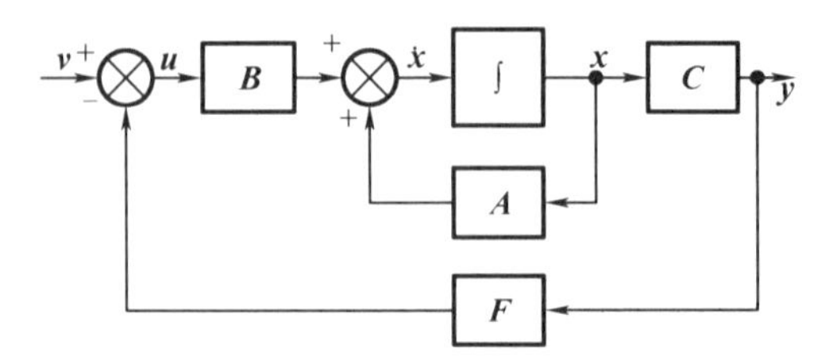

Figure 5.1.2 Output feedback

5.1.2 Comparison between state feedback and output feedback

State feedback has a better ability than output feedback in changing system's structural attributes and improving its performance. From eq. (5.1.3) and eq. (5.1.6), in order to finish the same task of synthesizing systems, the state feedback and output feedback have the following relationship:

$$FC = K \tag{5.1.8}$$

The above equation shows that: if output feedback can meet the requirements of the closed-loop system, it must find corresponding state feedback also. But given a K, we can not find the solution of $FC = K$ in general, i. e. the state feedback can achieve the requirements, output feedback is generally impossible to achieve.

5.2 Effects of state feedback and output feedback on controllability and observability

In this section, we discuss the effects on controllability and observability of the closed-loop system after introducing the state feedback and the output feedback. To better understand, we discuss this problem from single-input systems first.

5.2.1 Effect of state feedback on controllability of single-input systems

The controllability is invariant under state feedback for single-input systems. We discuss it in three situations.

1. If the original open-loop system is completely controllable, then the closed-loop system, (A_k, b_k), is still controllable after introducing the state feedback k.

Proof Consider a controllable LTI system described by

$$\dot{\boldsymbol{x}} = \boldsymbol{A}\boldsymbol{x} + \boldsymbol{b}u \tag{5.2.1}$$

where, $\boldsymbol{x}$ is an n-dimensional state vector, u is 1-dimensional input variable. Since the system state is completely controllable, the rank of the controllability matrix $\boldsymbol{Q}_c$ is n, i. e. rank $\boldsymbol{Q}_c = n$. For convenience, we let

$$\boldsymbol{A} = \left[\begin{array}{c|cccc} 0 & 1 & 0 & \cdots & 0 \\ 0 & 0 & 1 & \cdots & 0 \\ \vdots & 0 & \ddots & \ddots & 0 \\ 0 & 0 & \cdots & 0 & 1 \\ \hline -a_0 & -a_1 & -a_2 & \cdots & -a_{n-1} \end{array}\right], \boldsymbol{b} = \begin{bmatrix} 0 \\ 0 \\ \vdots \\ 0 \\ 1 \end{bmatrix} \tag{5.2.2}$$

i. e. $(\boldsymbol{A}, \boldsymbol{b})$ is controllable canonical form. If the matrix $(\boldsymbol{A}, \boldsymbol{b})$ is not controllable canonical form, it can be transformed into the above controllable canonical form by linear nonsingular transformation. According to conclusion 3. 7. 1, this transformation does not change the controllability of the system.

After introducing state feedback $\boldsymbol{k}$, the state equation of the closed-loop system is

$$\dot{\boldsymbol{x}} = \boldsymbol{A}_k\boldsymbol{x} + \boldsymbol{b}_k u \tag{5.2.3}$$

where $\boldsymbol{k} = [k_1 \quad k_2 \quad \cdots \quad k_n]$, and

$$\boldsymbol{A}_k = \boldsymbol{A} - \boldsymbol{b}\boldsymbol{k} = \begin{bmatrix} 0 & 1 & 0 & \cdots & 0 \\ 0 & 0 & 1 & \cdots & 0 \\ \vdots & 0 & \ddots & \ddots & 0 \\ 0 & 0 & \cdots & 0 & 1 \\ -k_1 - a_0 & -k_2 - a_1 & -k_3 - a_2 & \cdots & -k_n - a_{n-1} \end{bmatrix}, \boldsymbol{b}_k = \begin{bmatrix} 0 \\ 0 \\ \vdots \\ 0 \\ 1 \end{bmatrix} \tag{5.2.4}$$

Because $(\boldsymbol{A}_k, \boldsymbol{b}_k)$ is controllable canonical form, the closed-loop system after introducing the state feedback $\boldsymbol{k}$ is also controllable.

2. If the original open-loop system is completely uncontrollable, then the closed-loop system, $(\boldsymbol{A}_k, \boldsymbol{b}_k)$, is still completely uncontrollable after introducing the state feedback $\boldsymbol{k}$.

Proof Let the state space description of the system is

$$\begin{aligned} \dot{\boldsymbol{x}}_{\bar{c}} &= \boldsymbol{A}_{\bar{c}}\boldsymbol{x}_{\bar{c}} \\ \boldsymbol{y} &= \boldsymbol{C}_{\bar{c}}\boldsymbol{x}_{\bar{c}} \end{aligned} \tag{5.2.5}$$

If $(\boldsymbol{A}_{\bar{c}}, \boldsymbol{b}_{\bar{c}})$ is not the above form, the system can be transformed into the above form by linear nonsingular transformation. According to conclusion 3. 7. 1, this transformation does not change the controllability of the system.

After introducing state feedback $\boldsymbol{k}$, the state equation of the closed-loop system is

$$\dot{\boldsymbol{x}}_{\bar{c}} = \boldsymbol{A}_{\bar{c}}\boldsymbol{x}_{\bar{c}} \tag{5.2.6}$$

Hence the closed-loop system after introducing the state feedback $\boldsymbol{k}$ is also completely uncontrollable.

3. If the original open-loop system is not completely controllable, then the closed-loop

system, $(\boldsymbol{A}_k, \boldsymbol{b}_k)$, is still not completely controllable after introducing the state feedback $\boldsymbol{k}$.

Proof Let the state space description of the system is

$$\begin{cases} \dot{\boldsymbol{x}} = \boldsymbol{A}\boldsymbol{x} + \boldsymbol{b}u \\ y = \boldsymbol{C}\boldsymbol{x} \end{cases} \tag{5.2.7}$$

We introduce linearly nonsingular transformation $\bar{\boldsymbol{x}} = \boldsymbol{P}\boldsymbol{x}$ to decompose the system according to the controllability. The decomposed system is described by

$$\begin{bmatrix} \dot{\bar{\boldsymbol{x}}}_c \\ \dot{\bar{\boldsymbol{x}}}_{\bar{c}} \end{bmatrix} = \begin{bmatrix} \bar{\boldsymbol{A}}_c & \bar{\boldsymbol{A}}_{12} \\ \boldsymbol{0} & \bar{\boldsymbol{A}}_{\bar{c}} \end{bmatrix} \begin{bmatrix} \bar{\boldsymbol{x}}_c \\ \bar{\boldsymbol{x}}_{\bar{c}} \end{bmatrix} + \begin{bmatrix} \bar{\boldsymbol{b}}_c \\ \boldsymbol{0} \end{bmatrix} \boldsymbol{u}$$

$$y = \begin{bmatrix} \bar{\boldsymbol{C}}_c & \bar{\boldsymbol{C}}_{\bar{c}} \end{bmatrix} \begin{bmatrix} \bar{\boldsymbol{x}}_c \\ \bar{\boldsymbol{x}}_{\bar{c}} \end{bmatrix} \tag{5.2.8}$$

According to conclusion 3.7.1, this transformation does not change the controllability of the system. So we analyze controllability of the closed-loop system after introducing the state feedback $\boldsymbol{k}$ based on eq. (5.2.8). From eq. (5.2.8), the system are decomposed into controllable subsystem and uncontrollable subsystem. The controllable subsystem is described by

$$\begin{cases} \dot{\bar{\boldsymbol{x}}}_c = \bar{\boldsymbol{A}}_c \bar{\boldsymbol{x}}_c + \bar{\boldsymbol{A}}_{12} \bar{\boldsymbol{x}}_{\bar{c}} + \bar{\boldsymbol{b}}_c \boldsymbol{u} \\ y_1 = \bar{\boldsymbol{C}}_c \bar{\boldsymbol{x}}_c \end{cases} \tag{5.2.9}$$

The uncontrollable subsystem is described by

$$\begin{cases} \dot{\bar{\boldsymbol{x}}}_{\bar{c}} = \bar{\boldsymbol{A}}_{\bar{c}} \bar{\boldsymbol{x}}_{\bar{c}} \\ y_2 = \bar{\boldsymbol{C}}_{\bar{c}} \bar{\boldsymbol{x}}_{\bar{c}} \end{cases} \tag{5.2.10}$$

According to the previous proof, the closed-loop system is still controllable after introducing state feedback to the controllable subsystem. The closed-loop system is still completely uncontrollable after introducing state feedback to the uncontrollable subsystem. The diagram of closed-loop system with state feedback is shown in Figure 5.2.1.

Through the proof of this conclusion, we can have a deeper understanding of the content of canonical decomposition of systems. For the uncontrollable subsystem, state feedback cannot change the system matrix of the corresponding subsystem. Next, we will give proof that state feedback does not affect the controllability of multiple-input systems.

5.2.2 Effect of state feedback on controllability of multiple-input systems

For multiple-input systems, the controllability is also invariant under state feedback.

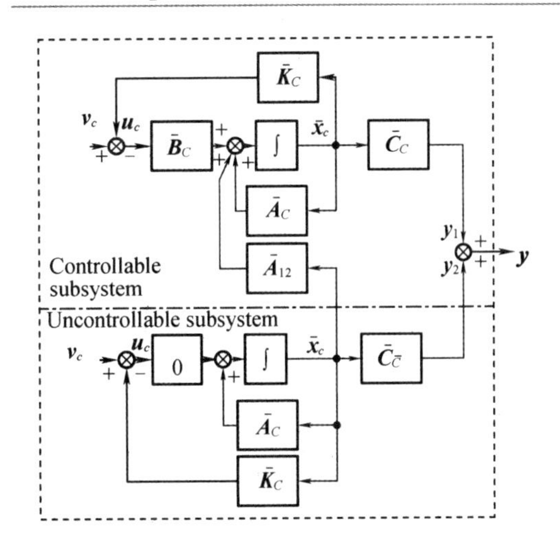

Figure 5.2.1 Diagram of a state feedback system

Conclusion 5.2.1 Consider a MIMO LTI system described by

$$\begin{cases} \dot{\boldsymbol{x}} = \boldsymbol{A}\boldsymbol{x} + \boldsymbol{B}\boldsymbol{u} \\ \boldsymbol{y} = \boldsymbol{C}\boldsymbol{x} \end{cases} \tag{5.2.11}$$

where, $\boldsymbol{x}$ is an $n \times 1$ state vector, $\boldsymbol{u}$ is a $p \times 1$ input vector, $\boldsymbol{y}$ is a $q \times 1$ output vector. Controllability of the closed-loop system is also invariant after introducing the state feedback $\boldsymbol{K}$.

Proof After introducing the state feedback $\boldsymbol{K}$, the state equation of the closed-loop system is

$$\dot{\boldsymbol{x}} = \boldsymbol{A}_k\boldsymbol{x} + \boldsymbol{B}_k\boldsymbol{u} \tag{5.2.12}$$

where $\boldsymbol{A}_k = \boldsymbol{A} - \boldsymbol{B}\boldsymbol{K}$, $\boldsymbol{B}_k = \boldsymbol{B}$. Because

$$[s\boldsymbol{I} - \boldsymbol{A} \vdots \boldsymbol{B}] = [s\boldsymbol{I} - \boldsymbol{A} + \boldsymbol{B}\boldsymbol{K} \vdots \boldsymbol{B}]\begin{bmatrix} \boldsymbol{I}_{n\times n} & \boldsymbol{0} \\ -\boldsymbol{K} & \boldsymbol{I}_{p\times p} \end{bmatrix}$$

For any $\boldsymbol{K}$ and s, we have

$$\text{rank}[s\boldsymbol{I} - \boldsymbol{A} \vdots \boldsymbol{B}] = \text{rank}[s\boldsymbol{I} - \boldsymbol{A} + \boldsymbol{B}\boldsymbol{K} \vdots \boldsymbol{B}]\begin{bmatrix} \boldsymbol{I}_{n\times n} & \boldsymbol{0} \\ -\boldsymbol{K} & \boldsymbol{I}_{p\times p} \end{bmatrix} = \text{rank}[s\boldsymbol{I} - \boldsymbol{A} + \boldsymbol{B}\boldsymbol{K} \vdots \boldsymbol{B}]$$

Hence, according to the PBH rank criterion, state feedback does not change the controllability of single-input systems or multi-input systems.

5.2.3 Effect of state feedback on observability

Conclusion 5.2.2 The observability of systems may be changed after introducing state feedback.

Proof Here, we explain the reason by an example. Consider a LTI system

$$\dot{\boldsymbol{x}} = \begin{bmatrix} 1 & 2 \\ 0 & 3 \end{bmatrix}\boldsymbol{x} + \begin{bmatrix} 0 \\ 1 \end{bmatrix}\boldsymbol{u}$$

$$y = [1 \quad 1]\boldsymbol{x}$$

Calculate the rank of the observability matrix

$$\text{rank } \boldsymbol{Q}_o = \text{rank}\begin{bmatrix} \boldsymbol{c} \\ \boldsymbol{cA} \end{bmatrix} = \begin{bmatrix} 1 & 1 \\ 1 & 5 \end{bmatrix} = 2 = n$$

Thus the system is observable. The open-loop system's transfer function is

$$g(s) = \frac{s+1}{(s-1)(s-3)}$$

Introduce the state feedback $\boldsymbol{k} = [0 \quad 4]$, the state space description of the closed-loop system is

$$\dot{\boldsymbol{x}} = (\boldsymbol{A} - \boldsymbol{bk})\boldsymbol{x} + \boldsymbol{b}v = \begin{bmatrix} 1 & 2 \\ 0 & -1 \end{bmatrix}\boldsymbol{x} + \begin{bmatrix} 0 \\ 1 \end{bmatrix}\boldsymbol{v}$$

$$\boldsymbol{y} = [1 \quad 1]\boldsymbol{x}$$

The rank of the observability matrix of the closed-loop system is

$$\text{rank } \boldsymbol{Q}_{ko} = \text{rank}\begin{bmatrix} \boldsymbol{c} \\ \boldsymbol{c}(\boldsymbol{A} - \boldsymbol{bk}) \end{bmatrix} = \begin{bmatrix} 1 & 1 \\ 1 & 1 \end{bmatrix} = 1 < n$$

The open-loop system's transfer function is

$$g_k(s) = \frac{s+1}{(s-1)(s+1)} = \frac{1}{s-1}$$

Obviously, the closed-loop system is not observable. It indicates that state feedback may change the observability of systems. Moreover, when the cancellation between pole and zero takes place, the observability is not be preserved.

5.2.4 Effect of output feedback on controllability and observability

Conclusion 5. 2. 3 Controllability and observability are invariant under any output feedback.

Proof (1) Because

$$[(s\boldsymbol{I} - \boldsymbol{A}) \quad \boldsymbol{B}] = [(s\boldsymbol{I} - \boldsymbol{A} + \boldsymbol{BFC}) \quad \boldsymbol{B}]\begin{bmatrix} \boldsymbol{I}_n & 0 \\ -\boldsymbol{FC} & \boldsymbol{I}_p \end{bmatrix}$$

$$\text{rank } [(s\boldsymbol{I} - \boldsymbol{A} + \boldsymbol{BFC}) \quad \boldsymbol{B}] = \text{rank}[(s\boldsymbol{I} - \boldsymbol{A}) \quad \boldsymbol{B}], \ \forall s \in C$$

According to the PBH rank criterion, the output feedback does not change controllability.

(2) Because

$$\text{rank}\begin{bmatrix} \boldsymbol{C} \\ s\boldsymbol{I} - \boldsymbol{A} \end{bmatrix} = \text{rank}\begin{bmatrix} \boldsymbol{I}_{q \times q} & \boldsymbol{0} \\ -\boldsymbol{BF} & \boldsymbol{I}_{n \times n} \end{bmatrix}\begin{bmatrix} \boldsymbol{C} \\ s\boldsymbol{I} - \boldsymbol{A} + \boldsymbol{BFC} \end{bmatrix} = \text{rank}\begin{bmatrix} \boldsymbol{C} \\ s\boldsymbol{I} - \boldsymbol{A} + \boldsymbol{BFC} \end{bmatrix}, \ \forall s \in C$$

According to the PBH rank criterion, the output feedback does not change observability.

5.3 Pole placement of single-input systems

Pole placement problem is also called pole assignment problems. From Section 5. 1, introducing state feedback can change the system matrix of the system, $\boldsymbol{A}$, i. e. change the eigenvalues of the closed-loop system. According to eq. (5. 1. 4), the poles of the transfer matrix of the closed-loop system are also changed. When the system is completely observable and controllable, the eigenvalues of the system are the same as the poles of the transfer matrix which will be proved later in the following section. State feedback has the same effect in changing the eigenvalues and poles of systems. Thus, we often call eigenvalue placement pole assignment. By state feedback, the poles of the closed-loop system can be placed in the desired position, so that the stability, dynamic performance, and steady-state accuracy of the system can be improved. The selection of desired closed-loop poles is determined by performance indexes, such as transient time, overshoot, and steady-state error in the time domain, and gain margin and phase margin in the frequency domain.

In this section, we deal with the problem of pole placement in a single input system using state feedback.

5.3.1 Role of poles

The value of poles, i. e. the position of poles in the S-plane, affects the stability, transient, and steady-state performance of the system. Figure 5. 3. 1 shows the effect of pole position on system performance through the unit step response of SISO LTI systems.

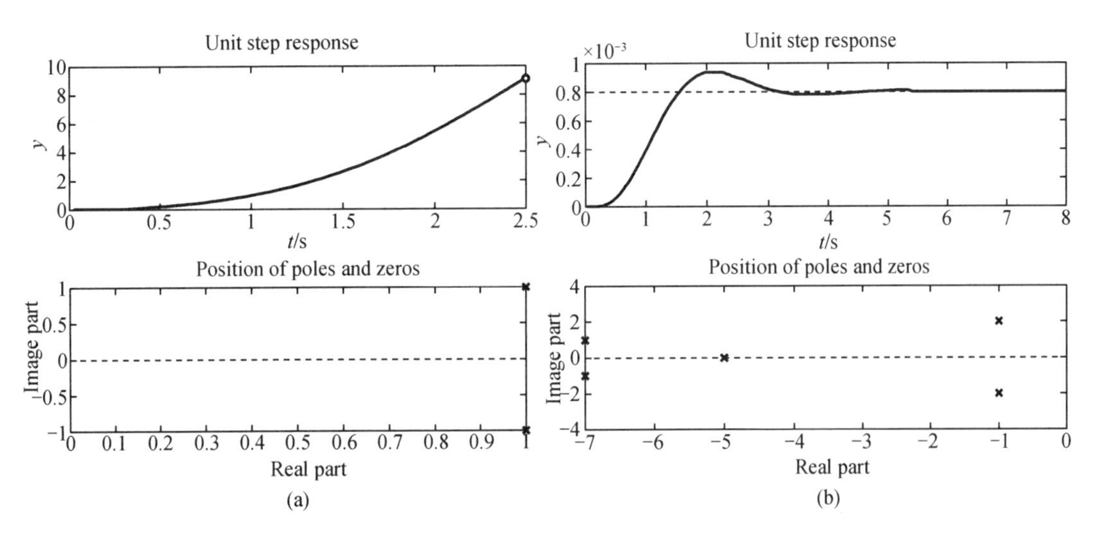

Figure 5.3.1 Role of poles

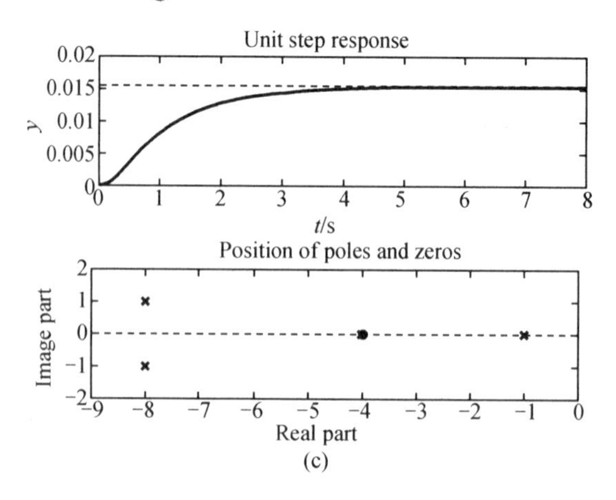

Figure 5.3.1 (Continued)

In Figure 5.3.1(a), the system has poles with positive real parts. So, it is unstable. In Figure 5.3.1(b), all poles have negative real parts and the performance of the system depends on the leading poles, $\lambda_{1,2} = -1 \pm 2j$. Because the damping ratio decided by the leading poles is less than one, the system overshoots. In Figure 5.3.1(c), all poles have negative real parts. So, the system is asymptotically stable. The performance mainly depends on the pole $\lambda_1 = -1$, and the corresponding damping ratio is one. Therefore, the system has no overshoot.

5.3.2 Conditions to place poles

We discuss the conditions to place all poles of closed-loop systems by state feedback for a single-input LTI systems.

Conclusion 5.3.1 Consider a plant described by the n-dimensional one input state equation

$$\begin{cases} \dot{\boldsymbol{x}} = \boldsymbol{A}\boldsymbol{x} + \boldsymbol{b}u \\ y = \boldsymbol{C}\boldsymbol{x} \end{cases} \tag{5.3.1}$$

if and only if it is completely controllable, then all poles of the system can be assigned arbitrarily by state feedback.

Proof Sufficiency. Prove all poles can be assigned arbitrarily if it is completely controllable.

By $\bar{\boldsymbol{x}} = \boldsymbol{P}^{-1}\boldsymbol{x}$, we transform $(\boldsymbol{A}, \boldsymbol{b})$ into the controllable canonical form:

$$\boldsymbol{A}_c = \boldsymbol{P}^{-1}\boldsymbol{A}\boldsymbol{P} = \left[\begin{array}{c:cccc} 0 & 1 & 0 & \cdots & 0 \\ 0 & 0 & 1 & \cdots & 0 \\ & & & \ddots & \\ 0 & 0 & \cdots & & 1 \\ \hdashline -a_0 & -a_1 & -a_2 & \cdots & -a_{n-1} \end{array}\right], \quad \boldsymbol{b}_c = \boldsymbol{P}^{-1}\boldsymbol{b} = \begin{bmatrix} 0 \\ 0 \\ \vdots \\ 0 \\ 1 \end{bmatrix} \tag{5.3.2}$$

The non-zero element of $\boldsymbol{A}_c$, $a_i (i = 0, 1, \cdots, n-1)$, comes from the characteristic polynomial of $\boldsymbol{A}_c$ or $\boldsymbol{A}$. Let the characteristic polynomial of $\boldsymbol{A}_c$ is

$$\det[s\boldsymbol{I}-\boldsymbol{A}_c]=\det[s\boldsymbol{I}-\boldsymbol{A}]=s^n+a_{n-1}s^{n-1}+\cdots+a_1s+a_0 \tag{5.3.3}$$

and state feedback matrix $\boldsymbol{k}_c=[k_{c1}\ k_{c2}\cdots k_{cn}]$. After introducing the state feedback, the obtained state equation of the closed-loop system is

$$\dot{\boldsymbol{x}}=\boldsymbol{A}_{ck}\boldsymbol{x}+\boldsymbol{b}_{ck}v \tag{5.3.4}$$

where,

$$\boldsymbol{A}_{ck}=\boldsymbol{A}_c-\boldsymbol{b}_c\boldsymbol{k}_c=\begin{bmatrix}0 & 1 & 0 & \cdots & 0\\ 0 & 0 & 1 & \cdots & 0\\ & & \cdots & & \\ -a_0-k_{c1} & -a_1-k_{c2} & -a_2-k_{c3} & \cdots & -a_{n-1}-k_{cn}\end{bmatrix},\boldsymbol{b}_{ck}=\boldsymbol{P}^{-1}\boldsymbol{b}=\begin{bmatrix}0\\ \vdots\\ 0\\ 1\end{bmatrix} \tag{5.3.5}$$

The characteristic polynomial of $\boldsymbol{A}_{ck}$ is:

$$\det[s\boldsymbol{I}-\boldsymbol{A}_{ck}]=s^n+(a_{n-1}+k_{cn})s^{n-1}+\cdots+(a_1+k_{c2})s+(a_0+k_{c1}) \tag{5.3.6}$$

Let the desired eigenvalues be $\lambda_1^*,\lambda_2^*,\cdots,\lambda_n^*$. So, we get the desired characteristic polynomial

$$D(s)=(s-\lambda_1^*)(s-\lambda_2^*)\cdots(s-\lambda_n^*)=s^n+a_{n-1}^*s^{n-1}+\cdots+a_1^*s+a_0^* \tag{5.3.7}$$

Comparing the coefficient of $s^i(i=0,1,\cdots,n-1)$ in eq. (6.3.6) and eq. (6.3.7), yields

$$\begin{cases}a_{n-1}+k_{cn}=a_{n-1}^*\\ \vdots\\ a_1+k_{c2}=a_1^*\\ a_0+k_{c1}=a_0^*\end{cases}\Rightarrow\begin{cases}k_{cn}=a_{n-1}^*-a_{n-1}\\ \vdots\\ k_{c2}=a_1^*-a_1\\ k_{c1}=a_0^*-a_0\end{cases}$$

So we get the state feedback matrix:

$$\boldsymbol{k}_c=[a_0^*-a_0\quad a_1^*-a_1\quad \cdots\quad a_{n-1}^*-a_{n-1}] \tag{5.3.8}$$

For the original system eq. (5.3.1), the state feedback matrix $\boldsymbol{k}$ which places the closed-loop eigenvalues at $\lambda_1^*,\lambda_2^*,\cdots,\lambda_n^*$, is calculated as the following equation

$$\boldsymbol{k}=\boldsymbol{k}_c\boldsymbol{P}^{-1} \tag{5.3.9}$$

This is because

$$\begin{aligned}\det[s\boldsymbol{I}-\boldsymbol{A}_c+\boldsymbol{b}_c\boldsymbol{k}_c]&=\det[s\boldsymbol{I}-\boldsymbol{P}^{-1}\boldsymbol{A}\boldsymbol{P}+\boldsymbol{P}^{-1}\boldsymbol{b}\boldsymbol{k}\boldsymbol{P}]\\ &=\det[\boldsymbol{P}^{-1}]\det[s\boldsymbol{I}-\boldsymbol{A}+\boldsymbol{b}\boldsymbol{k}]\det[\boldsymbol{P}]\\ &=\det[s\boldsymbol{I}-\boldsymbol{A}+\boldsymbol{b}\boldsymbol{k}]\\ &=(s-\lambda_1^*)(s-\lambda_2^*)\cdots(s-\lambda_n^*)=s^n+a_{n-1}^*s^{n-1}+\cdots+a_1^*s+a_0^*\end{aligned} \tag{5.3.10}$$

The above equation shows that the state feedback matrix $\boldsymbol{k}$ based on eq. (5.3.9) can place all eigenvalues at $\lambda_1^*,\lambda_2^*,\cdots,\lambda_n^*$.

Necessity. Prove system eq. (5.3.1) is completely controllable if all poles can be assigned arbitrarily. Adopt proof by contradiction to prove it. We suppose the system is not completely controllable. So, by introducing nonsingular transformation $\bar{x}=Px$, we decompose eq. (5.3.1) into:

$$\begin{bmatrix}\dot{\bar{x}}_c\\ \dot{\bar{x}}_{\bar{c}}\end{bmatrix}=\begin{bmatrix}\bar{A}_c & \bar{A}_{12}\\ 0 & \bar{A}_{\bar{c}}\end{bmatrix}\begin{bmatrix}\bar{x}_c\\ \bar{x}_{\bar{c}}\end{bmatrix}+\begin{bmatrix}\bar{b}_c\\ 0\end{bmatrix}u$$

$$y=[\bar{C}_c \quad \bar{C}_{\bar{c}}]\begin{bmatrix}\bar{x}_c\\ \bar{x}_{\bar{c}}\end{bmatrix} \tag{5.3.11}$$

Introducing the state feedback to eq. (5.3.11), we get the characteristic polynomial of the closed-loop system

$$\det\begin{bmatrix}\lambda I-\bar{A}_c+\bar{b}_c\bar{k}_c & -\bar{A}_{12}+\bar{b}_c\bar{k}_{\bar{c}}\\ 0 & \lambda I-\bar{A}_{\bar{c}}\end{bmatrix}=\det(\lambda I-\bar{A}_c+\bar{b}_c\bar{k}_c)\det(\lambda I-\bar{A}_{\bar{c}}) \tag{5.3.12}$$

It indicates that state feedback can not change all values of eigenvalues because the eigenvalues of the uncontrollable subsystem can not be changed. Similar to the proof of sufficiency, introducing state feedback to the original system eq. (5.3.1), $k=\bar{k}P$, then the characteristic polynomial of the obtained closed-loop system is also $\det(\lambda I-\bar{A}_c+\bar{b}_c\bar{k}_c)\det(\lambda I-\bar{A}_{\bar{c}})$. Thus, state feedback can not change all eigenvalues of the system eq. (5.3.1) which contradicts that all eigenvalues of the system can be placed at any position.

The procedure of proving the conclusion 5.3.1 is also the procedure of solving the state feedback gain matrix. The algorithm for solving the state feedback vector for single-input LTI systems is given below.

5.3.3 Pole assigning algorithm of single-input systems

Algorithm 6.3.1 Pole assigning algorithm. Given a controllable single-input system $\{A, b\}$ and desired closed-loop eigenvalues $\{\lambda_1^*, \lambda_2^*, \cdots, \lambda_n^*\}$, Find a $1\times n$ feedback gain vector k such that $\lambda_i(A-Bk)=\lambda_i^*$, $i=1,2,\cdots,n$, where $\lambda_i^*\ (i=1,2,\cdots,n)$ are n desired eigenvalues.

Step 1. Compute the characteristic polynomial of A, $\alpha(s)$:

$$\alpha(s)=\det(sI-A)=s^n+a_{n-1}s^{n-1}+\cdots+a_1s+a_0 \tag{5.3.13}$$

Step 2. Introducing $\bar{x}=P^{-1}x$ transforms $\{A, b\}$ into controllable canonical form $\{\bar{A}, \bar{b}\}$. Where we calculate P as the following formula:

$$P=[A^{n-1}b \ \cdots \ Ab \ b]\begin{bmatrix}1 & & & & \\ a_{n-1} & 1 & & & \\ a_{n-2} & a_{n-1} & 1 & & \\ \vdots & & \ddots & \ddots & \\ a_1 & \cdots & & a_{n-1} & 1\end{bmatrix} \tag{5.3.14}$$

Step 3. Compute the desired characteristic polynomial $\alpha^*(s)$ decided by desired eigenvalues $\{\lambda_1^*, \lambda_2^*, \cdots, \lambda_n^*\}$ of the closed-loop system.

$$\alpha^*(s)=\prod_{i=1}^{n}(s-\lambda_i^*)=s^n+a_{n-1}^*s^{n-1}+\cdots+a_1^*s+a_0^* \tag{5.3.15}$$

Step 4. From eq. (6. 3. 13) and eq. (6. 3. 15), compute the state feedback vector $\bar{\boldsymbol{k}}$ for controllable canonical form $\{\bar{\boldsymbol{A}},\bar{\boldsymbol{b}}\}$:

$$\bar{\boldsymbol{k}} = [a_0^* - a_0 \quad a_1^* - a_1 \quad \cdots \quad a_{n-1}^* - a_{n-1}] \tag{5.3.16}$$

We can obtain eq. (5,3,16) from eq. (6.3.2)—(6.3.8).

Step 5. Compute the state feedback vector $\boldsymbol{k}$ for system $\{\boldsymbol{A},\boldsymbol{b}\}$:

$$\boldsymbol{k} = \bar{\boldsymbol{k}}\boldsymbol{P}^{-1} \tag{5.3.17}$$

Example 5.3.1 Consider a single-input LTI system $\{\boldsymbol{A},\boldsymbol{b}\}$

$$\dot{\boldsymbol{x}} = \begin{bmatrix} 0 & 1 & 0 \\ 0 & 0 & 1 \\ 0 & -2 & -3 \end{bmatrix}\boldsymbol{x} + \begin{bmatrix} 0 \\ 0 \\ 1 \end{bmatrix}u$$

Try to determine the state feedback vector $\boldsymbol{k} = [k_1 \quad k_2 \quad k_3]$ such that the eigenvalues of the closed-loop system are placed at $s_1^* = -2$, $s_{2,3}^* = -1 \pm j$, and draw the state variable diagram of the system.

Answer

(1) Check the controllability of the system. Because $\{\boldsymbol{A},\boldsymbol{b}\}$ has controllable canonical form, the system is controllable .

(2) The desired characteristic polynomial of the closed-loop system is

$$D(s) = (s+2)(s+1+j)(s+1-j) = s^3 + 4s^2 + 6s + 4 \tag{5.3.18}$$

(3) The characteristic polynomial of $\boldsymbol{A} - \boldsymbol{bk}$ is

$$\det[s\boldsymbol{I} - \boldsymbol{A}_k] = \det(s\boldsymbol{I} - \boldsymbol{A} + \boldsymbol{bk}) = s^3 + (3+k_3)s^2 + (2+k_2)s + (0+k_1) \tag{5.3.19}$$

(4) Compare the coefficient of $s^i (i=0,1,2)$ in eq. (6.3.18) and eq. (6.3.19), we get

$$\begin{cases} k_1 = 4 \\ k_2 = 6 - 2 = 4 \\ k_3 = 4 - 3 = 1 \end{cases}$$

So, the state feedback gain vector $\boldsymbol{k} = [4\ 4\ 1]$.

(5) State variable diagram of the system is shown in Figure 5.3.2.

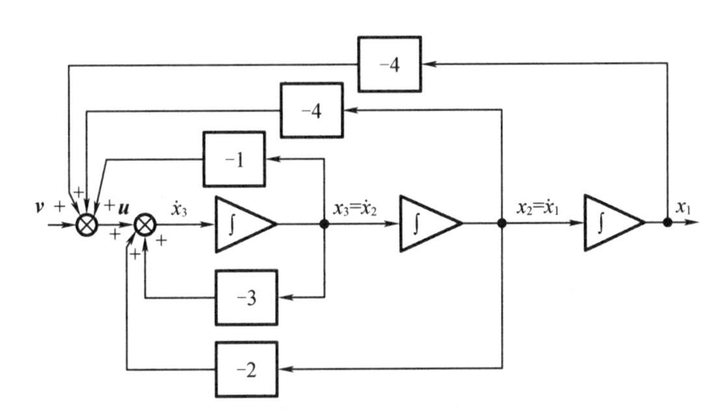

Figure 5.3.2 State variable diagram

Example 5.3.2 Given the transfer function of a SISO system

$$u \rightarrow \boxed{\frac{1}{s(s+1)}} \rightarrow y$$

try to calculate the state feedback gain vector $\boldsymbol{k}$, such that the closed-loop system has the following indices: overshoot $\sigma_p = 0.0432$, transient time $t_s = 4s$.

Answer The state equation of the open-loop system is

$$\dot{\boldsymbol{x}} = \begin{bmatrix} 0 & 1 \\ 0 & -1 \end{bmatrix} \boldsymbol{x} + \begin{bmatrix} 0 \\ 1 \end{bmatrix} u$$

Because it is a controllable canonical form, the system is controllable. The all eigenvalues can placed at any positions by using state feedback.

Let the state feedback gain vector $\boldsymbol{k} = [k_1 \quad k_2]$. The system matrix $\boldsymbol{A}_k$ of the closed-loop system is

$$\boldsymbol{A}_k = \boldsymbol{A} - \boldsymbol{bk} = \begin{bmatrix} 0 & 1 \\ -k_1 & -(k_2+1) \end{bmatrix}$$

The characteristic polynomial of the closed-loop system is

$$\det[s\boldsymbol{I} - \boldsymbol{A}_k] = s^2 + (1+k_2)s + k_1 \tag{5.3.20}$$

Form $\sigma_p = \mathrm{e}^{-\frac{\pi\zeta}{\sqrt{1-\zeta^2}}} = 0.0432$, the damping ratio of the system is $\zeta = \frac{1}{\sqrt{2}} = 0.707$. From $t_s = \frac{4}{\zeta\omega_n} = 4$, undamped oscillation frequency is $\omega_n = \sqrt{2} = 1.414$. So, the desired characteristic polynomial of the closed-loop system is

$$\alpha^*(s) = s^2 + 2\zeta\omega_n s + \omega_n^2 \tag{5.3.21}$$

From eq. (6.3.20) and eq. (6.3.21), we have

$$\begin{cases} k_1 = 2 \\ k_2 = 1 \end{cases}$$

Thus, the state feedback matrix $\boldsymbol{k} = [k_1 \quad k_2] = [2 \quad 1]$.

Example 5.3.3 Pole placement of a one-stage inverted pendulum system. For the one-stage inverted pendulum system shown in Figure 5.3.3, use state feedback to realize the control of the car moving to the desired position and making the swing Angle zero. Where, the car mass $M = 2$ kg, the length of the stick $l = 0.5$ m, the mass of the pendant $m = 0.1$ kg.

Answer First establish the mathematical model of the inverted pendulum system. According to Section 1.2, the state space description of the system is

$$\begin{bmatrix} \dot{x}_1 \\ \dot{x}_2 \\ \dot{x}_3 \\ \dot{x}_4 \end{bmatrix} = \begin{bmatrix} 0 & 1 & 0 & 0 \\ \frac{M+m}{Ml}g & 0 & 0 & 0 \\ 0 & 0 & 0 & 1 \\ -\frac{mg}{M} & 0 & 0 & 0 \end{bmatrix} \begin{bmatrix} x_1 \\ x_2 \\ x_3 \\ x_4 \end{bmatrix} + \begin{bmatrix} 0 \\ -\frac{1}{Ml} \\ 0 \\ \frac{1}{M} \end{bmatrix} u$$

$$y=[0 \quad 0 \quad 1 \quad 0]\boldsymbol{x} \tag{5.3.22}$$

Where $x_1=\theta$, $x_2=\dot{x}_1=\dot{\theta}$, $x_3=y$, $x_4=\dot{x}_3=\dot{y}$.

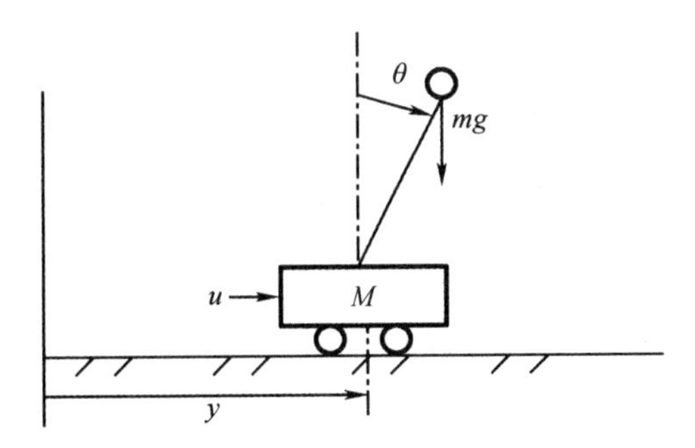

Figure 5.3.3 State feedback of a one-stage inverted pendulum system

Second, check controllability of the system. Because

$$\operatorname{rank}[\boldsymbol{b} \quad \boldsymbol{Ab} \quad \boldsymbol{A}^2\boldsymbol{b} \quad \boldsymbol{A}^3\boldsymbol{b}]=4$$

the inverted pendulum system is controllable and we can use state feedback to place all poles at any desired position.

Let $\dot{\boldsymbol{x}}=\boldsymbol{0}$, we get the equilibrium of the system eq. (5.3.22), i. e. $\theta=0$ and $y=$ any value. In order to move the car to a desired position, introduce a new state variable

$$\dot{x}_5=v-y \tag{5.3.23}$$

Where, v is the desired position of the car. Let

$$\boldsymbol{e}=\begin{bmatrix} x_1(t)-x_1(\infty) \\ x_2(t)-x_2(\infty) \\ x_3(t)-x_3(\infty) \\ x_4(t)-x_4(\infty) \\ x_5(t)-x_5(\infty) \end{bmatrix}$$

We get

$$\dot{\boldsymbol{e}}=\hat{\boldsymbol{A}}\boldsymbol{e}+\hat{\boldsymbol{b}}u_e \tag{5.3.24}$$

where,

$$\hat{\boldsymbol{A}}=\begin{bmatrix} 0 & 1 & 0 & 0 & 0 \\ \frac{M+m}{Ml}g & 0 & 0 & 0 & 0 \\ 0 & 0 & 0 & 1 & 0 \\ -\frac{mg}{M} & 0 & 0 & 0 & 0 \\ 0 & 0 & 1 & 0 & 0 \end{bmatrix}, \quad \hat{\boldsymbol{b}}=\begin{bmatrix} 0 \\ -\frac{1}{Ml} \\ 0 \\ 1 \\ 0 \end{bmatrix}$$

Use state feedback to system (5.3.24) and let

$$u_e(t)=-[k_1 \quad k_2 \quad k_3 \quad k_4 \quad k_5]\boldsymbol{e}=-\boldsymbol{ke}$$

The state equation of the closed-loop system is

$$\dot{e} = (\hat{A} - \hat{b}k)e \tag{5.3.35}$$

Let $\dot{e} = \mathbf{0}$, so $\theta = 0$ and $y = v$ is a equilibrium of the system eq. (5.3.35). According to the performance index, chose suitable state feedback matrix k, such that the eigenvalues of $\hat{A} - \hat{b}k$ are placed at the certain positions on the left side of the S-plane. Hence, $e(t)$ approaches exponentially toward the equilibrium $\theta = 0, y = v$. We chose $-2 \pm j\sqrt{3}$, -6, -6, -6as the desired eigenvalues. Using the pole assigning algorithm of single-input systems, we get the state feedback vector k:

$$k = [\,-214.88 \;\; -48.32 \;\; -104.59 \;\; -58.64 \;\; 89.29\,]$$

Under the state feedback controller, the response curves of the system are shown in Figure 5.3.4.

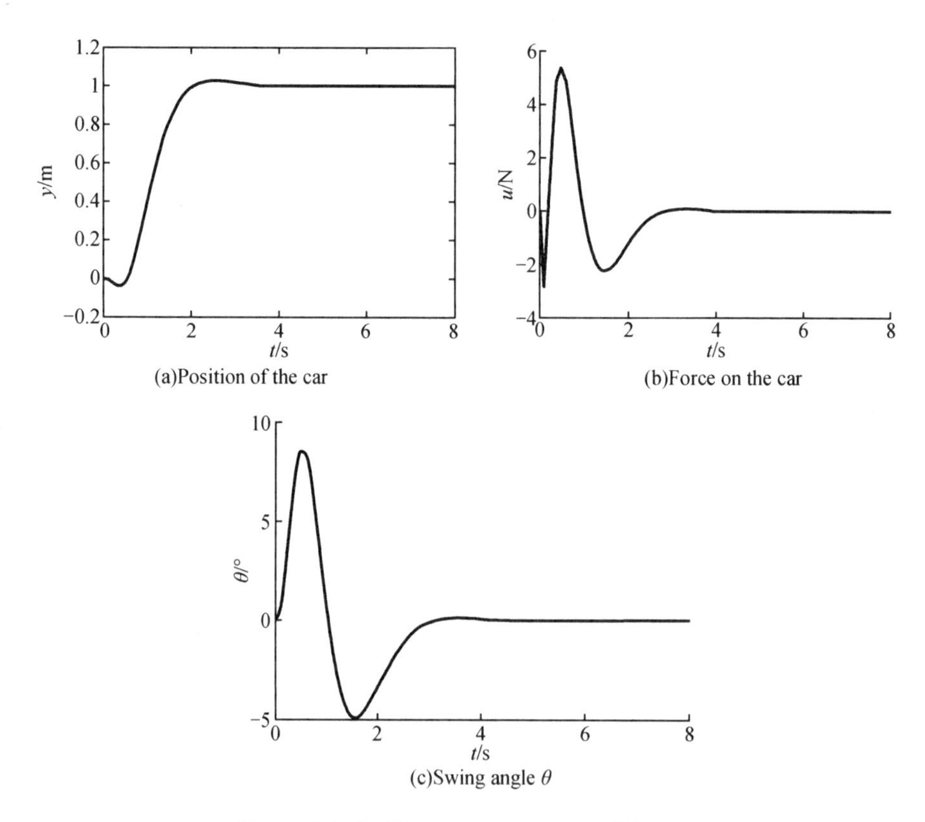

Figure 5.3.4 The response curves of the system

5.4 Pole placement of multiple-input systems

In this section, we discuss pole assignment by state feedback for multi-input systems. There are three pole assignment algorithms: Cyclic design method, Lyapunov equation method and Controllable canonical form method.

Similar to single-input systems, the sufficient and necessary condition for MIMO LTI systems to be able to arbitrarily assign all its poles by state feedback is that the system is completely controllable. The proof of this conclusion will be done in the Cyclic design method.

5.4.1 Cyclic design method

The cyclic design method is also called the direct method whose basic principle is to transform the pole assignment problem of a multi-input system into the pole assignment problem of a single-input system by a cyclic matrix. To this end, the concept of the cyclic matrix is introduced first.

Definition 5.4.1 (Cyclic matrix) A matrix $\boldsymbol{A}$ is called cyclic if its characteristic polynomial equals its minimal polynomial.

The cyclic matrix $\boldsymbol{A}$ has the following properties:

Property 1 The matrix $\boldsymbol{A}$ is cyclic if and only if the Jordan form of $\boldsymbol{A}$ has one and only one small Jordan block associated with each distinct eigenvalue.

Proof The eigenvalues of the matrix $\boldsymbol{A}$ are λ_1 (multiplicity σ_1), λ_2 (multiplicity σ_2), $\cdots\lambda_l$ (multiplicity σ_l) and $\sigma_1+\sigma_2+\cdots+\sigma_l=n$. There exists a invertible transformation matrix $\boldsymbol{Q}$. By introducing the transformation $\hat{\boldsymbol{x}}=\boldsymbol{Q}^{-1}\boldsymbol{x}$, we transform $\boldsymbol{A}$ into the Jordan form as follows:

$$\hat{\boldsymbol{A}}=\boldsymbol{Q}^{-1}\boldsymbol{A}\boldsymbol{Q}=\begin{bmatrix}\boldsymbol{J}_1 & & & & \\ & \ddots & & & \\ & & \boldsymbol{J}_i & & \\ & & & \ddots & \\ & & & & \boldsymbol{J}_l\end{bmatrix} \tag{5.4.1}$$

where, the $\sigma_i\times\sigma_i$ Jordan block $\boldsymbol{J}_i$ corresponding to λ_i has following form:

$$\boldsymbol{J}_i=\begin{bmatrix}\boldsymbol{J}_{i1} & & \\ & \ddots & \\ & & \boldsymbol{J}_{i\alpha_i}\end{bmatrix} \tag{5.4.2}$$

where, the $\sigma_{ik}\times\sigma_{ik}$ k^{th} small Jordan block $\boldsymbol{J}_{ik}$ has following form:

$$\boldsymbol{J}_{ik}=\begin{bmatrix}\lambda_i & 1 & & & \\ & \lambda_i & 1 & & \\ & & \ddots & \ddots & \\ & & & \lambda_i & 1 \\ & & & & \lambda_i\end{bmatrix} \tag{5.4.3}$$

and

$$\sigma_{i1}+\sigma_{i2}+\cdots+\sigma_{i\alpha_i}=\sigma_i \tag{5.4.4}$$

Let

$$\overline{\sigma}_i = \max\{\sigma_{i1}, \sigma_{i2}, \cdots, \sigma_{i\alpha_i}\} \tag{5.4.5}$$

From the matrix theory, the minimal polynomial of $\hat{A}$ or A is:

$$\boldsymbol{\Phi}(s) = \prod_{i=1}^{l} (s - \lambda_i)^{\overline{\sigma}_i} \tag{5.4.6}$$

The characteristic polynomial of $\hat{A}$ or A is

$$\det(s\boldsymbol{I} - \boldsymbol{A}) = \prod_{i=1}^{l} (s - \lambda_i)^{\sigma_i} \tag{5.4.7}$$

From eq. (5.4.6) and eq. (5.4.7), if only if

$$\overline{\sigma}_i = \sigma_i \tag{5.4.8}$$

i. e. when the Jordan form of A only has one small Jordan block for each different eigenvalue, the matrix A is cyclic.

Property 2 If the matrix A is cyclic, there exists a vector b, such that the vectors b Ab $\cdots$ $A^{n-1}b$ span n dimensional real linear space, i. e. $\{A, b\}$ is controllable.

Property 3 If all eigenvalues of A are distinct, then A is cyclic.

Proof Because all eigenvalues of A are distinct, A can be transformed into diagonal form, i. e. the Jordan form of A has one and only one small Jordan block associated with each distinct eigenvalue. According to Property 1, A is cyclic.

Conclusion 5.4.1 If $\{A_{n\times n}, B_{n\times p}\}$ is controllable and A is cyclic, then for almost any vector $\rho_{p\times 1}$, the single-input system $\{A, B\rho\}$ is controllable.

Here is an example to illustrate the basic idea of this conclusion. Let

$$\boldsymbol{A} = \left[\begin{array}{ccc|cc} 5 & 1 & 0 & 0 & 0 \\ 0 & 5 & 1 & 0 & 0 \\ 0 & 0 & 5 & 0 & 0 \\ \hline 0 & 0 & 0 & 1 & 1 \\ 0 & 0 & 0 & 0 & 1 \end{array}\right], \quad \boldsymbol{B} = \left[\begin{array}{cc} 0 & 1 \\ 1 & 1 \\ 1 & 3 \\ \hline 0 & 0 \\ 0 & 1 \end{array}\right]$$

$$\boldsymbol{B\rho} = \boldsymbol{B}\begin{bmatrix} \rho_1 \\ \rho_2 \end{bmatrix} = \left[\begin{array}{c} * \\ * \\ \rho_1 + 3\rho_2 \\ \hline * \\ \rho_2 \end{array}\right]$$

Because A has one and only one small Jordan block associated with each distinct eigenvalue, A is cyclic. According to the controllable canonical form criterion, $\{A, B\rho\}$ is controllable only if the row of $B\rho$, corresponding to the last row of each small Jordan block of A, is not zero, i. e.

$$\rho_1 + 3\rho_2 \neq 0 \text{ and } \rho_2 \neq 0$$

So, almost for any ρ other than $\rho_2 = 0$ and $\rho_1/\rho_2 = -3$, the single-input system $\{A, B\rho\}$ is controllable.

Conclusion 5.4.2 Cycling of an acyclic matrix.

If $\boldsymbol{A}$ is not cyclic, and $\{\boldsymbol{A}_{n\times n}, \boldsymbol{B}_{n\times p}\}$ is controllable, then for almost any real constant matrix $\boldsymbol{K}_{p\times n}$, the matrix $\boldsymbol{A}-\boldsymbol{BK}$ has only distinct eigenvalues and is, consequently, cyclic.

Proof Let

$$\boldsymbol{K}=[k_{ij}]_{p\times n} \quad i=1,2,\cdots,p \; j=1,2,\cdots,n$$

and the characteristic polynomial of $\boldsymbol{A}-\boldsymbol{BK}$ is

$$\alpha(s)=\det(s\boldsymbol{I}-\boldsymbol{A}+\boldsymbol{BK})=s^n+a_{n-1}s^{n-1}+\cdots+a_1 s+a_0$$

where, $a_i(i=1,2,\cdots,n-1)$ is function of the element k_{ij} of $\boldsymbol{K}$. Then take the derivative of $\alpha(s)$ with respect to s:

$$\frac{\mathrm{d}\alpha(s)}{\mathrm{d}s}=ns^{n-1}+(n-1)a_{n-1}s^{n-2}+\cdots+a_1$$

Only if $\alpha(s)=0$ has repeated roots, $\boldsymbol{A}-\boldsymbol{BK}$ is not cyclic. From the matrix theory, when $\alpha(s)=0$ has repeated roots, $\alpha(s)$ and $\frac{\mathrm{d}\alpha(s)}{\mathrm{d}s}$ are coprime, i. e.

$$\det\begin{bmatrix} a_0 & a_1 & \cdots & a_{n-1} & 1 & 0 & \cdots & 0 \\ 0 & a_0 & \cdots & a_{n-2} & a_{n-1} & 1 & \cdots & 0 \\ & & & \vdots & & & & \\ 0 & 0 & \cdots & a_0 & a_1 & a_2 & \cdots & 1 \\ a_1 & 2a_2 & \cdots & n & 0 & 0 & \cdots & 0 \\ 0 & a_1 & \cdots & (n-1)a_{n-1} & n & 0 & \cdots & 0 \\ & & & \vdots & & & & \\ 0 & 0 & \cdots & 0 & a_1 & 2a_2 & \cdots & n \end{bmatrix} \triangleq \gamma(k_{ij})=0 \qquad (5.4.9)$$

Note that $\boldsymbol{K}$ has $p\times n$ elements k_{ij}. When k_{ij} is arbitrary, $\{k_{ij}\}$ construct a $p\times n$ real vector space $R^{p\times n}$. The solution of eq. (5.4.9), $\{k_{ij}\}$, is only a lower dimensional subspace of $R^{p\times n}$. It indicates that for almost $\boldsymbol{K}$, $\gamma(k_{ij})\neq 0$, i. e. the roots of $\alpha(s)=0$ are distinct. Thus, for almost $\boldsymbol{K}$, $\boldsymbol{A}-\boldsymbol{BK}$ is cyclic.

Based on the cyclic matrix, we can get the condition of pole assignment for multi-input systems through state feedback. Consider a multiple-input system

$$\begin{cases} \dot{\boldsymbol{x}}=\boldsymbol{Ax}+\boldsymbol{Bu} \\ \boldsymbol{y}=\boldsymbol{Cx}+\boldsymbol{Du} \end{cases} \qquad (5.4.10)$$

where, $\boldsymbol{x}$ is an n-dimensional state vector, $\boldsymbol{u}$ is a p-dimensional input vector, $\boldsymbol{y}$ is q-dimensional output vector, $\boldsymbol{A}$ and $\boldsymbol{B}$ are $n\times n$ and $n\times p$ constant matrix, respectively, $\boldsymbol{C}$ and $\boldsymbol{D}$ are $q\times n$ and $q\times p$ constant matrix, respectively.

Conclusion 5.4.3 Pole assignment conditions for multiple input systems. For the system eq. (5.4.10), its poles can be placed arbitrarily if only if the system is controllable.

Proof The proof of necessity is the same as for single-input systems. Next to prove sufficiency, i. e. if the system (5.4.10) is controllable, then its poles can be placed arbitrarily by state feedback.

First, checkwhether $\boldsymbol{A}$ is cyclic. If $\boldsymbol{A}$ is cyclic, let $\overline{\boldsymbol{A}} = \boldsymbol{A}$. Else if A is not cyclic, chose a $p \times n$ real constant matrix $\boldsymbol{K}_1$, such that $\overline{\boldsymbol{A}} = \boldsymbol{A} - \boldsymbol{B}\boldsymbol{K}_1$ is cyclic. According to conclusion 5.4.2, the choice of $\boldsymbol{K}_1$ is almost arbitrary. For the cyclic matrix $\overline{\boldsymbol{A}}$, chose a $p \times 1$ real constant vector $\boldsymbol{\rho}$, such that the single input system $\{\overline{\boldsymbol{A}}, \boldsymbol{B}\boldsymbol{\rho}\}$ is controllable. According to conclusion 5.4.1, $\boldsymbol{\rho}$ is chosen almost arbitrarily. For the single input system $\{\overline{\boldsymbol{A}}, \boldsymbol{B}\boldsymbol{\rho}\}$, we can place all poles arbitrarily by introducing the state feedback $u = -\boldsymbol{k}\boldsymbol{x} + \bar{v}$. The state equation of the closed-loop system is

$$\dot{\boldsymbol{x}} = (\overline{\boldsymbol{A}} - \boldsymbol{B}\boldsymbol{\rho}k)\boldsymbol{x} + \boldsymbol{B}\boldsymbol{\rho}\bar{v} \tag{5.4.11}$$

Here, the following two cases are discussed:

(1) When $\boldsymbol{A}$ is cyclic, $\overline{\boldsymbol{A}} = \boldsymbol{A}$,

$$\det(s\boldsymbol{I} - \overline{\boldsymbol{A}} + \boldsymbol{B}\boldsymbol{\rho}k) = \det(s\boldsymbol{I} - \boldsymbol{A} + \boldsymbol{B}\boldsymbol{K}) \tag{5.4.12}$$

where

$$\boldsymbol{K} = \boldsymbol{\rho}\boldsymbol{k} \tag{5.4.13}$$

$\boldsymbol{K}$ is the state feedback matrix for the multiple-input system (5.4.10).

(2) When $\boldsymbol{A}$ is not cyclic, $\overline{\boldsymbol{A}} = \boldsymbol{A} - \boldsymbol{B}\boldsymbol{K}_1$,

$$\begin{aligned}\det(s\boldsymbol{I} - \overline{\boldsymbol{A}} + \boldsymbol{B}\boldsymbol{\rho}k) &= \det(s\boldsymbol{I} - \boldsymbol{A} + \boldsymbol{B}\boldsymbol{K}_1 + \boldsymbol{B}\boldsymbol{\rho}k) \\ &= \det(s\boldsymbol{I} - (\boldsymbol{A} - \boldsymbol{B}(\boldsymbol{K}_1 + \boldsymbol{\rho}\boldsymbol{k}))) \\ &= \det(s\boldsymbol{I} - \boldsymbol{A} + \boldsymbol{B}\boldsymbol{K})\end{aligned} \tag{5.4.14}$$

where

$$\boldsymbol{K} = \boldsymbol{K}_1 + \boldsymbol{\rho}\boldsymbol{k} \tag{5.4.15}$$

$\boldsymbol{K}$ is the state feedback matrix for the multiple-input system (5.4.10).

Because the state feedback $u = -\boldsymbol{k}\boldsymbol{x} + \bar{v}$, for the single input system, places all poles at any desired positions. So, from eq. (5.4.12) and eq. (5.4.13), or eq. (5.4.14) and eq. (5.4.15), the the state feedback for the multiple input system eq. (5.4.10) also places all poles at any desired positions.

In fact, the proof procedure of sufficiency of conclusion 5.4.3 is the solving procedure of the state feedback matrix of the multi-input system eq. (5.4.10). The pole assignment algorithm of multi-input systems is summarized as follows:

Algorithm 5.4.1 Direct method

Given a controllable multiple-input system eq. (5.4.10) and desired eigenvalues λ_1^*, $\lambda_2^*, \cdots, \lambda_n^*$ of the closed-loop system, by introducing the state feedback $\boldsymbol{u} = -\boldsymbol{K}\boldsymbol{x} + \boldsymbol{v}$($\boldsymbol{K}$ is a $p \times n$ constant matrix, $\boldsymbol{v}$ is a $p \times 1$ reference input), determine the state feedback matrix $\boldsymbol{K}$ such that the eigenvalues of the following closed-loop system

$$\begin{cases}\dot{\boldsymbol{x}} = (\boldsymbol{A} - \boldsymbol{B}\boldsymbol{K})\boldsymbol{x} + \boldsymbol{B}\boldsymbol{v} \\ \boldsymbol{y} = (\boldsymbol{C} - \boldsymbol{D}\boldsymbol{K})\boldsymbol{x} + \boldsymbol{D}\boldsymbol{v}\end{cases} \tag{5.4.16}$$

Satisfy $\lambda_i(\boldsymbol{A}-\boldsymbol{BK})=\lambda_i^*$, $i=1,2,\cdots,n$.

Step 1. Check whether $\boldsymbol{A}$ is cyclic. If it is not, we introduce a state feedback

$$\boldsymbol{u}=\boldsymbol{w}-\boldsymbol{K}_1\boldsymbol{x} \tag{5.4.17}$$

where, $\boldsymbol{K}_1$ is chosen almost arbitrarily, such that the system matrix of the following system

$$\dot{\boldsymbol{x}}=(\boldsymbol{A}-\boldsymbol{BK}_1)\boldsymbol{x}+\boldsymbol{Bw} \tag{5.4.18}$$

$\boldsymbol{A}-\boldsymbol{BK}_1$ is cyclic. Let

$$\overline{\boldsymbol{A}}=\begin{cases}\boldsymbol{A}-\boldsymbol{BK}_1\text{, if }\boldsymbol{A}\text{ is not cyclic}\\ \boldsymbol{A}\text{, if }\boldsymbol{A}\text{ is cyclic}\end{cases} \tag{5.4.19}$$

Because $\{\boldsymbol{A},\boldsymbol{B}\}$ is controllable, so $\{\overline{\boldsymbol{A}},\boldsymbol{B}\}$ is controllable. (Controllability is preserved under state feedback.)

Step 2. For the cyclic matrix $\overline{\boldsymbol{A}}$, choose a suitable real constant vector $\boldsymbol{\rho}_{p\times 1}$ such that $\{\overline{\boldsymbol{A}},\boldsymbol{b}\}$ is controllable. Where

$$\boldsymbol{b}_{n\times 1}=\boldsymbol{B}_{n\times p}\boldsymbol{\rho}_{p\times 1} \tag{5.4.20}$$

Step 3. For equivalent single input system $\{\overline{\boldsymbol{A}},\boldsymbol{b}\}$, we introduce another state feedback

$$\boldsymbol{w}=\boldsymbol{v}-\boldsymbol{kx}$$

such that closed-loop eigenvalues be assigned at desired positions. Where $\boldsymbol{v}$ is reference input, $\boldsymbol{k}$ is $1\times n$ real constant vector. After getting the state gain vector $\boldsymbol{k}_{1\times n}$, we go to the step 4.

Step 4. When $\boldsymbol{A}$ is cyclic, the final feedback gain matrix $\boldsymbol{K}$ is

$$\boldsymbol{K}_{p\times n}=\boldsymbol{\rho}_{p\times 1}\cdot\boldsymbol{k}_{1\times n} \tag{5.4.21}$$

when $\boldsymbol{A}$ is not cyclic, the final feedback gain matrix $\boldsymbol{K}$ is

$$\boldsymbol{K}_{p\times n}=\boldsymbol{\rho}_{p\times 1}\cdot\boldsymbol{k}_{1\times n}+\boldsymbol{K}_1 \tag{5.4.22}$$

In the above algorithm, because the selection of $\boldsymbol{\rho}$ and $\boldsymbol{K}_1$ is not unique, from eq. (5.2.21) and eq. (5.4.22), the feedback matrix $\boldsymbol{K}$ is not unique. In practice, the appropriate $\boldsymbol{\rho}$ and $\boldsymbol{K}_1$ are usually selected to make the absolute value of each element of $\boldsymbol{K}$ as small as possible.

Figure 5.4.1 shows a diagram of state feedback pole placement for multi-input systems.

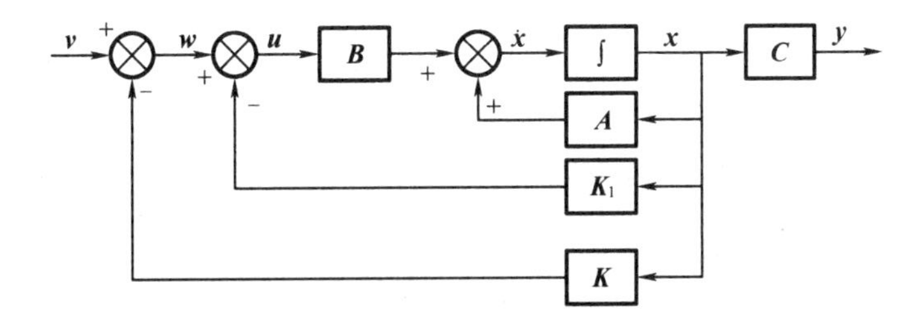

Figure 5.4.1 State feedback pole placement for multi-input systems

Example 5.4.1 Given a 2-input LTI system

$$\dot{\boldsymbol{x}}=\begin{bmatrix}3&1&0\\0&3&0\\0&0&2\end{bmatrix}\boldsymbol{x}+\begin{bmatrix}0&0\\1&0\\0&1\end{bmatrix}\boldsymbol{u}$$

compute the state feedback gain matrix $\boldsymbol{K}$ to place the closed-loop eigenvalues at -1, -2, and -3.

Answer According to Jordan form criterion, the system is controllable. So all poles can arbitrarily be assigned by using state feedback. Because $\boldsymbol{A}$ has one Jordan block associated with each distinct eigenvalue, so $\boldsymbol{A}$ is cyclic. Let

$$\boldsymbol{\rho} = \begin{bmatrix} 1 \\ 1 \end{bmatrix}$$

So,

$$\boldsymbol{B\rho} = \begin{bmatrix} 0 & 0 \\ 1 & 0 \\ 0 & 1 \end{bmatrix} \begin{bmatrix} 1 \\ 1 \end{bmatrix} = \begin{bmatrix} 0 \\ 1 \\ 1 \end{bmatrix} = \boldsymbol{b}$$

Because

$$\operatorname{rank}[\boldsymbol{b} \quad \boldsymbol{Ab} \quad \boldsymbol{A}^2\boldsymbol{b}] = \operatorname{rank}\begin{bmatrix} 0 & 1 & 6 \\ 1 & 3 & 9 \\ 1 & 2 & 4 \end{bmatrix} = 3$$

So the single-input system $\{\boldsymbol{A}, \boldsymbol{b}\}$ is controllable. For $\{\boldsymbol{A}, \boldsymbol{b}\}$, we design state feedback controller so that closed-loop poles be assigned at -1, -2, and -3.

First transform $\{\boldsymbol{A}, \boldsymbol{b}\}$ into controllable canonical form. The characteristic polynomial of $\boldsymbol{A}$ is:

$$\det(s\boldsymbol{I} - \boldsymbol{A}) = (s-3)^2(s-2) = s^3 - 8s^2 + 21s - 18$$

Through linearly nonsingular transformation $\bar{\boldsymbol{x}} = \boldsymbol{P}^{-1}\boldsymbol{x}$, we get controllable form of the system:

$$\boldsymbol{A}_c = \boldsymbol{P}^{-1}\boldsymbol{AP} = \begin{bmatrix} 0 & 1 & 0 \\ 0 & 0 & 1 \\ 18 & -21 & 8 \end{bmatrix}, \quad \boldsymbol{b}_c = \begin{bmatrix} 0 \\ 0 \\ 1 \end{bmatrix}$$

where,

$$\boldsymbol{P} = [\boldsymbol{A}^2\boldsymbol{b} \quad \boldsymbol{Ab} \quad \boldsymbol{b}] \begin{bmatrix} 1 & 0 & 0 \\ -8 & 1 & 0 \\ 21 & -8 & 1 \end{bmatrix} = \begin{bmatrix} 6 & 1 & 0 \\ 9 & 3 & 1 \\ 4 & 2 & 1 \end{bmatrix} \begin{bmatrix} 1 & 0 & 0 \\ -8 & 1 & 0 \\ 21 & -8 & 1 \end{bmatrix} = \begin{bmatrix} -2 & 1 & 0 \\ 6 & -5 & 1 \\ 9 & -6 & 1 \end{bmatrix}$$

We introduce the state feed back $u = -\boldsymbol{k}_c\bar{\boldsymbol{x}}_c + v$ for the single-input system $\{\boldsymbol{A}_c, \boldsymbol{b}_c\}$, where $\boldsymbol{k}_c = [k_{c1} \quad k_{c2} \quad k_{c3}]$. Then characteristic polynomial of $(\boldsymbol{A}_c - \boldsymbol{b}_c\boldsymbol{k}_c)$ is

$$\det(s\boldsymbol{I} - \boldsymbol{A}_c + \boldsymbol{b}_c\boldsymbol{k}_c) = s^3 + (-8 + k_{c3})s^2 + (21 + k_{c2})s + (-18 + k_{c1}) \qquad (5.4.23)$$

The desired characteristic polynomial of the closed-loop system is

$$(s+1)(s+2)(s+3) = s^3 + 6s^2 + 11s + 6 \qquad (5.4.24)$$

Comparing eq. (5.4.23) and eq. (5.4.24), yields

$$\begin{cases} -8+k_{c3}=6 \\ 21+k_{c2}=11 \\ -18+k_{c1}=6 \end{cases} \Rightarrow \begin{cases} k_{c1}=24 \\ k_{c2}=-10 \\ k_{c3}=14 \end{cases}$$

Thus, $\boldsymbol{k}_c = [24 \quad -10 \quad 14]$. The state feedback gain matrix for $\{\boldsymbol{A},\boldsymbol{b}\}$, $\boldsymbol{k}_{1\times 3}$, is

$$\boldsymbol{k}=\boldsymbol{k}_c \cdot \boldsymbol{P}^{-1}=[24 \quad -10 \quad 14]\begin{bmatrix} 1 & -1 & 1 \\ 3 & -2 & 2 \\ 9 & -3 & 4 \end{bmatrix}=[120 \quad -46 \quad 60]$$

The final state feedback gain matrix for state equation $\{\boldsymbol{A},\boldsymbol{B}\}$, $\boldsymbol{K}_{2\times 3}$, is

$$\boldsymbol{K}=\boldsymbol{\rho k}=\begin{bmatrix} 1 \\ 1 \end{bmatrix} \cdot [120 \quad -46 \quad 60]=\begin{bmatrix} 120 & -46 & 60 \\ 120 & -46 & 60 \end{bmatrix}$$

So, the state equation of closed-loop system is

$$\dot{\boldsymbol{x}}=(\boldsymbol{A}-\boldsymbol{BK})\boldsymbol{x}+\boldsymbol{Bv}=\begin{bmatrix} 3 & 1 & 0 \\ -120 & 49 & -60 \\ -120 & 46 & -58 \end{bmatrix}\boldsymbol{x}+\begin{bmatrix} 0 & 0 \\ 1 & 0 \\ 0 & 1 \end{bmatrix}\boldsymbol{v}$$

5.4.2 Lyapunov-eqauion method

Next, we discuss the pole assignment algorithm for multi-input systems, the Lyapunov-equation method. This method is applicable to the case where the desired eigenvalues are different from the eigenvalues of $\boldsymbol{A}$.

Algorithms 5.4.2 Lyapunov-equation method. Given a controllable multiple-input system (5.4.10) and desired eigenvalues $\lambda_1^*, \lambda_2^*, \cdots, \lambda_n^*$ of the closed-loop system, by introducing the state feedback $\boldsymbol{u}=-\boldsymbol{Kx}+\boldsymbol{v}$ ($\boldsymbol{K}$ is a $p\times n$ constant matrix, $\boldsymbol{v}$ is a $p\times 1$ reference input), determine the state feedback matrix $\boldsymbol{K}$ such that $\lambda_i(\boldsymbol{A}-\boldsymbol{BK})=\lambda_i^*$, $i=1,2,\cdots,n$, as long as the desired eigenvalue does not contain any eigenvalue of $\boldsymbol{A}$, i. e.

$$\lambda_i(\boldsymbol{A}) \neq \lambda_i^*, \quad i=1,2,\cdots,n \tag{5.4.25}$$

Step 1. Select an $n\times n$ matrix $\boldsymbol{F}$ such that its eigenvalues equal the desired eigenvalues, i. e.

$$\lambda_i(\boldsymbol{F})=\lambda_i^*, \quad i=1,2,\cdots,n \tag{5.4.26}$$

Step 2. Select a $p\times n$ constant matrix $\overline{\boldsymbol{K}}$ so that $\{\boldsymbol{F},\overline{\boldsymbol{K}}\}$ observable. In general, the probability that $\{\boldsymbol{F},\overline{\boldsymbol{K}}\}$ is completely observable is almost equal to 1 when $\overline{\boldsymbol{K}}$ is chosen arbitrarily.

Step 3. Determine the unique $n\times n$ $\boldsymbol{T}$ by solving the the following Lyapunov equation

$$\boldsymbol{AT}-\boldsymbol{TF}=\boldsymbol{B}\overline{\boldsymbol{K}} \tag{5.4.27}$$

Step 4. If $\boldsymbol{T}$ is singular, select a different $\overline{\boldsymbol{K}}$ and repeat the above process. If $\boldsymbol{T}$ is nonsingular, we compute the state feedback matrix $\boldsymbol{K}$ as the following equation

$$\boldsymbol{K}=\overline{\boldsymbol{K}}\boldsymbol{T}^{-1} \tag{5.4.28}$$

Next, we verify the correctness of this algorithm. If $\boldsymbol{T}$ is nonsingular, then from eq. (5.4.27) and eq. (5.4.8), we have

$$\boldsymbol{A}-\boldsymbol{BK}=\boldsymbol{A}-\boldsymbol{B}\overline{\boldsymbol{K}}\boldsymbol{T}^{-1}=(\boldsymbol{AT}-\boldsymbol{B}\overline{\boldsymbol{K}})\boldsymbol{T}^{-1}=\boldsymbol{TF}\boldsymbol{T}^{-1} \tag{5.4.29}$$

It indicates that the eigenvalues of $\boldsymbol{A}-\boldsymbol{BK}$ are the same as the eigenvalues of $\boldsymbol{T}$.

There are two conclusions about Lyapunov eq. (5.4.27).

Conclusion 5.4.4 There exists a unique solution of the Lyapunov eq. (5.4.27) if the $\boldsymbol{F}$ and $\boldsymbol{A}$ have no common eigenvalues.

Conclusion 5.4.5 If the solution of the Lyapunov eq. (5.4.27), $\boldsymbol{T}$ is nonsingular, then $\{\boldsymbol{A},\boldsymbol{B}\}$ is controllable and $\{\boldsymbol{F},\overline{\boldsymbol{K}}\}$ is observable. For single-input systems, the solution of the Lyapunov eq. (5.4.27), $\boldsymbol{T}$ is nonsingular, if only if $\{\boldsymbol{A},\boldsymbol{B}\}$ is controllable and $\{\boldsymbol{F},\overline{\boldsymbol{K}}\}$ is observable.

Example 5.4.2 Given a 2-input LTI system

$$\dot{\boldsymbol{x}}=\begin{bmatrix}-1&0&0\\0&0&1\\2&0&3\end{bmatrix}\boldsymbol{x}+\begin{bmatrix}1&0\\0&0\\0&1\end{bmatrix}\boldsymbol{u}$$

Determine the state feedback gain matrix $\boldsymbol{K}$ to place the eigenvalues of the closed-loop system at -3, -4, and -5.

Answer **Method 1** The desired characteristic polynomial of the closed-loop system is

$$\alpha^*(s)=(s+3)(s+4)(s+5)=s^3+12s^2+47s+60$$

Chose a matrix $\boldsymbol{F}$ such that its eigenvalues are -3, -4, and -5:

$$\boldsymbol{F}=\begin{bmatrix}0&1&0\\0&0&1\\-60&-47&-12\end{bmatrix}$$

Chose $\overline{\boldsymbol{K}}$ such that $\{\boldsymbol{F},\overline{\boldsymbol{K}}\}$ is observable:

$$\overline{\boldsymbol{K}}=\begin{bmatrix}0&0&1\\1&0&0\end{bmatrix}$$

Because

$$\text{rank}\begin{bmatrix}\overline{\boldsymbol{K}}\\\overline{\boldsymbol{K}}\boldsymbol{F}\\\overline{\boldsymbol{K}}\boldsymbol{F}^2\end{bmatrix}=\text{rank}\begin{bmatrix}0&0&1\\1&0&0\\-60&-47&-12\\*&*&*\\*&*&*\\*&*&*\end{bmatrix}=3$$

So, $\{\boldsymbol{F},\overline{\boldsymbol{K}}\}$ is observable. Let

$$\boldsymbol{T}=\begin{bmatrix}t_{11}&t_{12}&t_{13}\\t_{21}&t_{22}&t_{23}\\t_{31}&t_{32}&t_{33}\end{bmatrix}$$

Solve the following Lyapunov equation

$$AT - TF = B\bar{K}$$

we get

$$T = \begin{bmatrix} 2.5000 & -0.5417 & 0.0417 \\ 0.9909 & 0.2443 & 0.0207 \\ -1.2440 & 0.0164 & -0.0045 \end{bmatrix}$$

Since T is nonsingular, from eq. (5.4.28), the final state feedback matrix K is:

$$K = \bar{K}T^{-1} = \begin{bmatrix} 0 & 0 & 1 \\ 1 & 0 & 0 \end{bmatrix} \begin{bmatrix} -0.0670 & -0.0813 & -1.0033 \\ -1.0014 & 1.9059 & -0.4943 \\ 15.0033 & 29.6572 & 53.7718 \end{bmatrix}$$

$$= \begin{bmatrix} 15.0033 & 29.6572 & 53.7718 \\ -0.0670 & -0.0813 & -1.0033 \end{bmatrix}$$

Method 2 The desired characteristic polynomial of the closed-loop system is

$$\alpha^*(s) = (s+3)(s+4)(s+5) = s^3 + 12s^2 + 47s + 60$$

Construct the following constant matrix such that its characteristic polynomial equals $\alpha^*(s)$.

$$\begin{bmatrix} 0 & 1 & 0 \\ 0 & 0 & 1 \\ -60 & -47 & -12 \end{bmatrix}$$

Then, let

$$A - BK = \begin{bmatrix} 0 & 1 & 0 \\ 0 & 0 & 1 \\ -60 & -47 & -12 \end{bmatrix}$$

So,

$$BK = A - (A - BK) = \begin{bmatrix} -1 & 0 & 0 \\ 0 & 0 & 1 \\ 2 & 0 & 3 \end{bmatrix} - \begin{bmatrix} 0 & 1 & 0 \\ 0 & 0 & 1 \\ -60 & -47 & -12 \end{bmatrix} = \begin{bmatrix} -1 & -1 & 0 \\ 0 & 0 & 0 \\ 62 & 47 & 15 \end{bmatrix}$$

Let

$$K = \begin{bmatrix} k_{11} & k_{12} & k_{13} \\ k_{21} & k_{22} & k_{23} \end{bmatrix}$$

Then,

$$BK = \begin{bmatrix} k_{11} & k_{12} & k_{13} \\ 0 & 0 & 0 \\ k_{21} & k_{22} & k_{23} \end{bmatrix} = \begin{bmatrix} -1 & -1 & 0 \\ 0 & 0 & 0 \\ 62 & 47 & 15 \end{bmatrix}$$

Thus, the final state feedback gain matrix is

$$K = \begin{bmatrix} -1 & -1 & 0 \\ 62 & 47 & 15 \end{bmatrix}$$

5.4.3 Canonical-form method

We discuss the third method of solving state-feedback for multi-input systems, the controllable canonical form method, which is based on the Luenberger controllable canonical form.

Algorithm 5.4.3 Controllable canonical form method.

Consider a controllable multi-input LTI system eq. (5.4.10), determine state feedback $\boldsymbol{u} = -\boldsymbol{Kx} + \boldsymbol{v}$ ($\boldsymbol{K}$ is $p \times n$, $\boldsymbol{v}$ is $p \times 1$) such that eigenvalues of closed-loop system be assigned at desired positions $\lambda_1^*, \lambda_2^*, \cdots, \lambda_n^*$, i.e.

$$\lambda_i(\boldsymbol{A} - \boldsymbol{BK}) = \lambda_i^*, \quad i = 1, 2, \cdots, n$$

To simplify the discussion, we assume that the system has 9 state variables and three inputs, i.e. $n = 9$ and $p = 3$.

Step 1. Transform $\{\boldsymbol{A}, \boldsymbol{B}\}$ into Luenberger controllable canonical form:

$$\overline{\boldsymbol{A}} = \boldsymbol{S}^{-1}\boldsymbol{AS} = \left[\begin{array}{ccc:cc:cccc} 0 & 1 & 0 & 0 & 0 & 0 & 0 & 0 & 0 \\ 0 & 0 & 1 & 0 & 0 & 0 & 0 & 0 & 0 \\ -a_{10} & -a_{11} & -a_{12} & \beta_{14} & \beta_{15} & \beta_{16} & \beta_{17} & \beta_{18} & \beta_{19} \\ \hdashline 0 & 0 & 0 & 0 & 1 & 0 & 0 & 0 & 0 \\ \beta_{21} & \beta_{22} & \beta_{23} & -a_{20} & -a_{21} & \beta_{26} & \beta_{27} & \beta_{28} & \beta_{29} \\ \hdashline 0 & 0 & 0 & 0 & 0 & 0 & 1 & 0 & 0 \\ 0 & 0 & 0 & 0 & 0 & 0 & 0 & 1 & 0 \\ 0 & 0 & 0 & 0 & 0 & 0 & 0 & 0 & 1 \\ \beta_{31} & \beta_{32} & \beta_{32} & \beta_{34} & \beta_{35} & -a_{30} & -a_{31} & -a_{32} & -a_{33} \end{array}\right]$$

$$\overline{\boldsymbol{B}} = \boldsymbol{S}^{-1}\boldsymbol{B} = \left[\begin{array}{ccc} 0 & 0 & 0 \\ 0 & 0 & 0 \\ 1 & \gamma & 0 \\ \hdashline 0 & 0 & 0 \\ 0 & 1 & 0 \\ \hdashline 0 & 0 & 0 \\ 0 & 0 & 0 \\ 0 & 0 & 0 \\ 0 & 0 & 1 \end{array}\right]$$

Step 2. Group $\lambda_1^*, \lambda_2^*, \cdots, \lambda_9^*$ into 3 groups according to the number and dimension of the blocks in the diagonal. Then compute

$$\alpha_1^*(s) = (s - \lambda_1^*)(s - \lambda_2^*)(s - \lambda_3^*) = s^3 + a_{12}^* s^2 + a_{11}^* s + a_{10}^*$$

$$\alpha_2^*(s) = (s - \lambda_4^*)(s - \lambda_5^*) = s^2 + a_{21}^* s + a_{20}^*$$

$$\alpha_3^*(s) = (s - \lambda_6^*)(s - \lambda_7^*)(s - \lambda_8^*)(s - \lambda_9^*) = s^4 + a_{33}^* ss^3 + a_{32}^* s^2 + a_{31}^* s + a_{30}^*$$

Step 3. For $\{\overline{\boldsymbol{A}}, \overline{\boldsymbol{B}}\}$, compute $p \times n$ matrix $\overline{\boldsymbol{K}}$:

$$\overline{K}=\left[\begin{array}{ccc|cc|cccc} a_{10}^*-a_{10} & a_{11}^*-a_{11} & a_{12}^*-a_{12} & \beta_{14}-\gamma(a_{20}^*-a_{20}) & \beta_{15}-\gamma(a_{21}^*-a_{21}) & \beta_{16}-\gamma\beta_{26} & \beta_{17}-\gamma\beta_{27} & \beta_{18}-\gamma\beta_{28} & \beta_{19}-\gamma\beta_{29} \\ 0 & 0 & 0 & a_{20}^*-a_{20} & a_{21}^*-a_{21} & \beta_{26} & \beta_{27} & \beta_{28} & \beta_{29} \\ 0 & 0 & 0 & 0 & 0 & a_{30}^*-a_{30} & a_{31}^*-a_{31} & a_{32}^*-a_{32} & a_{33}^*-a_{33} \end{array}\right] \tag{5.4.30}$$

Step 4. Compute the state feedback gain matrix K for $\{A,B\}$:

$$K=\overline{K}S^{-1}$$

Next, we prove the correctness of the algorithm through the above example. Compute

$$\overline{A}-\overline{B}\,\overline{K}=\left[\begin{array}{ccc|cc|cccc} 0 & 1 & 0 & & & & & & \\ 0 & 0 & 1 & & & & & & \\ -a_{10}^* & -a_{11}^* & -a_{12}^* & & & & & & \\ \hline 0 & 0 & 0 & 0 & 1 & & & & \\ \beta_{21} & \beta_{22} & \beta_{23} & -a_{20}^* & -a_{21}^* & & & & \\ \hline & & & & & 0 & 1 & 0 & 0 \\ & & & & & 0 & 0 & 1 & 0 \\ & & & & & 0 & 0 & 0 & 1 \\ \beta_{31} & \beta_{32} & \beta_{33} & \beta_{34} & \beta_{35} & -a_{30}^* & -a_{31}^* & -a_{32}^* & -a_{33}^* \end{array}\right]$$

So, the eigenvalues of $\overline{A}-\overline{B}\,\overline{K}$ are $\lambda_1^*,\lambda_2^*,\cdots,\lambda_n^*$. Because

$$A-BK=S\overline{A}S^{-1}-S\overline{B}\,\overline{K}S^{-1}=S(\overline{A}-\overline{B}\,\overline{K})S^{-1}$$

According to conclusion 1.5.2, $A-BK$ has the same eigenvalues as $\overline{A}-\overline{B}\,\overline{K}$.

Algorithm 5.4.3 has two advantages. First is the normalization of the calculation process. The main work is the calculation of transformation matrix S^{-1} and the getting Luenberger controllable canonical form. Second is the element of the obtained state feedback matrix, K, is much smaller than the results of the first two algorithms.

In general, the larger the number of diagonal block matrices and the smaller the dimension of each block matrix in $\overline{A}$, the smaller the element of the obtained K.

Example 5.4.3 Consider a MIMO multiple-input

$$\dot{x}=\left[\begin{array}{ccc|cc} 0 & 1 & 0 & 0 & 0 \\ 0 & 0 & 1 & 0 & 0 \\ 3 & 1 & 0 & 1 & 2 \\ \hline 0 & 0 & 0 & 0 & 1 \\ 4 & 3 & 1 & -1 & -4 \end{array}\right]x+\left[\begin{array}{cc} 0 & 0 \\ 0 & 0 \\ 1 & 2 \\ \hline 0 & 0 \\ 0 & 1 \end{array}\right]u$$

Determine the state feedback gain matrix $K_{2\times 5}$ such that the eigenvalues be placed at $\lambda_1^*=-1,\lambda_{2,3}^*=-2\pm j,\lambda_{4,5}^*=-1\pm 2j$.

Answer Group the desired eigenvalues into two groups: $\{\lambda_1^*,\lambda_2^*,\lambda_3^*\}$ and $\{\lambda_4^*,$

$\lambda_5^* \}$. Compute

$$\alpha_1^*(s) = (s+1)(s+2-j)(s+2+j) = s^3+5s^2+9s+5$$

$$\alpha_2^*(s) = (s+1)(s+1-2j)(s+1+2j) = s^2+2s+5$$

From eq. (5.4.30), we get $\boldsymbol{K}$

$$\boldsymbol{K} = \begin{bmatrix} a_{10}^* - a_{10} & a_{11}^* - a_{11} & a_{12}^* - a_{12} & \beta_{14} - \gamma(a_{20}^* - a_{20}) & \beta_{15} - \gamma(a_{21}^* - a_{21}) \\ 0 & 0 & 0 & a_{20}^* - a_{20} & a_{21}^* - a_{21} \end{bmatrix}$$

$$= \left[\begin{array}{ccc|cc} 8 & 10 & 5 & -7 & 6 \\ \hline 0 & 0 & 0 & 4 & -2 \end{array}\right]$$

The state equation of the closed-loop system is

$$\dot{\boldsymbol{x}} = \left[\begin{array}{ccc|cc} 0 & 1 & 0 & 0 & 0 \\ 0 & 0 & 1 & 0 & 0 \\ -5 & -9 & -5 & 0 & 0 \\ \hline 0 & 0 & 0 & 0 & 1 \\ 4 & 3 & 1 & -5 & -2 \end{array}\right] \boldsymbol{x} + \left[\begin{array}{cc} 0 & 0 \\ 0 & 0 \\ 1 & 2 \\ \hline 0 & 0 \\ 0 & 1 \end{array}\right] \boldsymbol{v}$$

5.5 Effect of state feedback on transfer matrices

In this section, we discuss the influence of state feedback on transfer function matrices including the single-input case and multi-input case.

5.5.1 Effect of state feedback on transfer matrices for SISO systems

Consider a controllable SISO LTI system described by

$$\begin{cases} \dot{\boldsymbol{x}} = \boldsymbol{A}\boldsymbol{x} + \boldsymbol{b}u \\ y = \boldsymbol{c}\boldsymbol{x} \end{cases} \tag{5.5.1}$$

where, $\boldsymbol{x}$ is an n-dimensional state vector, u is a scale input, y is scale output. for the SISO LTI system, its transfer function matrix degenerates into a scalar function $g(s)$ about s. The following conclusions are drawn about the influence of state feedback on the transfer function.

Conclusion 5.5.1 For the single-variable system (5.5.1), state feedback can shift the poles of $g(s)$ to any positions and has no effect on the zeros.

Proof For the system (5.5.1), we introduce nonsingular transformation $\bar{\boldsymbol{x}} = \boldsymbol{P}^{-1}\boldsymbol{x}$ to transform system (5.5.1) into controllable canonical form:

$$\bar{\boldsymbol{A}} = \left[\begin{array}{c|cccc} 0 & 1 & 0 & \cdots & 0 \\ 0 & 0 & 1 & \cdots & 0 \\ & & & \ddots & \\ 0 & 0 & \cdots & & 1 \\ \hline -a_0 & -a_1 & -a_2 & \cdots & -a_{n-1} \end{array}\right], \bar{\boldsymbol{b}} = \left[\begin{array}{c} 0 \\ \vdots \\ 0 \\ \hline 1 \end{array}\right], \bar{\boldsymbol{c}} = [\bar{c}_0\ \bar{c}_1 \cdots \bar{c}_{n-1}] \tag{5.5.2}$$

The transfer function of the open-loop system(5.5.2)is

$$\bar{g}(s)=\bar{c}(s\boldsymbol{I}-\bar{\boldsymbol{A}})^{-1}\bar{\boldsymbol{b}}=\bar{\boldsymbol{c}}\begin{bmatrix} s & -1 & 0 & \cdots & 0 \\ 0 & s & -1 & \cdots & 0 \\ & & \vdots & & \\ 0 & 0 & 0 & \cdots & -1 \\ a_0 & a_1 & a_2 & \cdots & s+a_{n-1} \end{bmatrix}^{-1}\bar{\boldsymbol{b}}$$

where

$$(s\boldsymbol{I}-\bar{\boldsymbol{A}})^{-1}=\frac{\operatorname{adj}(s\boldsymbol{I}-\bar{\boldsymbol{A}})}{\det(s\boldsymbol{I}-\bar{\boldsymbol{A}})}=\frac{\begin{bmatrix} * & \cdots & * & 1 \\ * & \cdots & * & s \\ & \vdots & & \\ * & \cdots & * & s^{n-1} \end{bmatrix}}{s^n+a_{n-1}s^{n-1}+\cdots+a_1s+a_0}$$

So, the transfer function of the open-loop system (5.5.2)is

$$\bar{g}(s)=\bar{c}(s\boldsymbol{I}-\bar{\boldsymbol{A}})^{-1}\bar{\boldsymbol{b}}$$

$$=[\bar{c}_0\ \bar{c}_1\cdots\ \bar{c}_{n-1}]\frac{\begin{bmatrix} * & \cdots & * & 1 \\ * & \cdots & * & s \\ & \vdots & & \\ * & \cdots & * & s^{n-1} \end{bmatrix}}{s^n+a_{n-1}s^{n-1}+\cdots+a_1s+a_0}\cdot\begin{bmatrix} 0 \\ \vdots \\ 0 \\ 1 \end{bmatrix}$$

$$=[\bar{c}_0\ \bar{c}_1\cdots\ \bar{c}_{n-1}]\frac{\begin{bmatrix} 1 \\ s \\ \vdots \\ s^{n-1} \end{bmatrix}}{s^n+a_{n-1}s^{n-1}+\cdots+a_1s+a_0}$$

$$=\frac{\bar{c}_{n-1}s^{n-1}+\bar{c}_{n-2}s^{n-2}+\cdots+\bar{c}_1s+\bar{c}_0}{s^n+a_{n-1}s^{n-1}+\cdots+a_1s+a_0}$$

According to conclusion 1.5.2, The transfer function of the system (5.5.1)is

$$g(s)=\frac{\bar{c}_{n-1}s^{n-1}+\bar{c}_{n-2}s^{n-2}+\cdots+\bar{c}_1s+\bar{c}_0}{s^n+a_{n-1}s^{n-1}+\cdots+a_1s+a_0} \tag{5.5.3}$$

By introducing the state feedback $\bar{k}$ to the system (5.5.2), the closed-loop system is

$$\dot{\bar{\boldsymbol{x}}}=(\bar{\boldsymbol{A}}-\bar{\boldsymbol{b}}\,\bar{\boldsymbol{k}})\bar{\boldsymbol{x}}+\bar{\boldsymbol{b}}v=\bar{\boldsymbol{A}}_k\bar{\boldsymbol{x}}+\bar{\boldsymbol{b}}v$$

$$y=\overline{\boldsymbol{cx}} \tag{5.5.4}$$

where

$$\bar{\boldsymbol{k}}=[\bar{k}_1\ \bar{k}_2\cdots\bar{k}_n]$$

$$\bar{\boldsymbol{A}}_k=\bar{\boldsymbol{A}}-\bar{\boldsymbol{b}}\,\bar{\boldsymbol{k}}=\begin{bmatrix}0 & 1 & 0 & \cdots & 0\\ 0 & 0 & 1 & \cdots & 0\\ & & \vdots & & \\ -a_0-\bar{k}_1 & -a_1-\bar{k}_2 & -a_2-\bar{k}_3 & \cdots & -a_{n-1}-\bar{k}_n\end{bmatrix},$$

$$(s\boldsymbol{I}-\bar{\boldsymbol{A}}_k)^{-1}=\frac{\operatorname{adj}(s\boldsymbol{I}-\bar{\boldsymbol{A}}_k)}{\det(s\boldsymbol{I}-\bar{\boldsymbol{A}}_k)}=\frac{\begin{bmatrix}* & \cdots & * & 1\\ * & \cdots & * & s\\ & \vdots & & \\ * & \cdots & * & s^{n-1}\end{bmatrix}}{s^n+(a_{n-1}+\bar{k}_n)s^{n-1}+\cdots+(a_1+\bar{k}_2)s+(a_0+\bar{k}_1)}$$

The transfer function of the closed-loop system eq. (5.5.4) is

$$\bar{g}_k(s)=\bar{c}(s\boldsymbol{I}-\bar{\boldsymbol{A}}_k)^{-1}\bar{\boldsymbol{b}}$$

$$=[\bar{c}_0\ \bar{c}_1\cdots\ \bar{c}_{n-1}]\frac{\begin{bmatrix}* & \cdots & * & 1\\ * & \cdots & * & s\\ & \vdots & & \\ * & \cdots & * & s^{n-1}\end{bmatrix}}{s^n+(a_{n-1}+\bar{k}_n)s^{n-1}+\cdots+(a_1+\bar{k}_2)s+(a_0+\bar{k}_1)}\begin{bmatrix}0\\ \vdots\\ 0\\ 1\end{bmatrix}$$

$$=[\bar{c}_0\ \bar{c}_1\cdots\ \bar{c}_{n-1}]\frac{\begin{bmatrix}1\\ s\\ \vdots\\ s^{n-1}\end{bmatrix}}{s^n+(a_{n-1}+\bar{k}_n)s^{n-1}+\cdots+(a_1+\bar{k}_2)s+(a_0+\bar{k}_1)}$$

$$=\frac{\bar{c}_{n-1}s^{n-1}+\bar{c}_{n-2}s^{n-2}+\cdots+\bar{c}_1s+\bar{c}_0}{s^n+(a_{n-1}+\bar{k}_n)s^{n-1}+\cdots+(a_1+\bar{k}_2)s+(a_0+\bar{k}_1)}\tag{5.5.5}$$

From eq. (3.5.5) and eq. (5.3.17), we have

$$\bar{g}_k(s)=\bar{\boldsymbol{c}}(s\boldsymbol{I}-\bar{\boldsymbol{A}}_k)^{-1}\bar{\boldsymbol{b}}=\boldsymbol{cP}(s\boldsymbol{I}-\boldsymbol{P}^{-1}\boldsymbol{AP}+\boldsymbol{P}^{-1}\boldsymbol{bkP})^{-1}\boldsymbol{P}^{-1}\boldsymbol{b}=\boldsymbol{c}(s\boldsymbol{I}-\boldsymbol{A}_k)^{-1}\boldsymbol{b}=g_k(s)\tag{5.5.6}$$

Thus, from eq. (5.5.5) and eq. (5.5.6), after introducing the state feedback $\boldsymbol{k}$ to the system (5.5.1), the transfer function of the closed-loop system is

$$g_k(s)=\frac{\bar{c}_{n-1}s^{n-1}+\bar{c}_{n-2}s^{n-2}+\cdots+\bar{c}_1s+\bar{c}_0}{s^n+(a_{n-1}+\bar{k}_n)s^{n-1}+\cdots+(a_1+\bar{k}_2)s+(a_0+\bar{k}_1)}\tag{5.5.7}$$

Comparing eq. (5.5.3) and eq. (5.5.7), we get the conclusion that state feedback to SISO LTI systems only changes the poles not the zeros of the transfer function.

Example 5.5.1 Given the transfer function of a system

$$g(s)=\frac{(s+2)(s+3)}{(s+1)(s-2)(s+4)}$$

Is there a state feedback matrix $\boldsymbol{k}$ so that the transfer function of the closed-loop system is

$$g_k(s) = \frac{s+3}{(s+2)(s+4)}$$

If so, find the state feedback vector $\boldsymbol{k}$.

Answer We rewrite the transfer function of the system as

$$g(s) = \frac{s^2+5s+6}{s^3+3s^2-6s-8}$$

The state space description of the above transfer function is

$$\dot{\boldsymbol{x}} = \begin{bmatrix} 0 & 1 & 0 \\ 0 & 0 & 1 \\ 8 & 6 & -3 \end{bmatrix} \boldsymbol{x} + \begin{bmatrix} 0 \\ 0 \\ 1 \end{bmatrix} u$$

$$y = [6 \quad 5 \quad 1] \boldsymbol{x}$$

Because

$$\text{rank}[\boldsymbol{b}\ \boldsymbol{Ab}\ \boldsymbol{A}^2\boldsymbol{b}] = \text{rank}\begin{bmatrix} 0 & 0 & 1 \\ 0 & 1 & -3 \\ 1 & -3 & -3 \end{bmatrix} = 3$$

So, the system is controllable and we can place the eigenvalues at any desired positions. According conclusion 5.5.1, state feedback does not change the zeros of the closed-loop system. So, the desired transfer function of the closed-loop system is

$$g_k(s) = \frac{(s+2)(s+3)}{(s+2)(s+4)(s+2)}$$

And the desired characteristic polynomial of the closed-loop system is

$$\alpha^*(s) = (s+2)(s+4)(s+2) = s^3+8s^2+20s+16 \tag{5.5.8}$$

By introducing the state feedback $k = [k_1 \quad k_2 \quad k_3]$, the characteristic polynomial of the $\boldsymbol{A} - \boldsymbol{bk}$ is

$$\det[s\boldsymbol{I} - \boldsymbol{A} + \boldsymbol{bk}] = s^3 + (3+k_3)s^2 + (-6+k_2)s + (-8+k_1) \tag{5.5.9}$$

Comparing eq. (5.5.8) and eq. (5.5.9), yields

$$\begin{cases} -8+k_1 = 16 \\ -6+k_2 = 20 \\ 3+k_3 = 8 \end{cases} \Rightarrow \begin{cases} k_1 = 24 \\ k_2 = 26 \\ k_3 = 5 \end{cases}$$

Thus the state feedback gain vector $\boldsymbol{k} = [24 \quad 26 \quad 6]$.

5.5.2 Effect of state feedback on transfer matrices for MIMO systems

Consider a controllable MIMO LTI system described by

$$\begin{cases} \dot{\boldsymbol{x}} = \boldsymbol{Ax} + \boldsymbol{Bu} \\ \boldsymbol{y} = \boldsymbol{Cx} \end{cases} \tag{5.5.10}$$

where, $\boldsymbol{x}$ is an n-dimensional state vector, u is a $p \times 1$ input vector, y is $q \times 1$ output vector. Its transfer matrix is:

$$G(s) = C(sI - A)^{-1}B \tag{5.5.11}$$

In general, the zeros of $G(s)$ can be defined in various ways. The following is the definition of the zeros of a system when the system is completely controllable and completely observable: the value of s is the zero if

$$\text{rank}\begin{bmatrix} sI - A & B \\ -C & 0 \end{bmatrix} < n + \min(p,q) \tag{5.5.12}$$

Conclusion 5. 5. 2 For the controllable n-order time-continuous MIMO LTI system (5.5.10), the state feedback generally does not affect the zeros of $G(s)$ when all n poles are placed.

Proof Omitted.

Note that the above conclusion does not mean that the numerator polynomial of the transfer function of each element of $G(s)$ is not affected by state feedback. In the following, we give an example of this.

Example 5.5.2 Consider a two-input LTI system, $\{A, B, C\}$ in its state space description are

$$A = \begin{bmatrix} 1 & 0 & 0 \\ 0 & 2 & 0 \\ 0 & 0 & 3 \end{bmatrix}, \quad B = \begin{bmatrix} 1 & 0 \\ 0 & 1 \\ 1 & 1 \end{bmatrix}, \quad C = \begin{bmatrix} 1 & 0 & 2 \\ 2 & 1 & 0 \end{bmatrix}$$

The transfer matrix of the system, $G(s)$, is

$$G(s) = \begin{bmatrix} \dfrac{3s-5}{(s-3)(s-1)} & \dfrac{2}{s-3} \\ \dfrac{2}{s-1} & \dfrac{1}{s-2} \end{bmatrix}$$

The poles of $G(s)$ are $\lambda_1 = 1, \lambda_2 = 2, \lambda_3 = 3$, and the zero of $G(s)$ is 3. By introducing the state feedback matrix

$$K = \begin{bmatrix} -6 & -15 & 15 \\ 0 & 3 & 0 \end{bmatrix}$$

We get the parameter matrices of the closed-loop system:

$$A - BK = \begin{bmatrix} 7 & 15 & -15 \\ 0 & -1 & 0 \\ 6 & 12 & -12 \end{bmatrix}, \quad B = \begin{bmatrix} 1 & 0 \\ 0 & 1 \\ 1 & 1 \end{bmatrix}, \quad C = \begin{bmatrix} 1 & 0 & 2 \\ 2 & 1 & 0 \end{bmatrix}$$

The transfer matrix of the closed-loop system, $G_K(s)$, is

$$G_K(s) = \begin{bmatrix} \dfrac{3s-5}{(s+2)(s+3)} & \dfrac{2s^2+12s-17}{(s+1)(s+2)(s+3)} \\ \dfrac{2(s-3)}{(s+2)(s+3)} & \dfrac{(s-3)(s+8)}{(s+1)(s+2)(s+3)} \end{bmatrix}$$

Comparing $G(s)$ and $G_K(s)$, the state feedback places poles at

$$\lambda_1^* = -1, \lambda_2^* = -2, \lambda_3^* = -3$$

Meanwhile, it affects the numerator polynomials of the elements of $\boldsymbol{G}_K(s)$. But the zero of $\boldsymbol{G}(s)$ is also 3.

5.5.3 Effect of state feedback on observability

Based on the above discussion, the influence of state feedback on observability is further discussed.

Conclusion 5.5.3 For a SISO LTI system, it is either uncontrollable, unobservable or both uncontrollable and unobservable when zero-pole cancellation occurs in its transfer function.

Proof We assume that $\boldsymbol{A}$ of the SISO LTI system described by

$$\begin{cases} \dot{\boldsymbol{x}} = \boldsymbol{A}\boldsymbol{x} + \boldsymbol{b}u \\ y = \boldsymbol{c}\boldsymbol{x} \end{cases}$$

The matrix $\boldsymbol{A}$ has n distinct eigenvalues and the diagonal canonical form of the system is

$$\boldsymbol{A} = \begin{bmatrix} s_1 & & \\ & \ddots & \\ & & s_n \end{bmatrix}, \boldsymbol{b} = \begin{bmatrix} \alpha_1 \\ \alpha_2 \\ \vdots \\ \alpha_n \end{bmatrix}, \boldsymbol{c} = [\beta_1 \quad \beta_2 \quad \cdots \quad \beta_n]$$

The transfer function is

$$\begin{aligned} g(s) &= \frac{y(s)}{u(s)} \\ &= c(s\boldsymbol{I} - \boldsymbol{A}) - 1\boldsymbol{b} \\ &= [\beta_1 \quad \beta_2 \quad \cdots \quad \beta_n] \begin{bmatrix} \frac{1}{s - s_1} & & \\ & \ddots & \\ & & \frac{1}{s - s_n} \end{bmatrix} \begin{bmatrix} \alpha_1 \\ \alpha_1 \\ \vdots \\ \alpha_n \end{bmatrix} \\ &= \frac{\alpha_1 \beta_1}{s - s_1} + \cdots + \frac{\alpha_n \beta_n}{s - s_n} \end{aligned}$$

when zero-pole cancellation occurs, for example, if the pole s_2 is canceled, then we have

$$\alpha_2 \beta_2 = 0$$

So, there exist three situations:

(1) If $\alpha_2 = 0, \beta_2 \neq 0$, then the system is uncontrollable and observable;

(2) If $\alpha_2 \neq 0, \beta_2 = 0$, then the system is controllable but unobservable;

(3) if $\alpha_2 = 0, \beta_2 = 0$, then the system is uncontrollable and unobservable.

Because for SISO LTI systems, state feedback only changes poles of the transfer function and doesn't change the zeros. Therefore, state feedback may lead the pole-zero cancellation. According to conclusion 5.5.3, for SISO LTI systems, state feedback may destroy the observability.

The same is true for multiple-input systems.

Conclusion 5. 5. 4 For a controllable linear time-invariant system, the controllability remains unchanged and the observability may be destroyed after the state feedback is introduced.

5.6 Pole placement of not completely controllable systems

Consider an n-order not complete controllable LTI system described by

$$\begin{cases} \dot{\boldsymbol{x}} = \boldsymbol{A}\boldsymbol{x} + \boldsymbol{B}\boldsymbol{u} \\ \boldsymbol{y} = \boldsymbol{C}\boldsymbol{x} \end{cases} \tag{5.6.1}$$

where, $\boldsymbol{x}$ is an $n \times 1$ state vector, $\boldsymbol{u}$ is a $p \times 1$ input vector, $\boldsymbol{y}$ is a $q \times 1$ output vector. Because system eq. (5.6.1) is not complete controllable, the rank of its controllability matrix must be less than n, i. e.

$$\text{rank}\, \boldsymbol{Q}_c = \text{rank}[\boldsymbol{B} \vdots \boldsymbol{AB} \vdots \cdots \vdots \boldsymbol{A}^{n-1}\boldsymbol{B}] = n_c < n \tag{5.6.2}$$

Decompose not completely system (5.6.1) according to controllability, i. e. introduce the linearly non-singular transformation

$$\bar{\boldsymbol{x}} = \boldsymbol{P}\boldsymbol{x} \tag{5.6.3}$$

Then the state space description of thedecomposed system is

$$\begin{bmatrix} \dot{\bar{\boldsymbol{x}}}_c \\ \dot{\bar{\boldsymbol{x}}}_{\bar{c}} \end{bmatrix} = \begin{bmatrix} \bar{\boldsymbol{A}}_c & \bar{\boldsymbol{A}}_{12} \\ \boldsymbol{0} & \bar{\boldsymbol{A}}_{\bar{c}} \end{bmatrix} \begin{bmatrix} \bar{\boldsymbol{x}}_c \\ \bar{\boldsymbol{x}}_{\bar{c}} \end{bmatrix} + \begin{bmatrix} \bar{\boldsymbol{B}}_c \\ \boldsymbol{0} \end{bmatrix} \boldsymbol{u} \tag{5.6.4}$$

where, $\bar{\boldsymbol{x}}_c$ is an $n_c \times 1$ controllable state vector, $\bar{\boldsymbol{x}}_{\bar{c}}$ us an $(n - n_c) \times 1$ uncontrollable state vector.

The state space equation of the controllable subsystem is

$$\dot{\bar{\boldsymbol{x}}}_c = \bar{\boldsymbol{A}}_c \bar{\boldsymbol{x}}_c + \bar{\boldsymbol{A}}_{12} \bar{\boldsymbol{x}}_{\bar{c}} + \bar{\boldsymbol{B}}_c \boldsymbol{u} \tag{5.6.5}$$

The state space equation of the uncontrollable subsystem is

$$\dot{\bar{\boldsymbol{x}}}_{\bar{c}} = \bar{\boldsymbol{A}}_{\bar{c}} \bar{\boldsymbol{x}}_{\bar{c}} \tag{5.6.6}$$

Introducing the state feedback $\boldsymbol{K}_c (p \times n_c)$ to subsystem eq. (5.6.5), yields the closed-loop system described by

$$\dot{\bar{\boldsymbol{x}}}_c = (\bar{\boldsymbol{A}}_c - \bar{\boldsymbol{B}}_c \boldsymbol{K}_c) \bar{\boldsymbol{x}}_c + \bar{\boldsymbol{A}}_{12} \bar{\boldsymbol{x}}_{\bar{c}} + \bar{\boldsymbol{B}}_c \boldsymbol{v} \tag{5.6.7}$$

And its n_c eigenvalues can be placed at any positions.

Introducing the state feedback $\boldsymbol{K}_{\bar{c}}(p \times (n - n_c))$ to subsystem (5.6.6), yields the closed-loop system described by

$$\dot{\bar{\boldsymbol{x}}}_{\bar{c}} = \bar{\boldsymbol{A}}_{\bar{c}} \bar{\boldsymbol{x}}_{\bar{c}} \tag{5.6.8}$$

Obviously, the $n - n_c$ eigenvalues can not be changed by state feedback. So, we get the following conclusion:

Conclusion 5.6.1 The system eq. (5.6.1) can be stabilized by state feedback if and only if the eigenvalues of the uncontrollable subsystem are stable, i. e. the eigenvalues have a negative real-part.

Definition 5.6.1 Stabilization Problem. The problem of how to design a controller, such as by introducing state feedback, so that the original unstable system becomes stable, is called Stabilization Problem.

Conclusion 5.6.2 A system is stabilizable if

(1) the system is completely controllable. Or

(2) the eigenvalues of the uncontrollable state variables are stable.

Proof

(1) Obviously, case (1) is correct.

(2) We can prove case (2) from the above discussion, or from eq. (5.6.4) we get

$$\det(s\boldsymbol{I}-\boldsymbol{A}+\boldsymbol{BK})=\det(s\boldsymbol{I}-\overline{\boldsymbol{A}}+\overline{\boldsymbol{B}}\,\overline{\boldsymbol{K}})=\det(s\boldsymbol{I}_r-\overline{\boldsymbol{A}}_c+\overline{\boldsymbol{B}}_c\overline{\boldsymbol{K}}_c)\cdot\det(s\boldsymbol{I}_{n-r}-\overline{\boldsymbol{A}}_{\bar{c}})$$

where,

$$\overline{\boldsymbol{A}}=\begin{bmatrix}\overline{\boldsymbol{A}}_c & \overline{\boldsymbol{A}}_{12}\\ \boldsymbol{0} & \overline{\boldsymbol{A}}_{\bar{c}}\end{bmatrix},\quad \overline{\boldsymbol{B}}=\begin{bmatrix}\overline{\boldsymbol{B}}_c\\ \boldsymbol{0}\end{bmatrix},\quad \overline{\boldsymbol{K}}=[\overline{\boldsymbol{K}}_c\quad \overline{\boldsymbol{K}}_{\bar{c}}],\quad \boldsymbol{K}=\overline{\boldsymbol{K}}\boldsymbol{P}$$

The above equation indicates that state feedback don't change poles of the uncontrollable subsystem. In order to ensure the system is stable, the uncontrollable system must be stable.

Algorithm 5.6.1 Pole assigning algorithm of not completely uncontrollable systems.

Consider the not completely uncontrollable system eq. (5.6.1) and assume it is stabilizable. Find a $p\times n$ feedback gain matrix $\boldsymbol{K}$ such that the eigenvalues of the closed-loop system are placed at $\lambda_1^*,\lambda_2^*,\cdots,\lambda_{n_c}^*$.

Step 1. Check whether $\{\boldsymbol{A},\boldsymbol{B}\}$ is controllable. If it is not completely controllable, go to step 2, else go to step 5.

Step 2. Decompose system eq. (5.6.1) according to controllability and get eq. (5.6.4) and $\boldsymbol{P}$.

Step 3. For $\{\overline{\boldsymbol{A}}_c,\overline{\boldsymbol{B}}_c\}$, by using the pole placement algorithm of multiple-input systems, design a $p\times n_c$ state feedback matrix $\boldsymbol{K}_c$, such that the eigenvalues of the closed-loop system are placed at $\lambda_1^*,\lambda_2^*,\cdots,\lambda_{n_c}^*$.

Step 4. Compute the $p\times n$ state feedback matrix $\boldsymbol{K}$ by using the following formula

$$\boldsymbol{K}=[\boldsymbol{K}_c\quad \boldsymbol{0}]\boldsymbol{P} \tag{5.6.9}$$

such that the closed-loop system of eq. (5.6.1) are placed at $\lambda_1^*,\lambda_2^*,\cdots,\lambda_{n_c}^*,\lambda_{n_c+1},\cdots,\lambda_n$, where $\lambda_{n_c+1},\cdots,\lambda_n$ are the eigenvalues of unstable subsystem of (5.6.1). Stop computing.

Step 5. For $\{\boldsymbol{A},\boldsymbol{B}\}$, by using the pole placement algorithm of multiple-input systems, design a $p\times n$ state feedback matrix $\boldsymbol{K}$, such that the eigenvalues of the closed-loop system are placed at $\lambda_1^*, \lambda_2^*, \cdots, \lambda_{n_c}^*, \lambda_{n_c+1}, \cdots, \lambda_n$. Stop computing.

5.7 Pole placement using output feedback

There are the following conclusions about pole placement of output feedback:

Conclusion 5.7.1 In general, all poles of LTI systems cannot be arbitrarily placed using the output feedback $\boldsymbol{u} = -\boldsymbol{F}\boldsymbol{y} + \boldsymbol{v}$($\boldsymbol{v}$ is the reference input).

Proof It can be seen from the previous discussion that all poles of a controllable LTI system can be arbitrarily placed using state feedback. From Section 5.1.2, output feedback cannot achieve all functions of state feedback. So, output feedback cannot arbitrarily place all poles of the system.

Output feedback does not change the controllability and observability of systems. Therefore, output feedback must not place the poles of the system to the position of zeros.

Conclusion 5.7.2 Consider an n-order controllable SISO LTI system described by

$$\begin{cases} \dot{\boldsymbol{x}} = \boldsymbol{A}\boldsymbol{x} + \boldsymbol{b}u \\ y = \boldsymbol{c}\boldsymbol{x} \end{cases} \tag{5.7.1}$$

The output feedback

$$\boldsymbol{u} = \boldsymbol{v} - fy = \boldsymbol{v} - f\boldsymbol{c}\boldsymbol{x} \tag{5.7.2}$$

can place closed-loop poles on the root locus, not outside the root locus.

Proof The transfer function of system eq. (5.7.1) is

$$g(s) = \boldsymbol{c}(s\boldsymbol{I} - \boldsymbol{A})^{-1}\boldsymbol{b} = \frac{\beta(s)}{\alpha(s)} \tag{5.7.3}$$

The characteristic polynomial of system eq. (5.7.1) is

$$\alpha(s) = \det(s\boldsymbol{I} - \boldsymbol{A}) \tag{5.7.4}$$

Introducing the output feedback eq. (5.7.2) to system eq. (5.7.1), the transfer function of the closed-loop system is

$$g_f(s) = c(s\boldsymbol{I} - \boldsymbol{A} + \boldsymbol{b}f\boldsymbol{c})^{-1}\boldsymbol{b} \tag{5.7.5}$$

The characteristic polynomial of the closed-loop system is

$$\alpha_f(s) = \det(s\boldsymbol{I} - \boldsymbol{A} + \boldsymbol{b}f\boldsymbol{c}) \tag{5.7.6}$$

Because

$$(s\boldsymbol{I} - \boldsymbol{A} + \boldsymbol{b}f\boldsymbol{c}) = (s\boldsymbol{I} - \boldsymbol{A})(I + (s\boldsymbol{I} - \boldsymbol{A})^{-1}\boldsymbol{b}f\boldsymbol{c}) \tag{5.7.7}$$

So, we get

$$\alpha_f(s) = \det(s\boldsymbol{I} - \boldsymbol{A})\det(\boldsymbol{I} + (s\boldsymbol{I} - \boldsymbol{A})^{-1}\boldsymbol{b}fc) \tag{5.7.8}$$

Based on properties of matrix determinant, we have

$$\det(\boldsymbol{I} + \boldsymbol{G}_1\boldsymbol{G}_2) = \det(\boldsymbol{I} + \boldsymbol{G}_2\boldsymbol{G}_1) \tag{5.7.9}$$

where, $\boldsymbol{G}_1$ and $\boldsymbol{G}_2$ are the matrices with corresponding dimension. This is because

$$\det(\boldsymbol{I} + \boldsymbol{G}_1\boldsymbol{G}_2) = \det \boldsymbol{G}_2^{-1}(\boldsymbol{I} + \boldsymbol{G}_2\boldsymbol{G}_1)\boldsymbol{G}_2$$

$$= \det G_2^{-1} \cdot \det(I + G_2 G_1) \cdot \det G_2$$
$$= \det(I + G_2 G_1) \quad (5.7.10)$$

From eq. (5.7.8) and eq. (5.7.7), we have

$$\alpha_f(s) = \alpha(s)\det(1 + fc(sI - A)^{-1}b) = \alpha(s)(1 + f\frac{\beta(s)}{\alpha(s)}) = \alpha(s) + f\beta(s) \quad (5.7.11)$$

It indicates that the poles of transfer function of the closed-loop system after introducing output feedback is the roots of the following equation

$$\alpha(s) + f\beta(s) = 0 \quad (5.7.12)$$

And the roots of $\alpha(s) = 0$ and $\beta(s) = 0$ are the poles and zeros of transfer function eq. (5.7.3) of open-loop system eq. (5.7.1), respectively. According to the root locus method of classical control theory, for the output feedback system, the closed-loop poles can only be distributed on the root loci which start from open-loop poles and end at open-loop zeros on the complex plane when $f = 0 \to \infty$, but not outside the root loci.

The output feedback has great limitations in pole assignment. One way to expand the ability of its pole assignment is to use output feedback plus a feedforward compensator which is shown in Figure 5.7.1. It can be proved that all poles of the dynamic output feedback system can be arbitrarily placed by selecting the compensator reasonably.

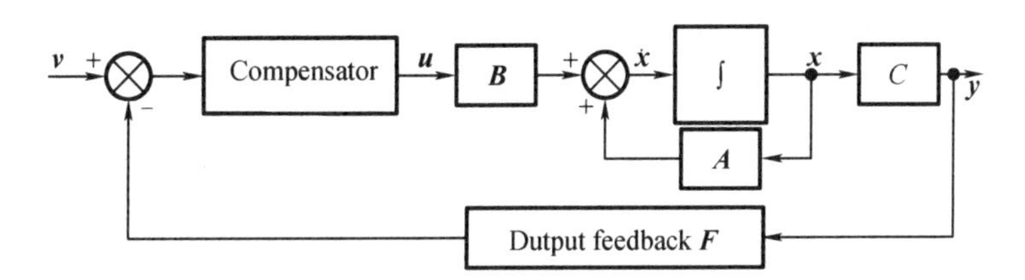

Figure 5.7.1 Dynamic output feedback

5.8 The decoupling of multivariable systems by state feedback

Decoupling control has greater theoretical importance and wider application potential. Decoupling control refers to a multiple-input multiple-output system, that uses a control algorithm such that each input of the system only individually controls single output and each output is controlled by only one input, i. e. there is a one-to-one relationship between the inputs and outputs by means of a decoupling algorithm.

In this section, we focus on conditions and algorithms for dynamic decoupling using state feedback.

5.8.1 Dynamic decoupling using state feedback

Consider a MIMO LTI system described by

$$\begin{cases} \dot{\boldsymbol{x}} = \boldsymbol{Ax} + \boldsymbol{Bu} \\ \boldsymbol{y} = \boldsymbol{Cx} \end{cases} \tag{5.8.1}$$

where $\boldsymbol{x}$ is an n-dimensional state vector, $\boldsymbol{u}$ is a p-dimensional input vector, $\boldsymbol{y}$ is a q-dimensional output vector, $p \leqslant n$.

There are three hypotheses:

(1) The dimension of the input vector of system eq. (5.8.1) is equal to the dimension of the output vector, i. e. $p = q$. In this case, the transfer function matrix of the system is a square rational fraction matrix.

(2) The decoupling control algorithm adopts state feedback combined with input transformation, i. e.

$$\boldsymbol{u} = -\boldsymbol{Kx} + \boldsymbol{Lv} \tag{5.8.2}$$

where, $\boldsymbol{K}$ is a $p \times n$ state feedback gain matrix, $\boldsymbol{L}$ is a $p \times p$ input transformation matrix, $\boldsymbol{v}$ is a $p \times 1$ reference input vector.

(3) The input transformation matrix $\boldsymbol{L}$ is non-singular, i. e. $\det \boldsymbol{L} \neq 0$.

Under the above three assumptions, the decoupling control algorithm of state feedback combined with input transformation (5.8.2) can realize the dynamic decoupling of systems, and the state space description of the closed-loop system obtained is described by

$$\begin{cases} \dot{\boldsymbol{x}} = (\boldsymbol{A} - \boldsymbol{BK})\boldsymbol{x} + \boldsymbol{BLv} \\ \boldsymbol{y} = \boldsymbol{Cx} \end{cases} \tag{5.8.3}$$

The diagram is shown in Figure 8.1. The transfer function matrix of the closed loop system, $\boldsymbol{G}_{KL}(s)$, is

$$\boldsymbol{G}_{KL}(s) = \boldsymbol{C}(s\boldsymbol{I} - \boldsymbol{A} + \boldsymbol{BK})^{-1}\boldsymbol{BL} \tag{5.8.4}$$

The decoupled transfer function matrix $\boldsymbol{G}_{KL}(s)$ is a non-singular diagonal rational fraction matrix:

$$\boldsymbol{G}_{KL}(s) = \begin{bmatrix} \bar{g}_{11}(s) & & \\ & \ddots & \\ & & \bar{g}_{pp}(s) \end{bmatrix} \tag{5.8.5}$$

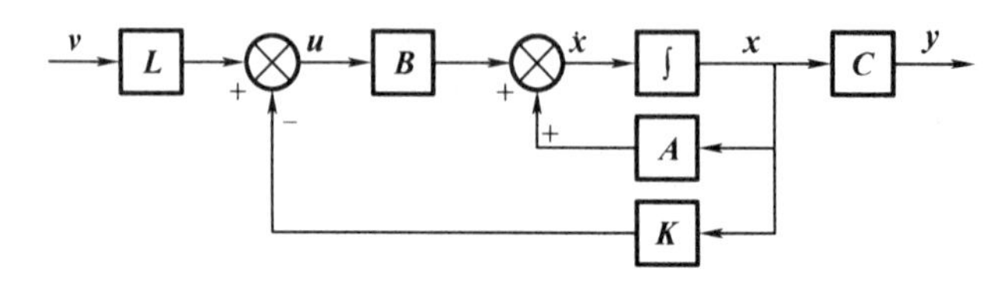

Figure 5.8.1 The diagram of the dynamic decoupling system

5.8.2 d_i and $\boldsymbol{E}_i$

In order to discuss the decupling conditions and algorithms for dynamic decoupling, d_i and $\boldsymbol{E}_i$ of the transfer function matrix of LTI systems, namely the structural characteristic index and the structural characteristic vector, are first introduced.

Consider a $p \times p$ transfer function matrix $\boldsymbol{G}(s)$:

$$\boldsymbol{G}(s) = \begin{bmatrix} \boldsymbol{g}_1(s) \\ \vdots \\ \boldsymbol{g}_p(s) \end{bmatrix} \tag{5.8.6}$$

where $\boldsymbol{g}_i(s)$ is i^{th} row of $\boldsymbol{G}(s)$:

$$\boldsymbol{g}_i(s) = [g_{i1}(s) \quad g_{i2}(s) \quad \cdots \quad g_{ip}(s)] \tag{5.8.7}$$

Let

$$\sigma_{ij} = \text{the degree in } s \text{ of the denominator of } g_{ij}(s) - \text{the degree in } s \text{ of the numerator of } g_{ij}(s) \tag{5.8.8}$$

The structural property index d_i is defined as

$$d_i = \min\{\sigma_{i1}, \sigma_{i2}, \cdots, \sigma_{ip}\} - 1 \tag{5.8.9}$$

where, $i = 1, \cdots, p$.

The structural characteristic vector $\boldsymbol{E}_i$ is defined as

$$\boldsymbol{E}_i = \lim_{s \to \infty} s^{d_i + 1} \boldsymbol{g}_i(s), i = 1, 2, \cdots, p \tag{5.8.10}$$

Example 5.8.1 Given a $\boldsymbol{G}(s)$ as

$$\boldsymbol{G}(s) = \begin{bmatrix} \dfrac{s+2}{s^2+s+1} & \dfrac{1}{s^2+s+2} \\ \dfrac{1}{s^2+2s+1} & \dfrac{3}{s^2+s+4} \end{bmatrix}$$

Compute its d_i and $\boldsymbol{E}_i$.

Answer

$$d_1 = \min\{1, 2\} - 1 = 0; \; d_2 = \min\{2, 2\} - 1 = 1$$

$$\boldsymbol{E}_1 = \lim_{s \to \infty} s \begin{bmatrix} \dfrac{s+2}{s^2+s+1} & \dfrac{1}{s^2+s+2} \end{bmatrix} = [1 \quad 0]$$

$$\boldsymbol{E}_2 = \lim_{s \to \infty} s^2 \begin{bmatrix} \dfrac{1}{s^2+s+1} & \dfrac{3}{s^2+s+4} \end{bmatrix} = [1 \quad 3]$$

Some basic properties of d_i and $\boldsymbol{E}_i$ are given below.

(1) For the system eq. (5.8.1), its structural characteristic index d_i is a non-negative integer and its value range is

$$0 \leqslant d_i \leqslant n - 1, \quad i = 1, 2, \cdots, p \tag{5.8.11}$$

(2) The parameter matrices $\{\boldsymbol{A}, \boldsymbol{B}, \boldsymbol{C}\}$ can be used to find d_i and $\boldsymbol{E}_i$ directly.

Let

$$C = \begin{bmatrix} c_1 \\ \vdots \\ c_p \end{bmatrix}$$

where, c_i is i^{th} row of E.

d_i is computed as the following formula:

$$d_i = \begin{cases} \mu, & \text{when } c_i A^k B = 0, k = 0, 1, \cdots, \mu - 1, \text{ and } c_i A^{\mu} B \neq 0 \\ n - 1, & \text{when } c_i A^k B = \mathbf{0}, k = 0, 1, \cdots, n - 1 \end{cases} \tag{5.8.12}$$

and

$$E_i = c_i A^{d_i} B \tag{5.8.13}$$

(3) For aproper rational fractional matrix $G(s)$, its structural characteristic vector E_i is a $1 \times p$ constant vector .

(4) For $\{L, K\}$, where $\det L \neq 0$, the ith row of the transfer function matrix $G_{KL}(s)$ of the closed-loop system after introducing state feedback is

$$g_{KLi}(s) = \frac{1}{\bar{\alpha}(s)}[c_i \bar{R}_{n-1} BLs^{n-1} + c_i \bar{R}_{n-2} BLs^{n-2} + \cdots + c_i \bar{R}_1 BLs + c_i \bar{R}_0 BL] \tag{5.8.14}$$

where,

$$\bar{\alpha}(s) = \det(sI - A + BK) = s^n + \bar{a}_{n-1} s^{n-1} + \cdots + \bar{a}_1 s + \bar{a}_0 \tag{5.8.15}$$

and

$$\begin{cases} \bar{R}_{n-1} = I_{n \times n} \\ \bar{R}_{n-2} = (A - BK) + \bar{a}_{n-1} I \\ \bar{R}_{n-3} = (A - BK)^2 + \bar{a}_{n-1}(A - BK) + \bar{a}_{n-2} I \\ \cdots \\ \bar{R}_0 = (A - BK)^{n-1} + \bar{a}_{n-1}(A - BK)^{n-2} + \cdots + \bar{a}_1 I \end{cases} \tag{5.8.16}$$

And $\bar{d}_i$ and $\bar{E}_i$ of the transfer function matrix $G_{KL}(s)$ of the corresponding closed-loop system are

$$\bar{d}_i = \begin{cases} \bar{\mu}, & \text{when } c_i (A - BK)^k BL = \mathbf{0}, k = 0, 1, \cdots, \bar{\mu} - 1, \text{ and } c_i (A - BK)^{\mu} B \neq \mathbf{0} \\ n - 1, & \text{when } c_i (A - BK)^k BL = \mathbf{0}, k = 0, 1, \cdots, n - 1 \end{cases} \tag{5.8.17}$$

and

$$\bar{E}_i = c_i (A - BK)^{\bar{d}_i} BL \tag{5.8.18}$$

where $i = 1, \cdots, p$.

(4) For $\{L, K\}$, where $\det L \neq 0$, the relationship between d_i, E_i of the open-loop system and $\bar{d}_i, \bar{E}_i$ of the closed-loop system is

$$\begin{cases} \bar{d}_i = d_i \\ \bar{E}_i = E_i L \end{cases} \tag{5.8.19}$$

5.8.3 Decouple condition

Conclusion 5.8.1 Decouple condition.

The LTI system eq. (5.8.1) can be decoupled by state feedback with input transformation by using eq. (5.8.2), if and only if the constant matrix

$$\boldsymbol{E} = \begin{bmatrix} \boldsymbol{E}_1 \\ \vdots \\ \boldsymbol{E}_p \end{bmatrix}_{p \times p} \tag{5.8.20}$$

is nonsingular, i. e. $\det(\boldsymbol{E}) \neq 0$.

Conclusion 5.8.2 Integral form decoupling.

Assuming the LTI system eq. (5.8.1) satisfies decoupling condition, we adopt $\boldsymbol{u} = -\boldsymbol{K}\boldsymbol{x} + \boldsymbol{L}\boldsymbol{v}$, where,

$$\begin{cases} \boldsymbol{L} = \boldsymbol{E}^{-1} \\ \boldsymbol{K} = \boldsymbol{E}^{-1}\boldsymbol{F} \end{cases} \tag{5.8.21}$$

$\boldsymbol{E}$ is computed by eq. (5.8.20) and

$$\boldsymbol{F} = \begin{bmatrix} \boldsymbol{c}_1 \boldsymbol{A}^{d_1+1} \\ \vdots \\ \boldsymbol{c}_p \boldsymbol{A}^{d_p+1} \end{bmatrix} \tag{5.8.22}$$

So the transfer matrix of decoupled closed-loop system is

$$\boldsymbol{G}_{KL}(s) = \begin{bmatrix} \frac{1}{s^{d_1+1}} & & \\ & \ddots & \\ & & \frac{1}{s^{d_p+1}} \end{bmatrix} \tag{5.8.23}$$

5.8.4 Algorithm of determining $\{\boldsymbol{K},\boldsymbol{L}\}$ to place all poles at any desire positions

Conclusion 5.8.3 Dynamic decoupling algorithm.

Consider a controllable LTI system described by

$$\begin{cases} \dot{\boldsymbol{x}} = \boldsymbol{A}\boldsymbol{x} + \boldsymbol{B}\boldsymbol{u} \\ \boldsymbol{y} = \boldsymbol{C}\boldsymbol{x} \end{cases} \tag{5.8.24}$$

where, $\boldsymbol{x}$ is an n-dimensional state vector, $\boldsymbol{u}$ is a p-dimensional input vector, $\boldsymbol{y}$ is a q-dimensional output vector. Determine $\{\boldsymbol{K},\boldsymbol{L}\}$ such that the closed-loop system is decoupled and all poles of each decoupled SISO system are assigned at desired positions.

Procedure:

Step 1. Compute

$$\{d_i, i=1,2,\cdots,p\}, \{\boldsymbol{E}_i = \boldsymbol{c}_i \boldsymbol{A}^{d_i}\boldsymbol{B}, i=1,2,\cdots,p\}$$

Step 2. Construct matrix $\boldsymbol{E}$:

$$\boldsymbol{E} = \begin{bmatrix} \boldsymbol{E}_1 \\ \vdots \\ \boldsymbol{E}_p \end{bmatrix}_{p\times p}$$

If $\boldsymbol{E}$ is nonsingular, the system can be decoupled and go to Step3. Else the system can not be decoupled and stop.

Step 3. Compute

$$\boldsymbol{E}^{-1}, \quad \boldsymbol{F} = \begin{bmatrix} \boldsymbol{c}_1 \boldsymbol{A}^{d_1+1} \\ \vdots \\ \boldsymbol{c}_p \boldsymbol{A}^{d_p+1} \end{bmatrix}$$

Step 4. Let

$$\overline{\boldsymbol{L}} = \boldsymbol{E}^{-1}, \quad \overline{\boldsymbol{K}} = \boldsymbol{E}^{-1}\boldsymbol{F}$$

We get integral form decoupled system described by

$$\begin{cases} \dot{\boldsymbol{x}} = \overline{\boldsymbol{A}}\boldsymbol{x} + \overline{\boldsymbol{B}}\boldsymbol{v} \\ \boldsymbol{y} = \overline{\boldsymbol{C}}\boldsymbol{x} \end{cases}$$

where,

$$\overline{\boldsymbol{A}} = \boldsymbol{A} - \boldsymbol{B}\boldsymbol{E}^{-1}\boldsymbol{F}, \quad \overline{\boldsymbol{B}} = \boldsymbol{B}\boldsymbol{E}^{-1}, \quad \overline{\boldsymbol{C}} = \boldsymbol{C}$$

and $\{\overline{\boldsymbol{A}}, \overline{\boldsymbol{B}}\}$ is controllable.

Step 5. Check whether $\{\overline{\boldsymbol{A}}, \overline{\boldsymbol{C}}\}$ is observable. If $\{\overline{\boldsymbol{A}}, \overline{\boldsymbol{C}}\}$ is not observable, compute

$$\text{rank}\boldsymbol{Q}_o = \text{rank}\begin{bmatrix} \overline{\boldsymbol{C}} \\ \overline{\boldsymbol{C}}\,\overline{\boldsymbol{A}} \\ \vdots \\ \overline{\boldsymbol{C}}\,\overline{\boldsymbol{A}}^{n-1} \end{bmatrix} = m$$

Step 6. Introducing nonsingular transformation $\tilde{\boldsymbol{x}} = \boldsymbol{T}^{-1}\boldsymbol{x}$ to transform the integral form decoupled system $\{\overline{\boldsymbol{A}}, \overline{\boldsymbol{B}}, \overline{\boldsymbol{C}}\}$ obtained at Step 4 into decoupling normal form:

$$\tilde{\boldsymbol{A}} = \boldsymbol{T}^{-1}\overline{\boldsymbol{A}}\boldsymbol{T}, \quad \tilde{\boldsymbol{B}} = \boldsymbol{T}^{-1}\overline{\boldsymbol{B}}, \quad \tilde{\boldsymbol{C}} = \overline{\boldsymbol{C}}\boldsymbol{T}$$

If $\{\overline{\boldsymbol{A}}, \overline{\boldsymbol{C}}\}$ is completely observable, the decoupling normal form is

$$\tilde{\boldsymbol{A}} = \begin{bmatrix} \tilde{\boldsymbol{A}}_1 & & \\ & \ddots & \\ & & \tilde{\boldsymbol{A}}_p \end{bmatrix}, \quad \tilde{\boldsymbol{B}} = \begin{bmatrix} \tilde{\boldsymbol{b}}_1 & & \\ & \ddots & \\ & & \tilde{\boldsymbol{b}}_p \end{bmatrix}, \quad \tilde{\boldsymbol{C}} = \begin{bmatrix} \tilde{\boldsymbol{c}}_1 & & \\ & \ddots & \\ & & \tilde{\boldsymbol{c}}_p \end{bmatrix} \tag{5.8.25}$$

where $\widetilde{\boldsymbol{A}}_i$ is $m_i \times m_i$, $\widetilde{\boldsymbol{b}}_i$ is $m_i \times 1$, $\widetilde{\boldsymbol{c}}_i$ is $1 \times m_i, i = 1, 2, \cdots, p$, $\sum_{i=1}^{p} m_i = n$.

If $\{\overline{\boldsymbol{A}}, \overline{\boldsymbol{C}}\}$ is not completely observable, the decoupling normal form is

$$\widetilde{\boldsymbol{A}} = \left[\begin{array}{ccc|c} \widetilde{\boldsymbol{A}}_1 & & & \boldsymbol{0} \\ & \ddots & & \vdots \\ & & \widetilde{\boldsymbol{A}}_p & \boldsymbol{0} \\ \hline \widetilde{\boldsymbol{A}}_{c1} & \cdots & \widetilde{\boldsymbol{A}}_{cp} & \widetilde{\boldsymbol{A}}_{p+1} \end{array}\right], \quad \widetilde{\boldsymbol{B}} = \left[\begin{array}{ccc} \widetilde{\boldsymbol{b}}_1 & & \\ & \ddots & \\ & & \widetilde{\boldsymbol{b}}_p \\ \hline \widetilde{\boldsymbol{b}}_{c1} & \cdots & \widetilde{\boldsymbol{b}}_{cp} \end{array}\right], \quad \widetilde{\boldsymbol{C}} = \left[\begin{array}{ccc|c} \widetilde{\boldsymbol{c}}_1 & & & \boldsymbol{0} \\ & \ddots & & \vdots \\ & & \widetilde{\boldsymbol{c}}_p & \boldsymbol{0} \end{array}\right] \tag{5.8.26}$$

where $\widetilde{\boldsymbol{A}}_i$ is $m_i \times m_i$, $\widetilde{\boldsymbol{b}}_i$ is $m_i \times 1$, $\widetilde{\boldsymbol{c}}_i$ is $1 \times m_i$, $i = 1, 2, \cdots, p$, $\sum_{i=1}^{p} m_i = n$.

When $m_i = d_i + 1$, we have

$$\underset{m_i \times m_i}{\widetilde{\boldsymbol{A}}_i} = \begin{bmatrix} 0 & 1 & & \\ \vdots & & \ddots & \\ 0 & & & 1 \\ 0 & 0 & \cdots & 0 \end{bmatrix}, \quad \underset{m_i \times 1}{\widetilde{\boldsymbol{b}}_i} = \begin{bmatrix} 0 \\ \vdots \\ 0 \\ 1 \end{bmatrix}, \quad \underset{1 \times m_i}{\widetilde{\boldsymbol{c}}_i} = [1 \quad 0 \quad \cdots \quad 0] \tag{5.8.27}$$

where, the $d_i + 1$ row of $\widetilde{\boldsymbol{A}}_i$, i. e. the lowest row of $\widetilde{\boldsymbol{A}}_i$, is zero vector, which reflects the characteristics of integral decoupling system.

When $m_i > d_i + 1$, we have

$$\underset{m_i \times m_i}{\widetilde{\boldsymbol{A}}_i} = \left[\begin{array}{c|ccc|c} 0 & & & & \\ \vdots & & \boldsymbol{I}_{d_i} & & \boldsymbol{0} \\ 0 & & & & \\ \hline 0 & 0 & \cdots & 0 & \\ \hline & & * & & * \end{array}\right] \begin{array}{l} \left.\vphantom{\begin{array}{c}0\\ \vdots\\ 0\\ 0\end{array}}\right\} (d_i + 1) \\ \} m_i - (d_i + 1) \end{array}, \quad \widetilde{\boldsymbol{b}}_i = \begin{bmatrix} 0 \\ \vdots \\ 0 \\ 1 \\ 0 \\ \vdots \\ 0 \end{bmatrix}$$

with column blocks of widths $(d_i + 1)$ and $m_i - (d_i + 1)$,

$$\underset{1 \times m_i}{\widetilde{\boldsymbol{c}}_i} = [1 \quad 0 \quad \cdots \quad 0] \tag{5.8.28}$$

where, the block denoted by * has no effect on the synthesis results.

Step 7. Compute $\boldsymbol{T}^{-1}$ from $\{\overline{\boldsymbol{A}}, \overline{\boldsymbol{B}}, \overline{\boldsymbol{C}}\}$ and $\{\widetilde{\boldsymbol{A}}, \widetilde{\boldsymbol{B}}, \widetilde{\boldsymbol{C}}\}$. If $\{\widetilde{\boldsymbol{A}}, \widetilde{\boldsymbol{B}}, \widetilde{\boldsymbol{C}}\}$ and $\{\overline{\boldsymbol{A}}, \overline{\boldsymbol{B}}, \overline{\boldsymbol{C}}\}$ are controllable and observable, there are

$$\widetilde{\boldsymbol{A}} = \boldsymbol{T}^{-1}\overline{\boldsymbol{A}}\boldsymbol{T}, \quad \widetilde{\boldsymbol{B}} = \boldsymbol{T}^{-1}\overline{\boldsymbol{B}}, \quad \widetilde{\boldsymbol{C}} = \overline{\boldsymbol{C}}\boldsymbol{T}$$

Let

$$\overline{\boldsymbol{Q}}_c = [\overline{\boldsymbol{B}}\ \overline{\boldsymbol{A}}\ \overline{\boldsymbol{B}} \cdots \overline{\boldsymbol{A}}^{n-1}\overline{\boldsymbol{B}}], \quad \widetilde{\boldsymbol{Q}}_c = [\widetilde{\boldsymbol{B}}\ \widetilde{\boldsymbol{A}}\ \widetilde{\boldsymbol{B}} \cdots \widetilde{\boldsymbol{A}}^{n-1}\widetilde{\boldsymbol{B}}]$$

$$\overline{Q}_o = \begin{bmatrix} \overline{C} \\ \overline{C}\,\overline{A} \\ \vdots \\ \overline{C}\ \overline{A}^{n-1} \end{bmatrix}, \quad \widetilde{Q}_o = \begin{bmatrix} \widetilde{C} \\ \widetilde{C}\ \widetilde{A} \\ \vdots \\ \widetilde{C}\ \widetilde{A}^{n-1} \end{bmatrix}$$

We have

$$T^{-1} = (\widetilde{Q}_o^T \widetilde{Q}_o)^{-1} \widetilde{Q}_o^T \overline{Q}_o, \quad T = \overline{Q}_c \widetilde{Q}_c^T (\widetilde{Q}_c \widetilde{Q}_c^T)^{-1} \tag{5.8.29}$$

Step 8. For decoupling canonical form $\{\widetilde{A}, \widetilde{B}, \widetilde{C}\}$, choose the $p \times n$ state feedback matrix $\widetilde{K}$, such that the system $\{\widetilde{A}, \widetilde{B}, \widetilde{C}\}$ can be dynamically decoupled. For eq. (5.8.25), $\widetilde{K}$ has the following form:

$$\widetilde{K} = \begin{bmatrix} \widetilde{k}_1 & & \\ & \ddots & \\ & & \widetilde{k}_p \end{bmatrix} \tag{5.8.30}$$

For eq. (5.8.26), $\widetilde{K}$ has the following form

$$\widetilde{K} = \begin{bmatrix} \widetilde{k}_1 & & & \mathbf{0} \\ & \ddots & & \vdots \\ & & \widetilde{k}_p & \mathbf{0} \end{bmatrix} \tag{5.8.31}$$

where, for eq. (5.8.27), we have

$$\underset{1 \times m_i}{\widetilde{k}_i} = [k_{i0} \quad k_{i1} \quad \cdots \quad k_{id_i}] \tag{5.8.32}$$

for eq. (5.8.28), we have

$$\underset{1 \times m_i}{\widetilde{k}_i} = [k_{i0} \quad k_{i1} \quad \cdots \quad k_{id_i} 0 \quad \cdots \quad 0] \tag{5.8.33}$$

And, $\widetilde{K}$ obtained from the above equations must make $\{\widetilde{A}, \widetilde{B}, \widetilde{C}\}$ dynamic decoupled, i. e.

$$\widetilde{C}(sI - \widetilde{A} + \widetilde{B}\widetilde{K})^{-1}\widetilde{B} = \begin{bmatrix} \widetilde{c}_1 (sI - \widetilde{A}_1 + \widetilde{b}_1 \widetilde{k}_1)^{-1} \widetilde{b}_1 & & \\ & \ddots & \\ & & \widetilde{c}_p (sI - \widetilde{A}_p + \widetilde{b}_p \widetilde{k}_p)^{-1} \widetilde{b}_p \end{bmatrix}$$

$$\widetilde{A}_i - \widetilde{b}_i \widetilde{k}_i = \begin{bmatrix} 0 & 1 & & \\ \vdots & & \ddots & \\ 0 & & & 1 \\ -k_{i0} & -k_{i1} & \cdots & -k_{id_i} \end{bmatrix}$$

or

$$\widetilde{A}_i - \widetilde{b}_i \widetilde{k}_i = \left[\begin{array}{c|ccc|c} 0 & & & & \\ \vdots & & I_{d_i} & & \mathbf{0} \\ 0 & & & & \\ \hline -k_{i0} & -k_{i1} & \cdots & -k_{id_i} & \\ \hline & & * & & * \end{array}\right]$$

Step 9. Specify the desired poles for each single-input single-output system after decoupling as

$$\{\lambda_{i_1}^*, \lambda_{i_2}^*, \cdots, \lambda_{i,d_i+1}^*\}, \quad i = 1, 2, \cdots, p$$

Determined the individual tuples of the state feedback matrix

$$\{k_{i0}, k_{i1}, \cdots, k_{id_i}\}, \quad i = 1, 2, \cdots, p$$

usingthe single-input pole placement algorithm.

Step 10. For the original system $\{A, B, C\}$ compute $\{L, K\}$ as eq. (5.8.34) such that the system is dynamic decoupled and all poles are placed at desired positions.

$$K = E^{-1}F + E^{-1}\widetilde{K}T^{-1}, \quad L = E^{-1} \tag{5.8.34}$$

One shortcoming of the above algorithm is that in Step 7, if $\{\overline{A}, \overline{B}, \overline{C}\}$ and $\{\widetilde{A}, \widetilde{B}, \widetilde{C}\}$ are completely controllable but not completely observable, the calculation of matrix T is not given.

Example 6.8.1 Given a two-input and two-output LTI system:

$$\dot{x} = \begin{bmatrix} -1 & 1 & 1 & 1 \\ 6 & 0 & -3 & 1 \\ -1 & 1 & 1 & 2 \\ 2 & -2 & -2 & 0 \end{bmatrix} x + \begin{bmatrix} 0 & 0 \\ 1 & 0 \\ 0 & 0 \\ 0 & 1 \end{bmatrix} u$$

$$y = \begin{bmatrix} 2 & 0 & -1 & 0 \\ -1 & 0 & 1 & 0 \end{bmatrix} x$$

determine $\{L, K\}$ such that the system is decoupled and the poles are placed at -2, -4, $-2+\mathrm{j}$, $-2-\mathrm{j}$.

Answer

(1) Computer structural characteristic index $\{d_1, d_2\}$ and structural characteristic vector $\{E_1, E_2\}$. From

$$c_1 B = [2 \quad 0 \quad -1 \quad 0] \begin{bmatrix} 0 & 0 \\ 1 & 0 \\ 0 & 0 \\ 0 & 1 \end{bmatrix} = [0 \quad 0]$$

$$c_1 AB = [2 \quad 0 \quad -1 \quad 0] \begin{bmatrix} -1 & 1 & 1 & 1 \\ 6 & 0 & -3 & 1 \\ -1 & 1 & 1 & 2 \\ 2 & -2 & -2 & 0 \end{bmatrix} \begin{bmatrix} 0 & 0 \\ 1 & 0 \\ 0 & 0 \\ 0 & 1 \end{bmatrix} = [1 \quad 0]$$

$$c_2B = [-1 \quad 0 \quad 1 \quad 0]\begin{bmatrix} 0 & 0 \\ 1 & 0 \\ 0 & 0 \\ 0 & 1 \end{bmatrix} = [0 \quad 0]$$

$$c_2AB = [-1 \quad 0 \quad 1 \quad 0]\begin{bmatrix} -1 & 1 & 1 & 1 \\ 6 & 0 & -3 & 1 \\ -1 & 1 & 1 & 2 \\ 2 & -2 & -2 & 0 \end{bmatrix}\begin{bmatrix} 0 & 0 \\ 1 & 0 \\ 0 & 0 \\ 0 & 1 \end{bmatrix} = [0 \quad 1]$$

We get

$$d_1 = 1, \quad d_2 = 1$$
$$E_1 = [1 \quad 0], \quad E_2 = [0 \quad 1]$$

(2) Check whether the system can be decoupled.

Compute

$$E = \begin{bmatrix} E_1 \\ E_2 \end{bmatrix} = \begin{bmatrix} 1 & 0 \\ 0 & 1 \end{bmatrix}$$

Because E is non-singular, the system can be decoupled.

(3) Derive integral form decoupling system.

First, compute

$$E^{-1} = \begin{bmatrix} 1 & 0 \\ 0 & 1 \end{bmatrix}, \quad F = \begin{bmatrix} c_1 A^2 \\ c_2 A^2 \end{bmatrix} = \begin{bmatrix} 6 & 0 & -3 & 2 \\ 2 & -2 & -2 & 0 \end{bmatrix}$$

The the state feedback matrix and input transformation matrix are computed as

$$\bar{L} = E^{-1} = \begin{bmatrix} 1 & 0 \\ 0 & 1 \end{bmatrix}, \quad \bar{K} = E^{-1}F = \begin{bmatrix} 6 & 0 & -3 & 2 \\ 2 & -2 & -2 & 0 \end{bmatrix}$$

So, we get the parameter matrices of the decoupling systemwith integral form are

$$\bar{A} = A - BE^{-1}F = \begin{bmatrix} -1 & 1 & 1 & 1 \\ 0 & 0 & 0 & -1 \\ -1 & 1 & 1 & 2 \\ 0 & 0 & 0 & 0 \end{bmatrix}$$

$$\bar{B} = BE^{-1} = \begin{bmatrix} 0 & 0 \\ 1 & 0 \\ 0 & 0 \\ 0 & 1 \end{bmatrix}$$

$$\bar{C} = C = \begin{bmatrix} 2 & 0 & -1 & 0 \\ -1 & 0 & 1 & 0 \end{bmatrix}$$

(4) Check whether $\{\bar{A}, \bar{C}\}$ is observable.

Based on the above parameter matrices, it is easy to get $\{\bar{A}, \bar{C}\}$ is observable.

(5) Derive the decoupling canonical form $\{\overline{A},\overline{B},\overline{C}\}$.

From $d_1 = 1, d_2 = 1$ and $n = 4$, we get $m_1 = d_1 + 1$, $m_2 = d_2 + 1$ and $m_1 + m_2 = 4$. Because $\{\overline{A},\overline{C}\}$ is observable, we get the decoupling norm form as:

$$\widetilde{A} = T^{-1}\overline{A}T = \left[\begin{array}{cc|cc} 0 & 1 & 0 & 0 \\ 0 & 0 & 0 & 0 \\ \hline 0 & 0 & 0 & 1 \\ 0 & 0 & 0 & 0 \end{array}\right]$$

$$\widetilde{B} = T^{-1}\overline{B} = \left[\begin{array}{c|c} 0 & 0 \\ 1 & 0 \\ \hline 0 & 0 \\ 0 & 1 \end{array}\right]$$

$$\widetilde{C} = \overline{C}T = \left[\begin{array}{cc|cc} 1 & 0 & 0 & 0 \\ \hline 0 & 0 & 1 & 0 \end{array}\right]$$

From $\{\widetilde{A},\widetilde{B},\widetilde{C}\}$ and $\{\overline{A},\overline{B},\overline{C}\}$, we get

$$T^{-1} = \begin{bmatrix} 2 & 0 & -1 & 0 \\ -1 & 1 & 1 & 0 \\ -1 & 0 & 1 & 0 \\ 0 & 0 & 0 & 1 \end{bmatrix}$$

$$T = \begin{bmatrix} 1 & 0 & 1 & 0 \\ 0 & 1 & -1 & 0 \\ 1 & 0 & 2 & 0 \\ 0 & 0 & 0 & 1 \end{bmatrix}$$

(6) Determine the state feedback $\widetilde{K}$ for the decoupling canonical form $\{\widetilde{A},\widetilde{B},\widetilde{C}\}$.

Based on the structure of $\{\widetilde{A},\widetilde{B},\widetilde{C}\}$, we get 2×4 feedback matrix $\widetilde{K}$

$$\widetilde{K} = \begin{bmatrix} k_{10} & k_{11} & 0 & 0 \\ 0 & 0 & k_{20} & k_{21} \end{bmatrix}$$

(7) Determine the state feedback $\widetilde{K}$ for the decoupled SISO system.

The decoupled SISO systems are two order. Specify the desired poles of two groups as

$$\lambda_{11}^* = -2, \quad \lambda_{12}^* = -4$$

$$\lambda_{21}^* = -2 + j, \quad \lambda_{22}^* = -2 - j$$

And two desired characteristic polynomials are

$$\alpha_1^*(s) = (s+2)(s+4) = s^2 + 6s + 8$$

$$\alpha_2^*(s) = (s+2-j)(s+2+j) = s^2 + 4s + 5$$

$$\tilde{A}-\tilde{B}\tilde{K}=\begin{bmatrix} 0 & 1 & & \\ -k_{10} & -k_{11} & & \\ & & 0 & 1 \\ & & -k_{20} & -k_{21} \end{bmatrix}$$

Using single input pole placement algorithm, we get

$$k_{10}=8,\quad k_{11}=6,\quad k_{20}=5,\quad k_{21}=4$$

Thus, the state feedback matrix which makes the system dynamic decoupling and all poles be placed at desired positions is:

$$\tilde{K}=\left[\begin{array}{cc:cc} 8 & 6 & 0 & 0 \\ \hdashline 0 & 0 & 5 & 4 \end{array}\right]$$

(8) Determine $\boldsymbol{L}$ and $\boldsymbol{K}$ for the original system $\{\boldsymbol{A},\boldsymbol{B},\boldsymbol{C}\}$.

$\boldsymbol{K}=\boldsymbol{E}^{-1}\boldsymbol{F}+\boldsymbol{E}^{-1}\tilde{\boldsymbol{K}}\boldsymbol{T}^{-1}$

$$=\begin{bmatrix} 6 & 0 & -3 & 2 \\ 2 & -2 & -2 & 0 \end{bmatrix}+\begin{bmatrix} 8 & 6 & 0 & 0 \\ 0 & 0 & 5 & 4 \end{bmatrix}\begin{bmatrix} 2 & 0 & -1 & 0 \\ -1 & 1 & 1 & 0 \\ -1 & 0 & 1 & 0 \\ 0 & 0 & 0 & 1 \end{bmatrix}$$

$$=\begin{bmatrix} 16 & 6 & -5 & 2 \\ -3 & -2 & 3 & 4 \end{bmatrix}$$

$$\boldsymbol{L}=\boldsymbol{E}^{-1}=\begin{bmatrix} 1 & 0 \\ 0 & 1 \end{bmatrix}$$

(9) The state space description of the closed-loop system after introducing state feedback and input transformation is

$$\dot{\boldsymbol{x}}=(\boldsymbol{A}-\boldsymbol{BK})\boldsymbol{x}+\boldsymbol{BLv}=\begin{bmatrix} -1 & 1 & 1 & 1 \\ -10 & -6 & 2 & -1 \\ -1 & 1 & 1 & 2 \\ 5 & 0 & -5 & -4 \end{bmatrix}\boldsymbol{x}+\begin{bmatrix} 0 & 0 \\ 1 & 0 \\ 0 & 0 \\ 0 & 1 \end{bmatrix}\boldsymbol{v}$$

$$\boldsymbol{y}=\boldsymbol{Cx}=\begin{bmatrix} 2 & 0 & -1 & 0 \\ -1 & 0 & 1 & 0 \end{bmatrix}\boldsymbol{x}$$

The transfer matrix of the closed-loop system is

$$\boldsymbol{G}_{KL}(s)=\boldsymbol{C}(s\boldsymbol{I}-\boldsymbol{A}+\boldsymbol{BK})^{-1}\boldsymbol{BL}=\begin{bmatrix} \frac{1}{s^2+6s+8} & \\ & \frac{1}{s^2+4s+5} \end{bmatrix}$$

5.9 Full-dimensional state estimator of linear systems

The state feedback discussed in previous sections has the implicit assumption that all state variables are available. This assumption may not hold in practice either sensing devices or transducers not available or very expensive. To apply state feedback in this case, we need to design a device or algorithm called a state estimator or observer, whose output will yield an estimate of the state. According to their structure, state estimators can be classified as full-dimensional or reduced-dimensional. The full-dimensional state estimator has the same dimension as the original system. In this section, we discuss the condition and algorithm of full-dimensional state estimators.

5.9.1 State estimator

1. Proposal of state estimator

The state feedback plays an significant role in improving the stability, steady-state error and dynamic performance of controlled systems. It can also decouple and stabilize systems. And the linear quadratic optimal control is fundamentally based on state feedback. As can be seen, state feedback is an important means for improving system performance. When designing a state feedback controller, it is always assumed that all of the system's state variables are measurable. However, in practice, the physical implementation of state feedback becomes impossible or extremely difficult because some state variables cannot be measured or they can be measured but at a costs. The state reconstruction, i. e. state estimator, is presented to solve the above mentioned problems.

2. The physical composition of state feedback using the state estimator

Reconstructing the state of the controlled system is one method of obtaining system state information to implement state feedback. That is, using theoretical analysis and corresponding algorithms, a rebuilt state equivalent to the original state is formed, and the reconstructed state is used to replace the real state to form the state feedback.

3. The essence of state reconstruction

The essence of state reconstruction is to construct a LTI system $\hat{\Sigma}$ with the same properties as the given LTI observed system Σ. The measurable output $\boldsymbol{y}$ and input $\boldsymbol{u}$ of the system Σ are used as inputs of $\hat{\Sigma}$ to construct the state $\hat{\boldsymbol{x}}$ of $\hat{\Sigma}$ or its transformation so that $\hat{\boldsymbol{x}}$ is equivalent to the state x of system Σ under a certain index. The index is usually taken as asymptotic equivalence, i. e.

$$\lim_{t\to\infty}\hat{\boldsymbol{x}}(t)=\lim_{t\to\infty}\boldsymbol{x}(t) \tag{5.9.1}$$

The $\hat{\boldsymbol{x}}$ is an estimate of x and the system $\hat{\Sigma}$ is said to be a state estimator of the observed system Σ. An illustration of the state reconstruction is shown in Figure 5. 9. 1. Further, if the observed system Σ is a stochastic linear system containing device noise and measurement noise, the system $\hat{\Sigma}$ that implements the state reconstruction needs to use a Kalman filter to estimate the state of the observed system Σ.

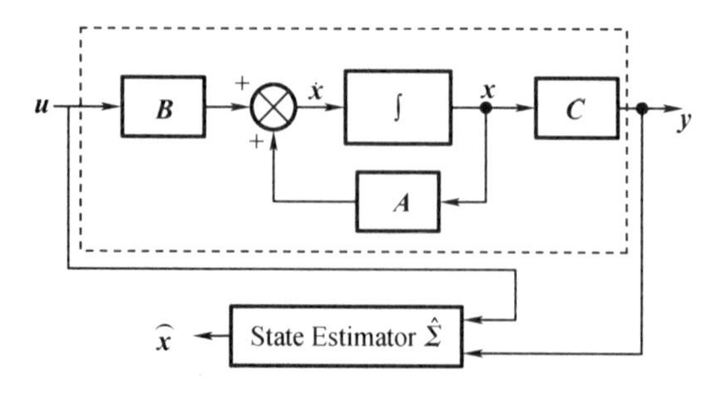

Figure 5.9.1 State Estimator

4. Classification of estimator

From the perspective of structure, state estimators can be classified into full-dimensional state estimators and reduced-dimensional state estimators. A state estimators whose dimension is equal to the dimension of the observed system is called full-dimensional state estimators, and state estimators whose dimension is less than the dimension of the observed system is called reduced-dimensional state estimators. From the perspective of function, estimators can be classified into state estimators and function estimators. The objective of state estimators is to reconstruct the state of the observed systems, and the asymptotic equivalence between the reconstructed state and the observed state, i. e. eq. (5.9.1), is taken as the equivalence index. The state of the estimator can be fully equivalent to the observed system state when $t \to \infty$.

The function estimator reconstructs a function of the observed system state, such as the feedback linear function $\boldsymbol{Kx}$, and takes asymptotically equivalent of the reconstructed output w to the observed state function, as equivalence index, i. e.

$$\lim_{t \to \infty} \boldsymbol{w}(t) = \lim_{t \to \infty} \boldsymbol{Kx}(t) \tag{5.9.2}$$

where $\boldsymbol{K}$ is a constant matrix. The the state of the state estimator can be fully equivalent to the function of observed system state, such as $\boldsymbol{Kx}$, when $t \to \infty$.

5.9.2 Conditions under which the state can be reconstructed

In the following, we discuss the conditions for state reconstruction using a single output system as an example. Consider a LTI system

$$\begin{cases} \dot{\boldsymbol{x}} = \boldsymbol{Ax} + \boldsymbol{Bu} \\ y = \boldsymbol{cx} \end{cases} \tag{5.9.3}$$

where, $\boldsymbol{x}$ is an n-dimensional state vector, $\boldsymbol{u}$ is a p-dimensional input vector, y is a output variable.

We have

$$\begin{cases} y = \boldsymbol{cx} \\ \dot{y} = \boldsymbol{c}\dot{\boldsymbol{x}} = \boldsymbol{cAx} + \boldsymbol{cBu}, \quad \dot{y} - \boldsymbol{cBu} = \boldsymbol{cAx} \\ y^{(2)} = \boldsymbol{cx}^{(2)} = \boldsymbol{cA}^2\boldsymbol{x} + \boldsymbol{cABu} + \boldsymbol{cBu}, \quad y^{(2)} - \boldsymbol{cB}\dot{\boldsymbol{u}} - \boldsymbol{cABu} = \boldsymbol{cA}^2\boldsymbol{x} \\ \cdots \\ y^{(n-1)} - \boldsymbol{cBu}^{(n-2)} - \boldsymbol{cABu}^{(n-3)} - \cdots - \boldsymbol{cABu} = \boldsymbol{cA}^{(n-1)}\boldsymbol{x} \end{cases} \tag{5.9.4}$$

From eq. (5.9.4), we get

$$\begin{bmatrix} y \\ \dot{y} - \boldsymbol{cBu} \\ y^{(2)} - \boldsymbol{cB}\dot{\boldsymbol{u}} - \boldsymbol{cABu} \\ \vdots \\ y^{(n-1)} - \boldsymbol{cBu}^{(n-2)} - \boldsymbol{cABu}^{(n-3)} - \cdots - \boldsymbol{cABu} \end{bmatrix} = \begin{bmatrix} \boldsymbol{c} \\ \boldsymbol{cA} \\ \boldsymbol{cA}^2 \\ \vdots \\ \boldsymbol{cA}^{(n-1)} \end{bmatrix} \boldsymbol{x} \tag{5.9.5}$$

So, as long as the system (5.9.3) is observable, i. e.

$$\text{rank}\ \boldsymbol{Q}_o = \text{rank} \begin{bmatrix} \boldsymbol{c} \\ \boldsymbol{cA} \\ \vdots \\ \boldsymbol{cA}^{n-1} \end{bmatrix} = n$$

we can obtain $\boldsymbol{x}$ from eq. (5.9.5). Hence, the state can be constructed if the system is observable.

5.9.3 Open-loop state estimator

Consider a MIMO LTI system described by

$$\begin{cases} \dot{\boldsymbol{x}} = \boldsymbol{Ax} + \boldsymbol{Bu} \\ \boldsymbol{y} = \boldsymbol{Cx} \end{cases} \tag{5.9.6}$$

where $\boldsymbol{x}$ is an n-dimensional state vector, $\boldsymbol{u}$ is a p-dimensional input vector, $\boldsymbol{y}$ is a q-dimensional output vector. We construct the open-loop state estimator as

$$\dot{\hat{\boldsymbol{x}}} = \boldsymbol{A}\hat{\boldsymbol{x}} + \boldsymbol{Bu} \tag{5.9.7}$$

From eq. (5.9.6) and eq. (5.9.7), we have

$$\dot{\boldsymbol{x}}(t) - \dot{\hat{\boldsymbol{x}}}(t) = \boldsymbol{A}(\boldsymbol{x}(t) - \hat{\boldsymbol{x}}(t)) \tag{5.9.8}$$

Eq. (5.9.8) is the state equation of a autonomous LTI system whose state motion or zero-input response is

$$\boldsymbol{x}(t) - \hat{\boldsymbol{x}}(t) = \mathrm{e}^{\boldsymbol{A}t}(\boldsymbol{x}(0) - \hat{\boldsymbol{x}}(0)) \tag{5.9.9}$$

where, $\boldsymbol{x}(0)$ is initial state of observed system (5.9.6), $\hat{\boldsymbol{x}}(0)$ is initial state of open-loop estimator (5.9.7). From eq. (5.9.9), whether the open-loop state observer (5.9.7) satisfies the asymptotic equivalence index (5.9.1) depends on the following conditions:

$$\text{Re}(\lambda_i(\boldsymbol{A})) < 0, \quad i = 1, 2, \cdots, n \tag{5.9.10}$$

That is, all the eigenvalues of $\boldsymbol{A}$ have negative real parts. Or

$$\boldsymbol{x}(0) = \hat{\boldsymbol{x}}(0) \tag{5.9.11}$$

Condition (5.9.10) depends on $\boldsymbol{A}$ of the controlled system, which is determined by the structure and parameters of the controlled system and cannot be changed by the designer, and condition (5.9.10) cannot be satisfied when $\boldsymbol{A}$ is unstable. In addition, the position of the eigenvalues of $\boldsymbol{A}$ determines the rate at which $\hat{\boldsymbol{x}}(t)$ approximates $\boldsymbol{x}(t)$, which is also not controlled by the designer. The condition (5.9.11), which requires the initial values of the open-loop state estimator to be exactly equal to the initial values of the controlled system, is also very difficult. Therefore, it is difficult for the open-loop state estimator to satisfy the asymptotic equivalence index (5.9.1).

5.9.4 closed-loop estimator

The basic idea of the closed-loop stateestimator is to introduce negative feedback to the open-loop state estimator to form a closed-loop system so that the distribution of the eigenvalues of the closed-loop state estimator is changed by changing the system matrix of the closed-loop system so that the asymptotic equivalence index equation (5.9.1) is satisfied. The schematic diagram of the closed-loop state estimator is shown in Figure 5.9.2.

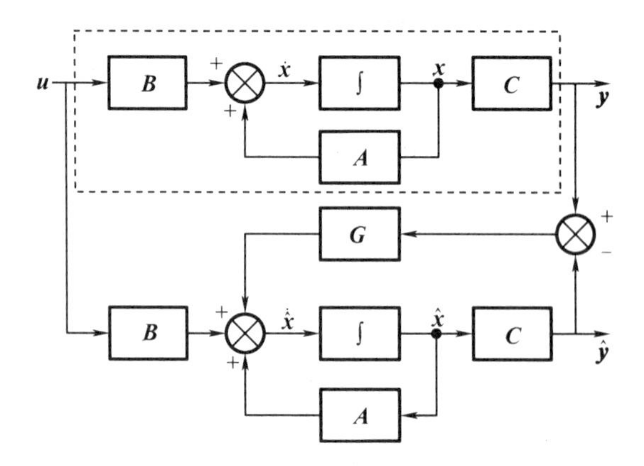

Figure 5.9.2 closed-loop state estimator

1. State space description of the full-dimensional closed-loop state estimator

Conclusion 5.9.1 For the closed-loop state estimator show in Figure 5.9.2, its state space description is

$$\begin{cases} \dot{\hat{\boldsymbol{x}}} = (\boldsymbol{A} - \boldsymbol{GC})\hat{\boldsymbol{x}} + \boldsymbol{Bu} + \boldsymbol{Gy}, & \hat{\boldsymbol{x}}(0) = \hat{\boldsymbol{x}}_0 \\ \hat{\boldsymbol{y}} = \boldsymbol{C}\hat{\boldsymbol{x}} \end{cases} \tag{5.9.12}$$

where, $\boldsymbol{G}$ is an $n \times q$ constant matrix.

Proof From Figure 5.9.2, we have:

$$\dot{\hat{\boldsymbol{x}}} = \boldsymbol{A}\hat{\boldsymbol{x}} + \boldsymbol{Bu} + \boldsymbol{Gc}(\boldsymbol{x} - \hat{\boldsymbol{x}}) = (\boldsymbol{A} - \boldsymbol{GC})\hat{\boldsymbol{x}} + \boldsymbol{Bu} + \boldsymbol{Gy}$$

2. Observation deviation of the full-dimensional closed-loop state estimator

Conclusion 5.9.2 State equation of observation error. For the closed-loop state estimator shown in Figure 5.9.2, where $\boldsymbol{x}$ is the state of the observed system and $\hat{\boldsymbol{x}}$ is the state of the closed-loop state estimator, the state equation of the observed deviation, $\tilde{\boldsymbol{x}} = \boldsymbol{x} - \hat{\boldsymbol{x}}$, is

$$\dot{\tilde{\boldsymbol{x}}} = (\boldsymbol{A} - \boldsymbol{GC})\tilde{\boldsymbol{x}}, \quad \tilde{\boldsymbol{x}}(0) = \tilde{\boldsymbol{x}}_0 = \boldsymbol{x}_0 - \hat{x}_0 \tag{5.9.13}$$

Proof From $\tilde{\boldsymbol{x}} = \boldsymbol{x} - \hat{\boldsymbol{x}}$, eq. (5.9.6) and eq. (5.9.12), we have

$$\begin{aligned}\dot{\tilde{\boldsymbol{x}}} &= \dot{\boldsymbol{x}} - \dot{\hat{\boldsymbol{x}}} \\ &= \boldsymbol{Ax} + \boldsymbol{Bu} - ((\boldsymbol{A} - \boldsymbol{GC})\hat{\boldsymbol{x}} + \boldsymbol{Bu} + \boldsymbol{Gy}) \\ &= (\boldsymbol{A} - \boldsymbol{GC})(\boldsymbol{x} - \hat{\boldsymbol{x}}) \\ &= (\boldsymbol{A} - \boldsymbol{GC})\tilde{\boldsymbol{x}}\end{aligned}$$

Conclusion 5.9.3 Analytic expressions of observation deviation. For the closed-loop state estimator shown in Figure 5.9.2, the analytical expression for the observation deviation $\tilde{\boldsymbol{x}} = \boldsymbol{x} - \hat{\boldsymbol{x}}$ is

$$\tilde{\boldsymbol{x}} = \mathrm{e}^{(\boldsymbol{A} - \boldsymbol{GC})t}\tilde{\boldsymbol{x}}_0 \tag{5.9.14}$$

Proof Solving the zero-input response of system (5.9.13), yields eq. (5.9.14).

For the case where $\boldsymbol{A} - \boldsymbol{GC}$ has distinct eigenvalues, there are

$$\boldsymbol{P}^{-1}(\boldsymbol{A} - \boldsymbol{GC})\boldsymbol{P} = \begin{bmatrix} \lambda_1 & & \\ & \ddots & \\ & & \lambda_n \end{bmatrix}$$

$$\boldsymbol{A} - \boldsymbol{GC} = \boldsymbol{P}\begin{bmatrix} \lambda_1 & & \\ & \ddots & \\ & & \lambda_n \end{bmatrix}\boldsymbol{P}^{-1}$$

$$\mathrm{e}^{(\boldsymbol{A} - \boldsymbol{GC})t} = \boldsymbol{P}\begin{bmatrix} \mathrm{e}^{\lambda_1 t} & & \\ & \ddots & \\ & & \mathrm{e}^{\lambda_n t} \end{bmatrix}\boldsymbol{P}^{-1}$$

where, $\lambda_i, i = 1, 2, \cdots, n$ is the eigenvalue of $\boldsymbol{A} - \boldsymbol{GC}$. It can be seen that in order to make $\hat{\boldsymbol{x}}$ converge to $\boldsymbol{x}$ as soon as possible, the eigenvalue λ_i should be located in the left half of the s-plane and as far away from the imaginary axis as possible. However, the dynamic performance of the estimator, such as the setting time, is inversely proportional to the cut off frequency ω_c. Too large ω_c will make the estimator less capable of suppressing high-frequency noise signals. Therefore, both of these factors should be considered when determining the position of the eigenvalue λ_i of $\boldsymbol{A} - \boldsymbol{GC}$.

3. The condition satisfying eq. (5.9.1) for full-dimensional closed-loop state estimators

Conclusion 5.9.4 For the full-dimensional closed-loop state estimator show in Figure 5.9.

2, there exists an $n \times q$ matrix $\boldsymbol{G}$ such that the following equation

$$\lim_{t\to\infty}\hat{\boldsymbol{x}}(t) = \lim_{t\to\infty}\boldsymbol{x}(t) \tag{5.9.15}$$

holds, if and only if the unobservable subsystem of the controlled system is asymptotically stable.

Proof Based on duality theorem, the observability of $\{\boldsymbol{A}, \boldsymbol{C}\}$ is equivalent to the controllability of $\{-\boldsymbol{A}^{\mathrm{T}}, \boldsymbol{C}^{\mathrm{T}}\}$, which in turn is equivalent to the controllability of $\{\boldsymbol{A}^{\mathrm{T}}, \boldsymbol{C}^{\mathrm{T}}\}$. And because

$$\det(s\boldsymbol{I} - \boldsymbol{A} + \boldsymbol{GC}) = \det(s\boldsymbol{I} - \boldsymbol{A} + \boldsymbol{GC})^{\mathrm{T}} = \det(s\boldsymbol{I} - \boldsymbol{A}^{\mathrm{T}} + \boldsymbol{C}^{\mathrm{T}}\boldsymbol{G}^{\mathrm{T}})$$

Therefore, designing an $n \times q$ state estimator's gain matrix $\boldsymbol{G}$ for $\{\boldsymbol{A}, \boldsymbol{C}\}$ is equivalent to designing a state feedback's gain matrix $\boldsymbol{G}^{\mathrm{T}}$ for $\{\boldsymbol{A}^{\mathrm{T}}, \boldsymbol{C}^{\mathrm{T}}\}$. This conclusion is proved by the related conclusions of the system stabilization problem.

Conclusion 5.9.5 Conditions under which the poles of full-dimensional closed-loop state estimators can be placed arbitrarily. For the estimator shown in Figure 5.9.2, there exists an $n \times q$ matrix $\boldsymbol{G}$ such that the poles of the state estimator can be placed arbitrarily if and only if $\{\boldsymbol{A}, \boldsymbol{C}\}$ of observed systems is observable.

Proof Referring to the proof process of conclusion 5.9.4, we can prove this conclusion.

4. Designing algorithm of full-dimensional closed-loop state estimators

Algorithm 5.9.1 Consider an observable MIMO LTI system described by

$$\begin{cases} \dot{\boldsymbol{x}} = \boldsymbol{Ax} + \boldsymbol{Bu} \\ \boldsymbol{y} = \boldsymbol{Cx} \end{cases}$$

where $\boldsymbol{x}$ is an n-dimensional state vector, $\boldsymbol{u}$ is a p-dimensional input vector, $\boldsymbol{y}$ is a q-dimensional output vector. Design a full-dimensional state estimator such that its eigenvalues are assigned at $\{\lambda_1^*, \lambda_2^*, \cdots, \lambda_n^*\}$.

Step 1. Compute the parameter matrices of its dual system, $\overline{\boldsymbol{A}} = \boldsymbol{A}^{\mathrm{T}}$, $\overline{\boldsymbol{B}} = \boldsymbol{C}^{\mathrm{T}}$.

Step 2. For $\{\overline{\boldsymbol{A}}, \overline{\boldsymbol{B}}\}$ and desired eigenvalues $\{\lambda_1^*, \lambda_2^*, \cdots, \lambda_n^*\}$, compute the $q \times n$ matrix $\overline{\boldsymbol{K}}$ using pole placement algorithm such that

$$\lambda_i(\overline{\boldsymbol{A}} - \overline{\boldsymbol{B}}\,\overline{\boldsymbol{K}}) = \lambda_i^*, \quad i = 1, 2, \cdots, n$$

Step 3. Let $\boldsymbol{G} = \overline{\boldsymbol{K}}^{\mathrm{T}}$.

Step 4. Compute $\boldsymbol{A} - \boldsymbol{GC}$.

Step 5. The state equation of obtained estimator is

$$\dot{\hat{\boldsymbol{x}}} = (\boldsymbol{A} - \boldsymbol{GC})\hat{\boldsymbol{x}} + \boldsymbol{Bu} + \boldsymbol{Gy}$$

Next we discuss designing full-dimensional state estimators for SISO systems.

Algorithm 5.9.2 Consider an observable SISO LTI system described by

$$\begin{cases} \dot{\boldsymbol{x}} = \boldsymbol{Ax} + \boldsymbol{b}u \\ \boldsymbol{y} = \boldsymbol{cx} \end{cases}$$

where $\boldsymbol{x}$ is an n-dimensional state vector, $\boldsymbol{u}$ is a 1-dimensional input variable, $\boldsymbol{y}$ is a 1-dimensional output variable. Design a full-dimensional state estimator such that its eigenvalues are assigned at $\{\lambda_1^*, \lambda_2^*, \cdots, \lambda_n^*\}$.

Step 1. Transform $\{\boldsymbol{A}, \boldsymbol{C}\}$ into observable canonical form. The characteristic polynomial of $\boldsymbol{A}$ is

$$\det(s\boldsymbol{I} - \boldsymbol{A}) = s^n + a_{n-1}s^{n-1} + \cdots + a_1 s + a_0$$

Introducing linearly nonsingular transformation $z = \boldsymbol{Q}\boldsymbol{x}$, yields the system's observable canonical form

$$\begin{cases} \dot{z} = \overline{\boldsymbol{A}}z + \overline{\boldsymbol{b}}u \\ y = \overline{\boldsymbol{c}}z \end{cases} \tag{5.9.16}$$

where

$$\overline{\boldsymbol{A}} = \boldsymbol{Q}\boldsymbol{A}\boldsymbol{Q}^{-1} = \begin{bmatrix} 0 & \cdots & 0 & -a_0 \\ 1 & & & -a_1 \\ & \ddots & & \vdots \\ & & 1 & -a_{n-1} \end{bmatrix}, \ \overline{\boldsymbol{b}} = \boldsymbol{Q}\boldsymbol{b}, \ \overline{\boldsymbol{c}} = \boldsymbol{c}\boldsymbol{Q}^{-1} = [0 \ \cdots \ 0 \ 1]$$

$$\boldsymbol{Q} = \begin{bmatrix} 1 & a_{n-1} & \cdots & a_1 \\ & \ddots & \ddots & \vdots \\ & & \ddots & a_1 \\ & & & 1 \end{bmatrix} \begin{bmatrix} \boldsymbol{c}\boldsymbol{A}^{n-1} \\ \vdots \\ \boldsymbol{c}\boldsymbol{A} \\ \boldsymbol{c} \end{bmatrix}$$

Step 2. For observable canonical form eq. (5.9.16), design gain matrix of state estimator $\overline{\boldsymbol{g}} = [\overline{g}_1 \ \ \overline{g}_2 \ \ \cdots \ \ \overline{g}_n]^T$ such that the eigenvalues of the closed-loop system are assigned at the desired positions $\{\lambda_1^*, \lambda_2^*, \cdots, \lambda_n^*\}$.

The characteristic polynomial of the closed-loop system is

$$\det[s\boldsymbol{I} - (\overline{\boldsymbol{A}} - \overline{\boldsymbol{g}}\,\overline{\boldsymbol{c}})] = s^n + (a_{n-1} + \overline{g}_n)s^{n-1} + \cdots + (a_1 + \overline{g}_2)s + (a_0 + \overline{g}_1) \tag{5.9.17}$$

The desired characteristic polynomial is

$$D(s) = (s - \lambda_1^*)(s - \lambda_2^*)\cdots(s - \lambda_n^*) = s^n + a_{n-1}^* s^{n-1} + \cdots + a_1^* s + a_0^* \tag{5.9.18}$$

Comparing eq. (5.9.17) and eq. (5.9.18), we have

$$\begin{cases} \overline{g}_1 = a_0^* - a_0 \\ \overline{g}_2 = a_1^* - a_1 \\ \vdots \\ \overline{g}_n = a_{n-1}^* - a_{n-1} \end{cases}$$

Step 3. Compute $\boldsymbol{g} = \boldsymbol{Q}^{-1}\overline{\boldsymbol{g}}$.

Step 4. The state equation of obtained estimator is

$$\dot{\hat{\boldsymbol{x}}} = (\boldsymbol{A} - \boldsymbol{g}\boldsymbol{c})\hat{\boldsymbol{x}} + \boldsymbol{b}u + \boldsymbol{g}y$$

Example 5.9.1 Given a SISO LTI system

$$\dot{\boldsymbol{x}} = \begin{bmatrix} 1 & 2 & 0 \\ 3 & -1 & 1 \\ 0 & 2 & 0 \end{bmatrix} \boldsymbol{x} + \begin{bmatrix} 2 \\ 1 \\ 1 \end{bmatrix} u$$

$$y = [0 \quad 0 \quad 1] \boldsymbol{x}$$

Design a 3-dimensional state estimator such that desired eigenvalues are assigned at $\lambda_1 = -3, \lambda_2 = -4, \lambda_3 = -5$.

Answer (1) Check whether the system is observable. Because

$$\text{rank}\begin{bmatrix} \boldsymbol{c} \\ \boldsymbol{cA} \\ \boldsymbol{cA}^2 \end{bmatrix} = \text{rank}\begin{bmatrix} 0 & 0 & 1 \\ 0 & 2 & 0 \\ 6 & -2 & 2 \end{bmatrix} = 3$$

The system is observable. We can design a full-dimensional state estimator to place all eigenvalues at any desired position.

(2) Get the observable canonical form of the system

$$\det(s\boldsymbol{I} - \boldsymbol{A}) = s^3 - 9s + 2$$

Introduce the linearly nonsingular transformation $\boldsymbol{z} = \boldsymbol{Qx}$, and

$$\boldsymbol{Q} = \begin{bmatrix} 1 & 0 & -9 \\ 0 & 1 & 0 \\ 0 & 0 & 1 \end{bmatrix} \begin{bmatrix} \boldsymbol{cA}^2 \\ \boldsymbol{cA} \\ \boldsymbol{c} \end{bmatrix} = \begin{bmatrix} 6 & -2 & 2 \\ 0 & 2 & 0 \\ 0 & 0 & 1 \end{bmatrix}$$

$$\boldsymbol{Q}^{-1} = \begin{bmatrix} 1/6 & 1/6 & 1/6 \\ 0 & 0.5 & 0 \\ 0 & 0 & 1 \end{bmatrix}$$

We get

$$\overline{\boldsymbol{A}} = \begin{bmatrix} 0 & 0 & -2 \\ 1 & 0 & 9 \\ 0 & 1 & 0 \end{bmatrix}, \overline{\boldsymbol{b}} = \boldsymbol{Qb} = \begin{bmatrix} 3 \\ 2 \\ 1 \end{bmatrix}, \overline{\boldsymbol{c}} = \boldsymbol{cQ}^{-1} = [0 \quad 0 \quad 1]$$

The desired characteristic polynomial is

$$(s+3)(s+4)(s+5) = s^3 + 12s^2 + 47s + 60 \tag{5.9.19}$$

The characteristic polynomial of $\overline{\boldsymbol{A}} - \overline{\boldsymbol{gc}}$ is

$$\det[s\boldsymbol{I} - (\overline{\boldsymbol{A}} - \overline{\boldsymbol{gc}})] = s^3 + \overline{g}_3 s^2 + (-9 + \overline{g}_2)s + (2 + \overline{g}_1) \tag{5.9.20}$$

Comparing eq. (5.9.19) and eq. (5.9.20), yields

$$\begin{cases} \overline{g}_1 = 58 \\ \overline{g}_2 = 56 \\ \overline{g}_3 = 12 \end{cases}$$

i.e. $\overline{\boldsymbol{G}} = [58 \quad 56 \quad 12]^{\mathrm{T}}$

(3) Compute

$$g = Q^{-1}\bar{g} = \begin{bmatrix} 33 \\ 28 \\ 12 \end{bmatrix}$$

(4) The state equation of obtained estimator is

$$\dot{\hat{x}} = (A - gc)\hat{x} + bu + gy$$

$$= \begin{bmatrix} 1 & 2 & -33 \\ 3 & -1 & -27 \\ 0 & 2 & -12 \end{bmatrix}\hat{x} + \begin{bmatrix} 2 \\ 1 \\ 1 \end{bmatrix}u + \begin{bmatrix} 33 \\ 28 \\ 12 \end{bmatrix}y$$

5.10 Feedback from estimated states

To enable the physical implementation of state feedback systems, the state estimator is introduced to constitute feedback control systems using reconstructed values $\hat{x}$ instead of the actual state x of the controlled system. In this section, we mainly discuss the introduction of the state estimator, its effect on the state feedback closed-loop system, and the basic characteristics of the estimator-based state feedback closed-loop system.

5.10.1 State space description of estimator-based state feedback systems

The estimator-based state feedback system consists of the controlled system, the state feedback and the state estimator.

Consider an n-dimensional time-continuous LTI system described by

$$\Sigma_0: \begin{cases} \dot{x} = Ax + Bu, & x(0) = x_0,\ t \geqslant 0 \\ y = Cx \end{cases} \tag{5.10.1}$$

where, u is a p-dimensional input vector, y is a q-dimensional output vector. Here we assume $\{A, C\}$ is observable and $\{A, B\}$ is controllable.

Introduce a full-dimensional closed-loop estimator

$$\Sigma_G: \begin{cases} \dot{\hat{x}} = (A - GC)\hat{x} + Bu + Gy \\ y = C\hat{x} \end{cases} \tag{5.10.2}$$

whose state $\hat{x}$ can approach the actual state x with any rate by selecting the $n \times q$ matrix G.

If x is not available, introduce the feedback from estimated states as

$$u = -K\hat{x} + v \tag{5.10.3}$$

as shown in Figure 5.10.1, where K is an $p \times n$ feedback matrix, v is a $p \times 1$ reference input.

Further, we establish the state space description of the estimator-based state feedback system.

Conclusion 5.10.1 Consider the estimator-based state feedback system $\sum_{KG}$ shown in Figure 5.10.1.

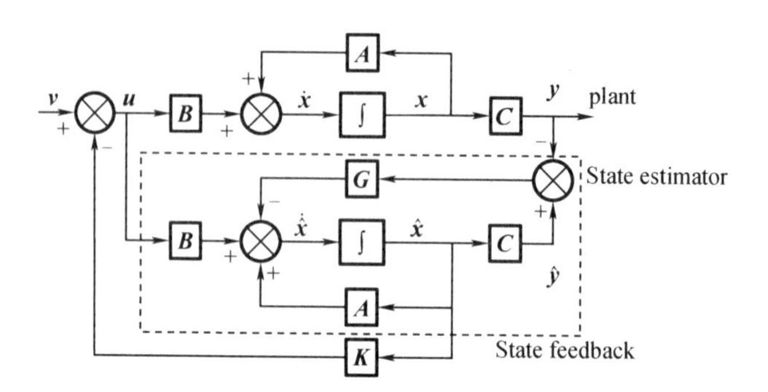

Figure 5.10.1 Feedback from estimated states

Take $\begin{bmatrix} x \\ \hat{x} \end{bmatrix}_{2n\times 1}$ as state, $v_{p\times 1}$ as input, $y_{q\times 1}$ as output. The state space description of the system is

$$\begin{cases} \begin{bmatrix} \dot{x} \\ \dot{\hat{x}} \end{bmatrix} = \begin{bmatrix} A & -BK \\ GC & A-GC-BK \end{bmatrix} \begin{bmatrix} x \\ \hat{x} \end{bmatrix} + \begin{bmatrix} B \\ B \end{bmatrix} v \\ y = \begin{bmatrix} C & 0 \end{bmatrix} \begin{bmatrix} x \\ \hat{x} \end{bmatrix} \end{cases} \tag{5.10.4}$$

Proof From eq. (5.10.1) to eq. (5.10.3), we have

$$\dot{x} = Ax + B(-K\hat{x} + v) = Ax - BK\hat{x} + Bv$$

and

$$\dot{\hat{x}} = (A - GC)\hat{x} + B(-K\hat{x} + v) + Gy = GCx + (A - GC - BK)\hat{x} + Bv$$

Rewriting the above two equations as matrix form, yields eq. (5.10.4).

5.10.2 Properties of estimator-based state feedback systems

The properties of the estimator-based state feedback system (5.10.4) are discussed below.

1. The dimension of the system Σ_{KG}

Conclusion 5.10.1 For the estimator-based state feedback system shown in Figure 5.10.1, we have

$$\dim(\Sigma_{KG}) = \dim(\Sigma_{K0}) + \dim(\Sigma_G) \tag{5.10.5}$$

The conclusion shows that after inserting the state estimator, the dimension of the system increases to 2n.

2. The eigenvalues of the system Σ_{KG}.

Conclusion 5.10.2 For the $2n$-dimensional estimator-based state feedback system Σ_{KG} shown in Figure 5.10.1, there exists

$$\lambda(\Sigma_{KG}) = \{\lambda(\Sigma_K),\lambda(\Sigma_G)\}$$
$$= \{\lambda_i(A-BK), i = 1,2,\cdots,n;\ \lambda_j(A-GC), j = 1,2,\cdots,n\} \tag{5.10.5}$$

where, $\lambda(\cdot)$ denotes the the set of eigenvalues of the matrix, $\lambda_i(\cdot)$ is i^{th} eigenvalues of the matrix, $\Sigma_K$denotes the original state feedback system, $\Sigma_G$denotes the state estimator system.

Proof For the system Σ_{KG}, introduce the following equivalence

$$\begin{bmatrix} x \\ e \end{bmatrix} = \begin{bmatrix} x \\ x-\hat{x} \end{bmatrix} = \begin{bmatrix} I & 0 \\ I & -I \end{bmatrix}\begin{bmatrix} x \\ \hat{x} \end{bmatrix} = P\begin{bmatrix} x \\ \hat{x} \end{bmatrix} \tag{5.10.6}$$

where

$$P = \begin{bmatrix} I_{n\times n} & 0_{n\times n} \\ -I_{n\times n} & I_{n\times n} \end{bmatrix},\quad P^{-1} = \begin{bmatrix} I_{n\times n} & 0_{n\times n} \\ I_{n\times n} & I_{n\times n} \end{bmatrix} \tag{5.10.7}$$

From eq. (5.10.6) and eq. (5.10.7), we have

$$P\begin{bmatrix} A & -BK \\ GC & A-GC-BK \end{bmatrix}P^{-1} = \begin{bmatrix} I_{n\times n} & 0_{n\times n} \\ -I_{n\times n} & I_{n\times n} \end{bmatrix}\begin{bmatrix} A & -BK \\ GC & A-GC-BK \end{bmatrix}\begin{bmatrix} I_{n\times n} & 0_{n\times n} \\ I_{n\times n} & I_{n\times n} \end{bmatrix}$$
$$= \begin{bmatrix} A-BK & -BK \\ 0 & A-GC \end{bmatrix}$$

And since the linear non-singular transformation does not change the eigenvalues of the matrix, we have

$$\lambda(\Sigma_{KG}) = \lambda\left(\begin{bmatrix} A & -BK \\ GC & A-GC-BK \end{bmatrix}\right) = \lambda\left(P\begin{bmatrix} A & -BK \\ GC & A-GC-BK \end{bmatrix}P^{-1}\right)$$
$$= \lambda\left(\begin{bmatrix} A-BK & -BK \\ 0 & A-GC \end{bmatrix}\right)$$
$$= \{\lambda_i(A-BK), i=1,2,\cdots,n;\ \lambda_j(A-GC), j=1,2,\cdots,n\}$$
$$= \{\lambda(\Sigma_K),\lambda(\Sigma_G)\}$$

Conclusion 5.10.2 shows that the eigenvalues of the estimator-based state feedback system are the union of the eigenvalues of the original state feedback system and the eigenvalues of the state estimator.

This conclusion leads to the following separation property.

3. The separation property of the system Σ_{KG}

Conclusion 5.10.3 The separation property. For the $2n$-dimensional estimator-based state feedback system Σ_{KG} shown in Figure 5.10.1, inserting the state estimator does not affect the eigenvalues of the original state feedback, nor the eigenvalues of the state estimator. Thus the design of state feedback and the design of state estimator can be carried out independently.

4. The transfer matrix of the system Σ_{KG}

Conclusion 5.10.4 The transfer matrix. The transfer matrix of the system $\Sigma_{KG}$is equal to the transfer matrix of the original state feedback system $\Sigma_K$without using a state estimator, i.e.

$$G_{KG}(s) = G_K(s) \tag{5.10.8}$$

where, $G_{KG}(s)$ and $G_K(s)$ denote the transfer matrices of the system Σ_{KG} and Σ_K, separately.

Proof Because

$$G_K(s) = C(sI - A + BK)^{-1}B$$

and

$$G_{KG}(s) = \bar{C}(sI - \bar{A})^{-1}\bar{B}$$

where

$$\bar{A} = \begin{bmatrix} A & -BK \\ GC & A - GC - BK \end{bmatrix}, \quad \bar{B} = \begin{bmatrix} B \\ B \end{bmatrix}, \quad \bar{C} = [C \quad 0]$$

Taking the equivalence(5.10.6) to $\{\bar{A}, \bar{B}, \bar{C}\}$, we have

$$\tilde{A} = P\bar{A}P^{-1} = P\begin{bmatrix} A & -BK \\ GC & A - GC - BK \end{bmatrix}P^{-1} = \begin{bmatrix} A - BK & -BK \\ 0 & A - GC \end{bmatrix} \tag{5.10.9}$$

$$\tilde{B} = P\bar{B} = \begin{bmatrix} I_{n\times n} & 0_{n\times n} \\ -I_{n\times n} & I_{n\times n} \end{bmatrix}\begin{bmatrix} B \\ B \end{bmatrix} = \begin{bmatrix} B \\ 0 \end{bmatrix} \tag{5.10.10}$$

$$\tilde{C} = \bar{C}P^{-1} = [C \quad 0]\begin{bmatrix} I_{n\times n} & 0_{n\times n} \\ I_{n\times n} & I_{n\times n} \end{bmatrix} = [C \quad 0] \tag{5.10.11}$$

Since the transfer matrix remains unchanged under linear non-singular transformation, it can be obtained from eq. (5.10.9) to eq. (5.10.11):

$$\begin{aligned} G_{KB}(s) &= \bar{C}(sI - \bar{A})^{-1}\bar{B} \\ &= \tilde{C}(sI - \tilde{A})^{-1}\tilde{B} \\ &= [C \quad 0]\begin{bmatrix} sI - A + BK & BK \\ 0 & sI - A + GC \end{bmatrix}^{-1}\begin{bmatrix} B \\ 0 \end{bmatrix} \\ &= [C \quad 0]\begin{bmatrix} (sI - A + BK)^{-1} & * \\ 0 & (sI - A + GC)^{-1} \end{bmatrix}\begin{bmatrix} B \\ 0 \end{bmatrix} \\ &= C(sI - A + BK)^{-1}B \\ &= G_K(s) \end{aligned}$$

5. Controllability and observability of the system Σ_{KG}

We discuss inserting a state estimator and how it affects on controllability and observability of the system.

Conclusion 5.10.5 The estimator-based state feedback system Σ_{KG} is not controllable and not observable after inserting a state estimator for a controllable and observable controlled system Σ_0. The controllable and observable part of the estimator-based state feedback system is $\{A - BK, B, C\}$.

Proof Since the transfer function matrix of the system Σ_{KG} only reflects the controllable and observable subsystem, it can be seen from conclusion 5.10.4 that the controllable and observable part of the system is $\{A - BK, B, C\}$, and its dimension is n. From eq. (5.10.9) to eq. (5.10.11) can be seen, the system Σ_{KG} also contains uncontrollable and unobservable states.

6. Design principles of state estimators of the system Σ_{KG}

For the estimator-based state feedback system Σ_{KG} shown in Figure 5. 10. 1, the general design principle of state estimator is to take the negative real-part of the eigenvalues of the estimator as 2—3 times of the negative real-part of the eigenvalues of $\boldsymbol{A}-\boldsymbol{BK}$, i. e.

$$\mathrm{Re}\lambda_i(\boldsymbol{A}-\boldsymbol{GC})=(2\text{—}3)\mathrm{Re}\lambda_i(\boldsymbol{A}-\boldsymbol{BK}) \tag{5.10.12}$$

Eq. (5. 10. 12) can rapidly reduce the errors betweenthe estimated state values and the actual state values that come from the initially estimated errors and ensure the system Σ_{KG} fast approach the desired input.

Example 5. 10. 1 Given a LTI system

$$\dot{\boldsymbol{x}}=\begin{bmatrix}0 & 1\\ -6 & -8\end{bmatrix}\boldsymbol{x}+\begin{bmatrix}0\\ 1\end{bmatrix}u, y=[4 \quad 0]\boldsymbol{x}$$

Based on the separation property, design an estimator-based state feedback system and determine the gain matrix $\boldsymbol{g}=[g_1 \quad g_2]^{\mathrm{T}}$ of the estimator such that its eigenvalues be assigned at -6 and -6. And determine the state feedback gain matrix $\boldsymbol{k}=[k_1 \quad k_2]^{\mathrm{T}}$ such that the other two eigenvalues of the system be assigned at -1, -3. Compute the transfer function of the estimator-based state feedback system.

Answer

(1) Check observability of the open-loop system. Because

$$\mathrm{rank}\ \boldsymbol{Q}_o=\mathrm{rank}\begin{bmatrix}\boldsymbol{c}\\ \boldsymbol{cA}\end{bmatrix}=\mathrm{rank}\begin{bmatrix}4 & 0\\ 0 & 4\end{bmatrix}=2$$

The open-loop is observable, we can place the eigenvalues arbitrarily.

The desired characteristic polynomial is $(s+6)(s+6)=s^2+12s+36$.

The characteristic polynomial of $\boldsymbol{A}-\boldsymbol{gc}$ is $\det(s\boldsymbol{I}-\boldsymbol{A}+\boldsymbol{gc})=s^2+(4l_1+8)s+(32l_1+4l_2+6)$.

Compare the above polynomials, we get $\boldsymbol{g}=[1 \quad 7.5]^{\mathrm{T}}$.

(2) Check controllability of the open-loop system. Because $\{\boldsymbol{A}, \boldsymbol{b}\}$ has controllable canonical form, the open-loop system is controllable. So we can place the other eigenvalues of the system arbitrarily.

The desired characteristic polynomial is $\alpha^*(s)=(s+1)(s+3)=s^2+4s+3$.

The characteristic polynomial of A is $\alpha^*(s)=(s+1)(s+3)=s^2+4s+3$.

And we get $\boldsymbol{k}=[\alpha_0^*-\alpha_0 \quad \alpha_1^*-\alpha_1]=[-3 \quad -4]$.

(3) The transfer function of the estimator-based state feedback system is

$$g_{KG}(s)=\boldsymbol{c}(s\boldsymbol{I}-\boldsymbol{A}+\boldsymbol{bk})^{-1}\boldsymbol{b}=\frac{4}{s^2+4s+3}$$

Problems

5.1 Determine whether the all eigenvalues of following systems can be placed arbitrarily using state feedback.

(1) $\dot{\boldsymbol{x}} = \begin{bmatrix} 1 & 2 \\ 3 & 2 \end{bmatrix}\boldsymbol{x} + \begin{bmatrix} 1 \\ 0 \end{bmatrix}\boldsymbol{u}$

(2) $\dot{\boldsymbol{x}} = \begin{bmatrix} 1 & 0 & 0 \\ 0 & -2 & 1 \\ 0 & 0 & -2 \end{bmatrix}\boldsymbol{x} + \begin{bmatrix} 1 & 0 \\ 0 & 2 \\ 0 & 0 \end{bmatrix}\boldsymbol{u}$

(3) $\dot{\boldsymbol{x}} = \begin{bmatrix} 0 & 1 & 0 & 0 \\ 0 & 0 & 1 & 0 \\ 0 & 0 & 0 & 1 \\ -2 & -4 & -3 & 0 \end{bmatrix}\boldsymbol{x} + \begin{bmatrix} 0 & 0 & 0 \\ 0 & 0 & 1 \\ 0 & 1 & 0 \\ 1 & 0 & 0 \end{bmatrix}\boldsymbol{u}$

5.2 Given a LTI system

$$\dot{\boldsymbol{x}} = \begin{bmatrix} 1 & 2 \\ 3 & 1 \end{bmatrix}\boldsymbol{x} + \begin{bmatrix} 1 \\ 0 \end{bmatrix}u$$

find the state feedback gain $\boldsymbol{x}$ so that the state feedback system has $\lambda_1^* = -3+j$ and $\lambda_2^* = -3-j$ as its eigenvalues.

5.3 Given a system with transfer function

$$g_0(s) = \frac{1}{s(s+4)(s+8)}$$

find the state feedback gain $\boldsymbol{k}$ so that the closed loop system has $\lambda_1^* = -1$, $\lambda_2^* = -4$ and $\lambda_3^* = -3$ as its eigenvalues.

5.4 For the controlled system in problem 5.3, find the state feedback gain $\boldsymbol{k}$ so that the unit step response of the closed loop system satisfies: overshoot $\sigma \leqslant 20\%$ and setting time $t_s \leqslant 0.4$s.

5.5 Give a LTI system

$$\dot{\boldsymbol{x}} = \begin{bmatrix} 1 & 1 \\ 0 & 1 \end{bmatrix}\boldsymbol{x} + \begin{bmatrix} 1 \\ 0 \end{bmatrix}u$$

$$\boldsymbol{y} = \begin{bmatrix} 2 & 0 \\ 0 & 1 \end{bmatrix}\boldsymbol{x}$$

find the output feedback gain $\boldsymbol{g}$ so that the closed-loop system has $\lambda_1^* = -2$ and $\lambda_2^* = -4$ as its eigenvalues.

5.6 Consider the following LTI system

$$\dot{x} = \begin{bmatrix} 2 & 1 & 0 & 0 \\ 0 & 2 & 0 & 0 \\ 0 & 0 & -4 & 0 \\ 0 & 0 & 0 & -4 \end{bmatrix} x + \begin{bmatrix} 0 \\ 1 \\ 1 \\ 1 \end{bmatrix} u$$

Is it possible to find a state feedback gain $\boldsymbol{k}$ so that the closed-loop system has eigenvalues $-4, -4, -4, -4$? Is it possible to have eigenvalues $-4, -4, -4, -2$? How about $-5, -5, -5, -5$? Is the system stabilizable?

5.7 Given a LTI system

$$\dot{x} = \begin{bmatrix} 1 & 1 & 0 \\ 0 & 1 & 0 \\ 0 & 0 & 2 \end{bmatrix} x + \begin{bmatrix} 0 & 0 \\ 1 & 0 \\ 0 & -1 \end{bmatrix} \boldsymbol{u}$$

find the two different state feedback gain $\boldsymbol{K}_1$ and $\boldsymbol{K}_2$ so that the closed-loop system has $\lambda_1^* = -2$, $\lambda_2^* = -1 + j2$ and $\lambda_3^* = -1 - j2$ as its eigenvalues.

5.8 Given a LTI system

$$\dot{x} = \begin{pmatrix} 0 & 2 & 0 & 0 \\ 0 & 0 & 1 & 0 \\ -3 & 1 & 2 & 3 \\ 2 & 1 & 0 & 0 \end{pmatrix} x + \begin{pmatrix} 0 & 0 \\ 0 & 0 \\ 1 & 2 \\ 0 & 2 \end{pmatrix} \boldsymbol{u}$$

find the two different state feedback gain $\boldsymbol{K}_1$ and $\boldsymbol{K}_2$ so that the closed-loop system has $\lambda_{1,2}^* = -2 \pm j3$ and $\lambda_{3,4}^* = -5 \pm j6$ as its eigenvalues.

5.9 Can the following systems be stabilized by state feedback?

(1) $\dot{x} = \begin{bmatrix} 1 & 3 \\ 2 & 1 \end{bmatrix} x + \begin{bmatrix} 0 \\ 1 \end{bmatrix} u$

(2) $\dot{x} = \begin{bmatrix} 4 & 2 \\ 0 & -2 \end{bmatrix} x + \begin{bmatrix} 1 \\ 0 \end{bmatrix} u$

(3) $\dot{x} = \begin{bmatrix} 1 & 0 & 0 \\ 0 & -2 & 1 \\ 0 & 0 & -2 \end{bmatrix} x + \begin{bmatrix} 1 & 0 \\ 0 & 1 \\ 0 & 0 \end{bmatrix} \boldsymbol{u}$

5.10 Can the following systems be stabilized by output feedback?

(1) $\dot{x} = \begin{bmatrix} 1 & 3 \\ 2 & 1 \end{bmatrix} x + \begin{bmatrix} 0 \\ 1 \end{bmatrix} u$

$\boldsymbol{y} = \begin{bmatrix} 0 & 2 \\ 1 & 0 \end{bmatrix} x$

(2) $\dot{x} = \begin{bmatrix} 4 & 2 \\ 0 & -2 \end{bmatrix} x + \begin{bmatrix} 1 \\ 0 \end{bmatrix} u$

$\boldsymbol{y} = \begin{bmatrix} 1 & 1 \\ 0 & 2 \end{bmatrix} x$

(3) $\dot{\boldsymbol{x}} = \begin{bmatrix} 4 & 0 & 0 \\ 0 & -1 & 1 \\ 0 & 0 & -1 \end{bmatrix} \boldsymbol{x} + \begin{bmatrix} 0 & 1 \\ 1 & 0 \\ 0 & 0 \end{bmatrix} \boldsymbol{u}$

$\boldsymbol{y} = \begin{bmatrix} 1 & 0 & 1 \\ 1 & 1 & 0 \\ 2 & 4 & 3 \end{bmatrix} \boldsymbol{x}$

5.11 Consider a system with transfer function

$$g_0(s) = \frac{(s+2)(s+3)}{(s+1)(s-2)(s+4)}$$

Is it possible to change the transfer function to

$$g(s) = \frac{(s+3)}{(s+2)(s+4)}$$

by state feedback? If it is, find the state feedback gain $\boldsymbol{k}$.

5.12 Given

$$\dot{\boldsymbol{x}} = \begin{bmatrix} 2 & 1 & 0 \\ 0 & 1 & 1 \\ 1 & 0 & 0 \end{bmatrix} \boldsymbol{x} + \begin{bmatrix} 0 \\ 1 \\ 0 \end{bmatrix} u$$

find the state feedback gain $\boldsymbol{k}$ so that $(\boldsymbol{A} - \boldsymbol{bk})$ is similar to

$$\boldsymbol{F} = \begin{bmatrix} -3 & 0 & 0 \\ 0 & -2 & 0 \\ 0 & 0 & -1 \end{bmatrix}$$

5.13 Can the following systems be decoupled by state feedback with input transformation?

(1) $\boldsymbol{G}_0(s) = \begin{bmatrix} \dfrac{3}{s^2+2} & \dfrac{2}{s^2+s+1} \\ \dfrac{4s+1}{s^3+2s+1} & \dfrac{1}{s} \end{bmatrix}$

(2) $\dot{\boldsymbol{x}} = \begin{bmatrix} 3 & 1 & 0 \\ 0 & 0 & -1 \\ 0 & 1 & -1 \end{bmatrix} \boldsymbol{x} + \begin{bmatrix} 0 & 0 \\ 1 & 0 \\ 0 & 1 \end{bmatrix} \boldsymbol{u}$

$\boldsymbol{y} = \begin{bmatrix} 2 & -1 & 1 \\ 0 & 1 & 1 \end{bmatrix} \boldsymbol{x}$

5.14 Consider the LTI system

$$\dot{\boldsymbol{x}} = \begin{bmatrix} -1 & 0 & 0 \\ 0 & -2 & -3 \\ 1 & 0 & 1 \end{bmatrix} \boldsymbol{x} + \begin{bmatrix} 1 & 0 \\ 0 & 1 \\ 0 & -1 \end{bmatrix} \boldsymbol{u}$$

$$\boldsymbol{y} = \begin{bmatrix} 1 & 2 & 0 \\ 0 & 1 & 1 \end{bmatrix} \boldsymbol{x}$$

(1) Can the system be decoupled?

(2) If it is, determine the $\{\boldsymbol{L},\boldsymbol{K}\}$ to realize integral form decoupling.

5.15 Design the full-dimensional state estimator by using two methods for the following system

$$\dot{\boldsymbol{x}} = \begin{bmatrix} 0 & 1 \\ 0 & 0 \end{bmatrix}\boldsymbol{x} + \begin{bmatrix} 0 \\ 1 \end{bmatrix}u$$

$$y = \begin{bmatrix} 1 & 0 \end{bmatrix}\boldsymbol{x}$$

Select the eigenvalues of the estimator from $\{-2, -4\}$.

5.16 Design a full-dimensional state estimator by using two methods for the following system

$$\dot{\boldsymbol{x}} = \begin{bmatrix} -1 & -2 & -2 \\ 0 & -1 & 1 \\ 1 & 0 & -1 \end{bmatrix}\boldsymbol{x} + \begin{bmatrix} 2 \\ 0 \\ 1 \end{bmatrix}u$$

$$y = \begin{bmatrix} 1 & 1 & 0 \end{bmatrix}\boldsymbol{x}$$

Select the eigenvalues of the estimator from $\{-3, -3, -4\}$.

References

[1] 郑大钟. 线性系统理论[M]. 北京:清华大学出版社,2005.

[2] 李保全,陈维远. 线性系统理论[M]. 北京:国防工业出版社,1997.

[3] 仝茂达. 线性系统理论和设计[M]. 2 版. 合肥:中国科学技术大学出版社, 2012.

[4] 陈啟宗. 线性系统理论与设计[M]. 北京:科学出版社,1998.

[5] 段广仁. 线性系统理论[M]. 哈尔滨:哈尔滨工业大学出版社,1996.

[6] 吕碧湖. 线性系统理论基础[M]. 合肥:中国科学技术大学出版社, 1990.

[7] 余贻鑫. 线性系统[M]. 天津:天津大学出版社,1991.

[8] T. 凯拉斯. 线性系统习题解答[M]. 李清泉,高龙,褚家晋,译. 北京:科学出版社,1985.

[9] 有本卓,高桥进一,滨田望. 线性系统理论、例题和习题[M]. 卢伯英,译. 北京:科学出版社,1982.

[10] 黄琳. 系统与控制理论中的线性代数[M]. 北京:科学出版社,1994.

[11] 韩京清,何关钰,许可康. 线性系统理论代数基础[M]. 沈阳:辽宁科学技术出版社,1985 .

[12] 吴沧浦. 最优控制的理论与方法[M]. 2 版. 北京:国防工业出版社,2000.

[13] 解学书. 最优控制理论与应用[M]. 北京:清华大学出版社,1986.

[14] 郑大钟. 一类 LQ 问题的最优控制和次优控制的综合方法[J]. 清华大学学报, 1989, 29(4):106 – 114.

[15] 中国矿业学院数学教研室. 数学手册[M]. 北京:科学出版社,1990.

[16] 楼顺天,于卫. 基于 MATLAB 的系统分析与设计:控制系统[M]. 西安:西安电子科技大学出版社,1999.

[17] 胡寿松. 自动控制原理[M]. 4 版. 北京:科学出版社,2001.

[18] D'Azzo J J , Houpis C H Linear Control System Analysis and Design[M]. New York: McGraw-Hill Book Company,1988.

[19] D'Souza A F. Design of Control Systems[M]. New Jersy: Prentice-Hall,1988.

[20] Kailatg T. Linear System[M]. Englewood Cliffs, New Jersy: Prentice-Hall,1980.

[21] Chi – T Chen. Linear System Theory and Design[M]. Oxford University Press, 1999 .

[22] Richard C Dorf, Robert H. Bishop. 现代控制系统[M]. 11 版. 北京:电子工业出版社,2009.

[23] Gene F. Franklin, J. David Powell, Abbas Emami-Naeini. 自动控制原理与设计[M]. 李中华,等译. 北京:电子工业出版社,2013.